Glaciers and Landscape
A Geomorphological Approach

Glaciers and Landscape
A Geomorphological Approach

David E. Sugden
Department of Geography, University of Aberdeen

and

Brian S. John
Formerly of Department of Geography, University of Durham

Edward Arnold

First published 1976 by
Edward Arnold (Publishers) Ltd
41 Bedford Square, London WC1B 3DQ

Reprinted 1977 with corrections and bibliographic additions.
Reprinted with additional references 1979
Reprinted with additional references 1982

ISBN 0 7131 5840 9 (paper)

Printed in Great Britain by
Butler & Tanner Ltd,
Frome and London

Contents

Preface

We are only too aware that much of this book deals with topics which others know far more about than we do. Among colleagues who have looked at parts of the text and tried to keep us on the right track we are particularly grateful to the following: Dr John Allen, Dr Geoffrey Boulton, Dr Parker Calkin, Dr George Denton, Dr Ian Evans, Dr John Glen, Professor Gunnar Hoppe, Dr Hal Lister, Dr N. Macmillan, Dr Mark Meier, Dr John Mercer, Professor John Nye, Dr Simon Ommanney, Professor Valter Schytt, Dr Ron Shreve, Dr Brian Sissons and Dr Charles Swithinbank. We have also received kind and helpful criticism from Dr Björn Anderson, Dr John Andrews, Dr Chalmers Clapperton, Patrick Hamilton, Dr Stig Jonsson, Wibjörn Karlén, Dr Jan Mangerud, and Dr Gunnar Østrem. We are indebted (far more than they realize) to all the student groups who have taken our option courses in polar geomorphology. Their cynicism has led to the weeding out of our wilder ideas, and their enthusiasm has encouraged the cultivation of many less wild ideas which now appear among the pages of the book. Others have provided stimulus through discussions, help in the field and comments on specific points from the book, and to all of them we offer our thanks. In spite of this expert help there will still be mistakes for which we are wholly to blame.

We value enormously the philosophy which ran through our own undergraduate and research training at the University of Oxford, under Dr Marjorie Sweeting. The late Professor David Linton never taught us formally, but he strongly influenced our thinking on the processes and patterns of glaciation. Through his persistent faith in the geographical approach and through his multitude of published ideas he was an inspiration. In recent years our own research projects have had the benefit of help from many colleagues and institutions. Individually or together we have received financial and material aid from many institutions, enabling us to undertake prolonged fieldwork in Iceland, Greenland and Antarctica and shorter visits to other Arctic areas. In particular we thank the British Antarctic Survey, the Royal Society, the Carnegie Trust, the British Natural Environment Research Council, the NATO Research Grants Programme, and the Universities of Aberdeen and Durham.

A large number of the drafts of maps and diagrams for this book were prepared in the drawing office of the Durham University Geography Department, and many more in the drawing office of Aberdeen University Geography Department by Bert Bremner and his assistants. Most of the typing and photographic work has been done by the secretarial and dark room staffs of both departments. We thank them all for their skill and patience, and we are happy to acknowledge the kind help of Professor K. Walton and Professor W. B. Fisher, who have allowed the use of many departmental facilities at various stages in the preparation of the book. Professor Gunnar Hoppe granted the full use of the facilities of the Physical Geography Department at Stockholm University while one of us (BSJ) spent an academic year there. Sven Stridsberg gave great help with photographic illustrations, and even managed to make some of our own photographic efforts quite presentable.

This book would never have been written but for the influence of three people. Mr E.

Paget, our tutor at Jesus College, Oxford, is a practitioner of human geography, but he influenced us more than he will ever know by his encouragement of criticial thought. We hope he still insists that ideas are as respectable as facts, and that they exist to be questioned. We hope, too, that he is not too disappointed that we turned into geomorphologists.

Our wives Britta and Inger have put up with glaciers for far too long, and we owe them the greatest debt of all. Britta has kept us alive during long working sessions, and Inger has typed much of the early manuscript. They have both helped with the drudgery of compiling the bibliography and index. They have kept small children at bay at times when we should have been taking our duties as fathers far more seriously, and they have rescued vital pieces of manuscript which were in danger of being covered with strawberry jam or turned into paper aeroplanes. They have encouraged, been critical where necessary, and above all they have been patient. They will be far more relieved than we are to see the book in print.

January 1976

DES
BSJ

1 Introduction

Glaciology and geomorphology

Viewed from outer space the earth can with some justification be described as a glacial planet (Figure 1.1, overleaf). During the northern hemisphere winter, over half of the world's land surface area and up to 30 per cent of its ocean area may be covered by a blanket of snow and ice. Approximately one tenth of the land surface is covered by glaciers and in the relatively recent history of the earth this figure was some three times higher. Furthermore, 75 per cent of the present fresh water resources of the world are contained in glaciers (Nace, 1969), and the overwhelming importance of glacial moisture storage as compared with river, lake and groundwater storage is shown in Figure 1.2 (p. 4). While all this is of profound significance for most sciences, it is quite crucial to geomorphology in several respects:

1 An understanding of the geomorphology of a considerable portion of the earth's surface is impossible without understanding the characteristics of glaciers.
2 Now, as always during the earth's history, the extent of glaciers exerts a profound control upon the amount of water available for release into rivers and oceans (Meier, 1967).
3 Glaciers contain in solid form three quarters of all water which is in transit on the land and therefore involved in geomorphological activity.

Historically, the popularity of glacial geomorphology can largely be attributed to the development of landform studies in those densely populated areas of the northern hemisphere middle latitudes which have been so drastically affected by glacial processes. However, the status and traditions of glacial geomorphology vary greatly between one country and another. In the United States it is traditionally studied in geology departments and forms part of glacial geology, while in western Europe, the USSR and Canada there is a much stronger tradition of study and research in geography departments. Inevitably, in countries where glacial deposits are thick and widespread, geologists have tended to become much involved with glacial stratigraphy and glacial chronology. Elsewhere the main focus of attention has been glacial geomorphology, and in particular glacial landscape evolution. One result of this focus has been to leave the study of glacial processes and forms somewhat in the cold and consequently poorly understood. At the same time and largely independently, the science of glaciology has developed rapidly, and currently there exists something of a gulf between those who study glaciology and those who study glacial landscapes and deposits. Insufficient modern glaciological theory has been applied to the study of glacial geomorphology and glaciologists have not been provided with reliable information on landforms. Perhaps there is a need for a more glaciological type of geomorphology and a more geomorphological type of glaciology. There is now a strong case for a realistic dialogue between those studying glacier dynamics and those studying forms. Until this occurs, there can be few spectacular advances such as those achieved recently in fluvial and slope geomorphology.

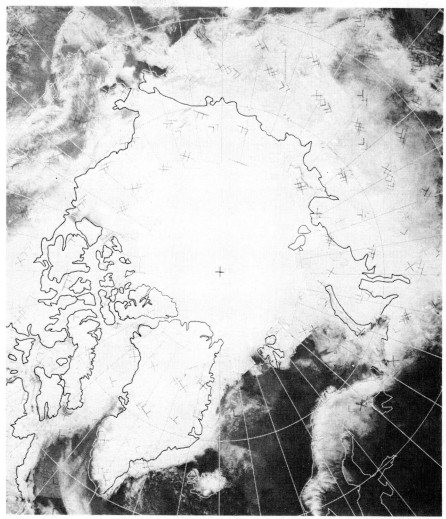

Figure 1.1
Composite minimum-brightness imagery of the north Polar regions *(above)*, and the south Polar regions *(opposite)*. Both images illustrate the early summer extent of the area covered by snow and ice. *(By permission of C.W.M. Swithinbank and the US Department of Commerce.)*

A wealth of evidence has recently become available from Greenland and the Antarctic as a result of the unprecedented research efforts directed at the world's two remaining ice sheets. Information is now also beginning to accumulate concerning the characteristics of the polar ice caps on Mars (Figure 1.3, p. 5), and some of this seems relevant to more worldly studies of cold-based ice sheets (Sharp, 1974). From North America and Scandinavia there is a mass of newly published evidence which has yet to be integrated into the main body of glacial geomorphology. Much of this evidence is concerned with the Pleistocene ice sheets, and it introduces a new dimension to the subject which is still dependent, often unsuspectingly, on ideas of Alpine glaciation

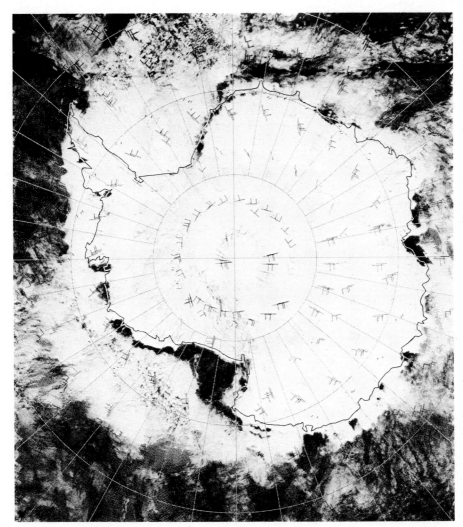

A brief report on glacial geomorphology in the 1960s and early 1970s could read:

1   overweight on chronology, thin on studies of process-form relationships;
2   thin on the links between glacial geomorphology and glaciological theory;
3   thin on the application of polar evidence to glacial problems.

This book attempts, in a limited way, to tackle these apparent shortcomings.

Glacier systems

In the following chapters there are many instances where complicated ideas are explained with the aid of a simple systems approach. This arises out of a firm belief in the value of a systems framework as a powerful explanatory tool. By focusing on links a systems approach is ideally suited to the elucidation of interrelationships between process and form. This in turn offers the opportunity of tackling such questions as why certain landforms have certain shapes and characteristics and why they occur in distinctive patterns. The approach is intended as an aid to explanation and should not be confused with explanation itself. Reality can be simplified and set within a systems framework in an attempt to demonstrate some of the complex relationships which operate in the real world (Chorley and Kennedy, 1971). It should not be expected that the systems and the linkages proposed in this book will be wholly adequate; no simplified model can be, and no lengthy analysis of a system can be expected to provide reliable or foolproof answers to problems of dynamics. But it is often helpful to use a simplified systems framework before going on to consider its inadequacy and eventually to replace it with a more complex and reliable framework. Hence systems are here used essentially as models for the demonstration of relationships in space and time.

There is a danger that the use of systems jargon can all too easily lead to confusion instead of enlightenment. In particular, it can become difficult to distinguish the genuinely useful

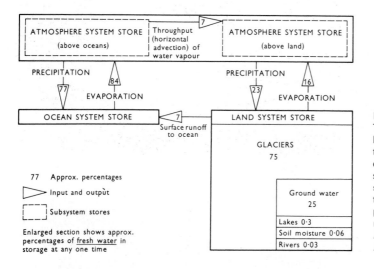

Figure 1.2
The major linkages in the hydrological cycle, showing the overwhelming importance of glaciers in the land system store. Note that the dimensions of the storage 'boxes' on the diagram are not proportional to their importance in reality. *(Nace, 1969, and Barry, 1969, Associated Book Publishers, London.)*

terms from those which are jargonese. This text is written with a deliberately restricted systems terminology and contains only those conceptual terms which are thought to be helpful for the understanding of glacial geomorphology.

There is also a danger in employing a purely mathematical approach to systems. In suitable situations mathematics provides the best means of expressing and working with complicated ideas. But there is more than an element of truth in the charge that in some geomorphological literature mathematics is used only to make a simple idea respectable. In the belief that mathematics has its main role as a problem solving tool and that it is not a substitute for insight and the ability to recognize a problem, equations are used sparingly throughout the book. However, it must be emphasized that the ideas which are discussed lean heavily on the results of a large number of mathematically based papers; many of these are referred to in the text.

There are two main types of glacier – the ice sheet or ice cap, and the valley glacier. Each can be recognized as a system having a set of internal characteristics in a distinctive combination. These characteristics or attributes are both morphological and functional or dynamic;

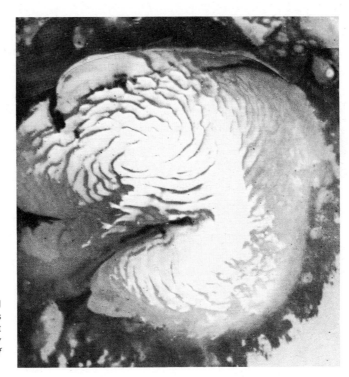

Figure 1.3
A photo mosaic of the residual north polar ice cap on Mars. Its maximum diameter is about 1000 km. *(Mariner 9 photography from 1972; by permission of NASA.)*

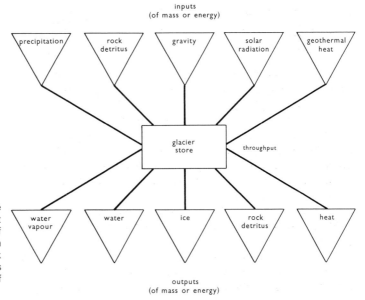

Figure 1.4
A simple model showing the relationships which exist between the various types of input and output within a glacier system. The link between input and output is provided by the throughput of material in the glacier store.

they can be measured and their relationships can be demonstrated. Seen as dynamic systems, glaciers are involved with the transfer of mass (ice, snow, water and rock detritus) in response to *input* of mass and energy (Miller, 1973). The movement of this material through the system can be termed *throughput* (Figure 1.4). As long as the material is held in the glacier, even if it is being transported slowly towards the snout, it can be said to be in *storage*.

The glacier system manifestly does not have a set of constant characteristics from the highest point in its collecting grounds to the lowest point reached by the snout. Part of the glacier receives net additions to its mass by input, and part experiences net loss of mass by

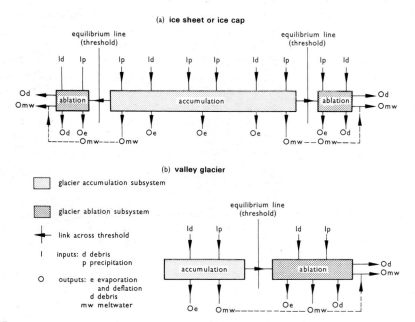

Figure 1.5
Models showing the input and ouput relationships on the two main types of glacier system. For convenience, inputs are shown on the top of each diagram and outputs on the bottom. Note that the ice sheet or ice cap system **(a)** may have its equilibrium line quite close to its periphery. In contrast, the valley glacier system **(b)** has its equilibrium line relatively further from the snout, separating accumulation and ablation subsystems of more equal dimensions.

output, so it is helpful to recognize two *subsystems* within the glacier, one called the accumulation subsystem and the other the ablation subsystem (Figure 1.5). They are separated by a conceptual line (the equilibrium line) which can be seen as a *threshold*. There are other clear thresholds surrounding the glacier system; on the bed there is a threshold between ice and bedrock, and on the glacier surface a threshold between ice and the atmosphere.

If the glacier experiences a change of input (usually as an increase or decrease in the amount of snow accumulating on the surface) it must respond accordingly. But because glaciers store material there is always a lag before the response becomes easily discernible at the snout. This is called the *response time*. The flow of ice through a glacier system varies according to how much material is being added or lost. Generally there is a tendency for some sort of equilibrium to be approached in which there is a balance between the rates of input, throughout and output. Hence, just as one can talk of glacier 'dynamics', one can refer to glacier 'economics'.

If the glacier is out of equilibrium, which it usually is on a short time scale, a *feedback mechanism* operates in order to restore equilibrium. Generally this is in the form of *negative feedback*, in which the effects of a change of input are damped down or eliminated. Sometimes, however, some instability in the system is exploited and the change is exaggerated or propagated. In such cases *positive feedback* can be recognized. This type of feedback is not common, but certain instances are discussed later in the book.

Such a systems framework can also be applied profitably to landforms created by glaciers. For example, certain landforms on a glacier bed can only be moulded effectively when a threshold is crossed which allows basal slipping to occur. The forms, especially if they are streamlined, may become relatively stable when they offer a minimum of resistance to the overrunning ice. Thus they can be regarded as achieving some sort of equilibrium. The response time describes the time taken for the stable form to be established. Negative feedback mechanisms may allow a glacier bed to change its shape constantly by localized deposition or erosion in response to fluctuating glacier conditions. In other circumstances, positive feedback mechanisms may cause a major change in the glacier bed to follow a relatively minor initial change. An example is the excavation of a rock basin beneath a glacier (King, 1970).

The awareness of scale

The use of a systems approach in glacier and landform studies is closely involved with problems of scale. The significance of scale in many different fields of study has been stressed by authors such as Schumm and Lichty (1965), Hanwell and Newson (1973), and Cooke and Warren (1973). These and other authors have attempted to provide logical and useful sets of terms to be used in the consideration of scale, for there is still much confusion about what is meant by such terms as 'small scale', 'large scale', 'short term', and 'long term'. Perhaps the most useful list of terms is that of Tricart (1965), modified for use in glacial geomorphology in Table 1.1 (overleaf). Wherever possible the terminology employed in the book conforms with usage in the table, although it should be realized that all the definitions are arbitrary.

In this book, scale is used as a major theme in the belief that the various types of glaciers and their effects must be considered in terms both of their areal extent and of the appropriate time scale over which they operate. The fundamental idea is that the most extensive features in space, such as *ice sheets* and land*scapes*, have also to be considered on the longest time scales. Smaller *valley glaciers* and land*forms* are best considered on shorter time scales. Only in this way is it possible to approach an understanding of the fundamental variables which are relevant and the way in which they are related to one another. Following the ideas of Schumm and Lichty (1965), these can be viewed as relatively *independent* variables, which are external to the system, or relatively *dependent* variables which are part of the system. The former can be referred to as *cause* variables, and the latter as *effect* variables. The status of any variable depends on the time scale employed, as illustrated in Table 1.2 (p. 9).

Often it is difficult to decide which variables really are the independent ones in a particular situation, and the feedback mechanisms which operate in respect of glacier systems are no less complex than those operating in fluvial systems. In general, however, the smaller the

Table 1.1 Definitions of different scales over space and time.
(Based partly upon the ideas of Tricart, 1965.)

Scale	Order	Units of earth's surface in km²	Characteristics, with glacial examples	Equivalent climatic units	Basic controlling mechanisms/processes	Time span of persistence
Macro- or large scale	1	10^7	Largest continents, ocean basins (No glacial examples yet)	Zonal systems, controlled by astronomic factors	Structural differentiation of earth's crust	10^9 or 10^8 years
Macro- or large scale	2	$>10^5$	Major ice sheets (East Antarctic; Laurentide; Scandinavian)	Broad climatic types	Major climatic or astronomic events; moutain-building	$>10^5$ years
Macro- or large scale	3	$>10^3$	Plateau and highland ice caps (alpine Pleistocene ice cap; Vatnajökull)	Subdivisions of broad climatic types	Interplay of long term climatic changes and relief effects	10^4 years
Meso- or medium scale	4 (Regional)	10^2	Highland ice fields, small ice caps (Jan Mayen ice cap)	Regional climates, influenced by regional upland topography etc.	Medium term climatic changes and relief effects	$<10^4$ years
Meso- or medium scale	5 (Local)	10	Valley glaciers (Blue glacier)	Meso climates, influenced by relief and aspect		10^3 years
Micro- or small scale	6	10^{-1}	Minor glacial features and landforms	Climate linked to land surface details	Short term climatic-changes and local relief. Processes linked with nature of glacier	10^2 years
Micro- or small scale	7	10^{-4}	Microforms	Microclimate, linked to the form	Glacial processes determined by nature of basal ice and bedrock	10^{-1} years
Micro- or small scale	8	10^{-6}	Microscopic forms	Micro-environment		10^{-3} years

Table 1.2 Glacier variables and their status during time spans of different durations. *(Modified after Schumm and Lichty, 1965.)*

Glacier variables	Status of variables during designated time spans		
	Long term ($>10^4$ years)	Medium term (10^4 and 10^3 years)	Short term ($<10^3$ years)
1 Time	Ind	Ind	Ind
2 Geology	Ind	Ind	Ind
3 Climate	(Ind)	Ind	Ind
4 Regional relief	Dep	Ind	Ind
5 Slope forms	Dep	Ind	Ind
6 Glacier morphology	Dep	(Ind)	Ind
7 Channel geometry		Dep	(Ind)
8 Sediment discharge			Dep
9 Ice discharge			Dep

Ind = Independent (cause) variables
Dep = Dependent (effect) variables

glacier system and the shorter its response time, the greater the number of variables which can be classified as independent. On the other hand, when a glacier is big and operating on a long time scale, the interdependence of a larger number of variables cannot be overlooked.

Models and hypotheses

Generalizations and models of various types are an essential part of this book. Again this is because of a conscious attempt to simplify reality as an aid to understanding and as a stimulus to critical thought. A major advantage of using models such as those in Figure 1.5 is that they are always inadequate and almost always wrong. The models in this book should be considered as working hypotheses. They will be criticized for being too simple or for omitting features which others would consider critical. This is right and proper. The reader will wish to think about models as representations of reality, and will proceed either to elaborate or reject them.

As the philosopher K. Popper (1972) has pointed out, science commonly advances by the creation of hypotheses which are, by their very nature, inadequate and available for falsification and rejection. The theories and models presented in this book are unlikely to survive prolonged scrutiny. They are used nevertheless in an attempt to encourage questioning and disbelief. At the same time there is a need to find reasonable and justifiable grounds for their rejection or modification. This latter point is a critical one, and it would be unacceptable to write a geomorphology text in which ideas and theory predominated to the extent that basic information was largely neglected. This information is vital as a foundation for argument and criticism. Each chapter of the book does attempt, therefore, to synthesize data from a wide range of glacial topics as well as proposing a large number of hypotheses and frameworks for study.

The layout of the book

Inevitably this book reflects the interests and experience of the authors. It is a geomorphology text written from a geographical viewpoint. However, it is also a book about glaciers, and there is a fundamental belief that glaciers need to be considered as forms in their own right. The morphologic and dynamic properties of glaciers have to be studied before their special appearance can be understood. Accordingly part I of the book is concerned with glaciers and glacier dynamics, including chapters on the properties of glacier ice, the manner in which ice movement is organized within the glacier system, and the relationship of both of these to glacier morphology. Part II is also concerned with glaciers *per se*, but here there is a discussion of how glaciers are distributed in space (chapter 5) and in time (chapters 6 and 7).

Parts III and IV are concerned with the erosional and depositional effects of glaciers. In glaciers, erosional and depositional processes often operate side by side, so it is somewhat artificial to treat them separately. Nevertheless, the convention of separate treatment is retained here as a means of simplifying reality and aiding understanding. In each part there is first a consideration of basic processes as far as they are known, with stress on both theoretical considerations and on deductions made on the basis of the interpretation of both small scale forms and sediments. The other chapters in each part are concerned with the relationships between these processes and forms. Landforms and landscapes are treated separately, according to the ideas of scale described above. For example, the landscape chapters (10 and 13) deal with large areas and long time scales.

Finally, part V of the book concentrates upon meltwater as a glacier subsystem. Meltwater merits special treatment because of its profound influence upon the characteristics of ice, upon ice movement, and upon the effects of ice on the landscape. In addition, it is capable itself of profound landscape modification. The treatment of the meltwater subsystem is a microcosm of the book as a whole. It begins with a consideration of the fundamental role of meltwater in glaciers and then goes on to discuss its erosional and depositional effects. However, as in the other parts of the book, there is no treatment of processes and landforms beyond the confines of the glacier.

SI-units

This book uses SI-units and those other units allowed to be used in conjunction with SI-units. There will perhaps be some problem in relating these expressions to those used in some of the older literature. In many places in the text the older equivalent is given alongside the SI-approved unit. In case of difficulty the reader is directed to a note: Anon. 1968: SI-units and glaciology. *J. Glaciol.* 7 (50), 151–3.

The main changes introduced by the SI system that are relevant to glacial geomorphology are:

1   The metre and the kilogramme replace the centimetre and gramme as basic units. Thus the density of ice, for example, is expressed in $Mg\ m^{-3}$ rather than in $g\ cm^{-3}$.

2   The unit of energy (including heat) is the joule and of power is the watt. These replace the various different calories.

3   Stresses and pressures are measured in multiples of newtons per square metre ($N\ m^{-2}$) or bars and not in $kg\ cm^{-2}$, since the kilogramme is a unit of mass, not force.

Table 1.3 lists some replaced units and their appropriate equivalents.

Table 1.3   Some examples of units which are not now used and their
permissible equivalents

Units not now used	Equivalent in SI or other allowed units
μ cal cm^{-2}s^{-1}	= $4 \cdot 1868 \times 10^{-2}$W m^{-2} = $41 \cdot 868$ mW m^{-2}
kg cm^{-2}	= $0 \cdot 98$ bar = $9 \cdot 8 \times 10^{4}$ N m^{-2}
calorie (IT)	= $4 \cdot 1868$ J
Langley	= $41 \cdot 868$ kJm^{-2}
g cm^{-2}	= 10 kg m^{-2}
g cm^{-3}	= Mg m^{-3}

In some figures in the text, and in paragraphs which refer specifically to a published paper using older units of measurement, the author's units are presented unchanged. Where possible a footnote draws attention to this fact.

Further reading

CHORLEY, R. J. and KENNEDY, B. A. 1971: *Physical geography: a systems approach.* Prentice-Hall, London (370 pp).

KING, C. A. M. 1970: Feedback relationships in geomorphology. *Geogr. Annlr* **52A** (3–4), 147–59.

MEIER, M. F. 1967: Why study glaciers? *Trans Am. Geophys. Un.* **48** (2), 798–802.

MILLER, M. M. 1973: Entropy and self-regulation of glaciers in Arctic and Alpine regions. In Fahey, B. D. and Thompson, R. D. (eds) Research in polar and alpine geomorphology, *3rd Guelph Symposium on geomorphology, 1973, Geo Abstracts,* Norwich, 136–58.

POPPER, K. 1972: *The logic of scientific discovery.* Hutchinson, London (480 pp).

SHARP, R. P. 1974: Ice on Mars. *J. Glaciol.* **13** (68), 173–86.

Part I
Glaciers and glacier dynamics

2 Glacier ice

A glacier consists of ice crystals, air, water and rock debris. Of these, ice crystals are obviously the fundamental component. This chapter is concerned with the characteristics and dynamic behaviour of glacier ice and, as such, it examines the basic independent variable relevant to all glaciers. The behaviour of glacier ice determines what is and what is not possible and hence defines the basic constraints of glacial geomorphology.

In the pressure and temperature ranges represented by glaciers, ice crystallizes on the hexagonal system. This structure is reflected to perfection in an undisturbed snowflake, sketches of which can be seen on the front cover of any issue of the *Journal of Glaciology*. It is beyond the scope of this book to discuss the structure of ice crystals, but two crystal characteristics are especially worthy of note. The first is that ice crystals are weak and can easily be made to slip on planes parallel to the basal plane. Fundamentally, it is this weakness which allows glaciers to deform readily under their own weight and thus to flow, although (as will be seen later) other processes are also involved. The second, and very unusual, characteristic is that water is a substance which is less dense in its solid form than in its liquid form. This means that liquid water may exist at the bottom of suitable glaciers and that in certain situations glaciers may float. The importance of this observation is easily appreciated if, for a moment, one tries to imagine a glacial geomorphology in which glacier ice is more dense than water!

Sources

Glacier ice is derived indirectly from the precipitation of snow or ice crystals from the atmosphere, or directly from liquid transformed to ice at the glacier surface. There is a wide variety of solid precipitation forms and even a simplified international classification lists ten basic types (International Association of Scientific Hydrology 1954)—usually the lower the temperature the smaller and simpler the crystal assemblage. Once on the surface, the characteristics of the snow vary considerably, according to the environment of deposition. In polar latitudes, for example, wind plays a crucial role and may transport snow for considerable distances before depositing it (Loewe, 1970b); in general, given a constant temperature, there is a correlation between wind speed and snow density, with higher wind speeds favouring higher densities. In mountainous areas avalanches may be an important means whereby fresh snow is moved from slopes on either side of a glacier to its surface (Vivian, 1975).

Water vapour which freezes on contact with the glacier surface forms several types of ice, the most important of which is rime. Rime ice is formed when supercooled water droplets strike a cold solid object and freeze on impact (Figure 2.1, overleaf). The ice is whitish in appearance as a result of entrapped air bubbles and is quite firmly attached to the receiving surfaces. Rime accumulates most rapidly in cool, humid conditions on surfaces which are most exposed to the wind (Mellor, 1964). Thus it is important in cool maritime glacial environments. For example, Koerner (1961) describes how cauliflower-shaped masses of rime

Figure 2.1
Rime ice on the summit of Snaefellsjökull, Iceland. *(Photograph by H. R. Bárðarson.)*

smother exposed rock summits in the Antarctic Peninsula area and build up to considerable sizes before breaking off and falling to the glacier below. On horizontal surfaces, the contribution of rime to the ice mass is less significant and in the Antarctic Peninsula area Koerner calculated that rime contributed only about 2·5 per cent of the total accumulation, while a lower figure of 0·1 per cent is mentioned by Rundle (1970).

Superimposed ice is formed when water comes into contact with a cold glacier surface and freezes (Figure 2.2). The water may be derived from rain or more commonly from meltwater from the summer melting of the previous winter's snow cover. Superimposed ice can only form when air temperatures are at or above freezing point and the ice is below freezing point. It is particularly important as an ice source in 'polar continental' areas like northern Canada and Arctic Siberia (Paterson, 1969; Grosval'd and Kotlyakov, 1969). Here winter

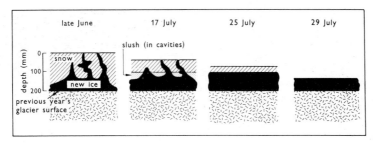

Figure 2.2
The growth of superimposed ice on Devon Island ice cap in June–July 1961. *(Koerner, J. Glaciol. 1970, by permission of the International Glaciological Society.)*

snowfall is relatively light and allows the winter cold to cool the ice effectively. In the following summer most of the snow is melted and, with the exception of meltwater flowing right off the glacier, refreezes as superimposed ice. As an example some 90 per cent of the ice of the Meighen ice cap is derived from superimposed ice (Koerner, 1968). In the Antarctic, the process occurs in zones subjected to summer melting, such as those near the continental periphery.

Figure 2.3
Layering is often apparent in firn and even at considerable depths in glacier ice (Gudmandsen, 1975). Here successive accumulations of firn are picked out by volcanic ash layers exposed in an ice wall at Grimsvötn, Vatnajökull. *(Photograph by Magnus Jóhannson.)*

Transformation

Assuming they are not removed in the following summer, snow and any associated ice layers accumulate and gradually undergo a change to glacier ice (Figure 2.3). The term *firn* is generally applied to snow which has survived a summer melt season and has begun this transformation. It consists of loosely consolidated, randomly orientated ice crystals with interconnecting air passages and a density generally greater than $0.4\ \mathrm{Mg\,m^{-3}}$. Transformation of firn to ice takes place by a variety of processes whose overall effect is to increase the crystal

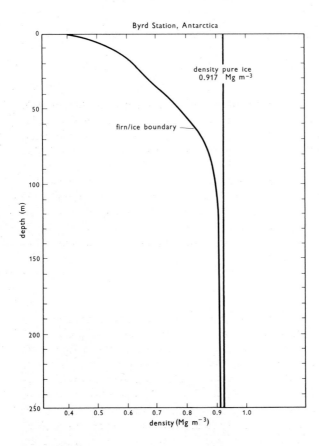

Figure 2.4
The increase in firn/ice density with depth at Byrd Station, Antarctica. *(Mellor, US Army CRREL Res. Rept 1964.)*

size and eliminate the air passages (Paterson, 1969). Eventually, crystals may surpass a football in size. When consolidation has proceeded sufficiently to isolate the air into separate bubbles the firn becomes *glacier ice*. This change to ice takes place at densities of between 0.8 and $0.85\ \mathrm{Mg\,m^{-3}}$. Further compression of the air bubbles increases the density of ice until it approaches the 'pure' value of around $0.9\ \mathrm{Mg\,m^{-3}}$ (Figure 2.4).

As in other solids these processes associated with the transition from firn to glacier ice operate most freely close to the melting point of the crystals; there is a close correlation between the temperature of the firn and the rate of crystal growth and therefore transformation to ice (Gow, 1969). This means that under cold conditions the firn/glacier ice boundary

Table 2.1
The thickness of the firn layer and the time taken for firn to be transformed to ice at four cold stations in Antarctica and Greenland. The colder the climate the longer the transformation takes. *(Gow, 1971).*

Location			Firn temperature °C	Depth firn/ice transition (m)	Time taken for transition to ice (yr)
Plateau Inge	79°15′S	40°30′E	−57	160	3500
Lehmann	77°57′N	39°11′W	−30	60	400
Byrd Camp	79°59′S	120°01′W	−28	65	200
Century	77°11′N	61°10′W	−24	68	125

will be deep and the rate of increase of crystal size and density with depth will be less than in temperate areas. Table 2.1 lists some data for localities on the Antarctic and Greenland ice sheets and at the coldest site (Plateau Station), it can be seen that the transition from firn to glacier ice is estimated to take place at a depth of 160 m and to take some 3,500 years, while at the warmer site of Camp Century the figures are 68 m and 125 years. At temperatures of about 0°C, meltwater derived from melting snow causes a dramatic increase in the rate of transformation from snow to ice. Under these circumstances firn is transformed into ice in a few years within a depth of a few metres. For example on Seward glacier in Alaska it takes 3–5 years and is accomplished within a depth of about 13 m (Sharp, 1951). In extreme cases where the entire snow cover is converted to superimposed ice, the conversion takes place close to the surface within a year.

Temperature characteristics

The behaviour of glacier ice is intimately related to its temperature. For this reason it is essential to consider how temperatures vary in glacier ice. Ice temperatures are influenced by heat derived from three sources – the surface, the base (geothermal heat flux) and internal friction. Heat from the surface may be derived from the direct incorporation of firn with a certain temperature, from conduction (direct influence of air temperature and radiation) and from the transfer of latent heat by refreezing of water. Geothermal heat flows into the base of a glacier at an average of 59·9 mW m^{-2} which is enough to melt 6 mm of ice at its pressure melting point each year (Paterson, 1969). Although for convenience this is often regarded as a constant source of heat for all glaciers, there is considerable variation from place to place depending on geological conditions (Table 2.2, overleaf). For example, the geothermal heat flux beneath Icelandic glaciers is about twice as high as that beneath the Greenland ice sheet. As will be developed later, these spatial variations can be of critical importance in certain threshold situations. Frictional heat is derived from differential movement within and at the base of the ice, and a rate of ice movement of 20 m per year at the stresses common in glaciers produces as much heat as the average geothermal heat flux (Paterson, 1969). A rate of flow of 100 m per year produces 5 times as much heat.

Two fundamental types of ice occur in response to these heat sources – *warm* ice and *cold* ice. Cold ice is below the pressure melting point while warm ice is sufficiently close to its

Table 2.2　Variations in the geothermal heat flux associated with major geological features. *(Stacey, 1969.)*

Major geological feature	Geothermal heat flux Mean and standard deviation mW m^{-2}
Precambrian shields	38·5± 7·1
Palaeozoic orogenic areas	62·0±23·5
Post precambrian non-orogenic areas	64·5±15·3
Continental shelves	71·6±44·0
Mesozoic and Cenozoic orogenic areas	73·7±24·3
Ocean ridges	76·3±65·4
Cenozoic volcanic areas	90·5±19·3

pressure melting point to contain water. The term *pressure melting point* is used because the temperature at which water freezes diminishes under additional pressure (by a rate of *c.* 1°C for every 140 bars[1]). This means that the melting point of ice under the pressures at the base of a glacier may be slightly below 0°C. For example, at Byrd Station in the Antarctic where the ice thickness is 2,164 m, the pressure melting point at the base of the ice sheet is −1·6°C, and represents a pressure of approximately 200 bars (Gow, 1970). As a result of the work of Ahlmann (1948) around the North Atlantic, these two types of ice have been termed 'polar' and 'temperate' in the past, but the terms can be confusing since both types of ice occur in polar and temperate regions. Also whole glaciers have been classified as temperate and polar and this is now thought to be unsatisfactory since both types of ice commonly occur in individual glaciers in polar and temperate areas.

Cold ice forms in two main situations. The first is where the firn accumulates at temperatures so low that there is little or no surface melting in summer. In such areas the temperature of the firn below the level of seasonal temperature variations is a close approximation to the mean annual air temperature at that site. Thus over most of the Antarctic the mean annual temperature can be obtained simply by digging a hole and measuring the snow temperature at a depth of *c.* 10 m. As with most sedimentary deposits on the earth's surface, the temperature of a glacier formed from cold firn increases with depth due to geothermal heat. This is clearly seen in Figure 2.5 which shows the temperature curves for boreholes at Byrd Station and Camp Century. There is a slight inversion near the top which is due to the presence of colder layers of firn and ice. These cold layers relate to the effect of movement within the ice sheet which brings firn and ice from colder areas at higher altitudes (Robin, 1955), and to the effect of climatic change, the upper layers representing a cooler period during the Middle Ages.

The rate of the temperature increase with depth is strongly influenced by the rate of accumulation of firn. Since the firn subsides into the ice sheet, heavy annual accumulation carries cold into the ice sheet more effectively than light accumulation. Heavy accumulation is thus better able to counteract the effect of geothermal heating and as a result the temperature gradient is decreased (Robin, 1955) (Figure 2.6). For example, assuming negligible horizontal flow, a 3,000 m thick ice sheet with an annual accumulation of 40 mm water equivalent should have a basal temperature *c.* 18°C warmer than one with three times that

[1] 1 bar=10^{5}N m^{-2} (1·02 kg cm^{-2})

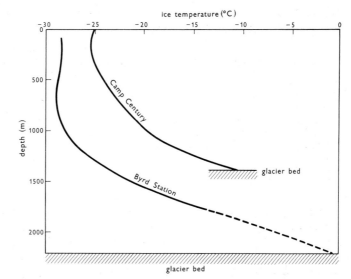

ice temperature (°C)

Figure 2.5
The rise in temperature with depth in cold ice, as illustrated by measurements from ice cores in the Greenland and Antarctic ice sheets. *(Ueda and Garfield, Antarct. J. US 1968; Hansen and Langway, Antarct. J. US 1966.)*

accumulation. It is important to note the assumption of zero horizontal flow. The numerical values of the model as given are thus likely to be realistic only near the centre of an ice sheet.

The second situation where cold ice forms is related to cooling of surface layers of a glacier by winter cold. Surface layers to a depth of 5–20 m are generally cooled below 0°C unless autumn snow insulates the ice from winter cold. Cold ice as defined above occurs on the surface of all glaciers in winter. If a greater thickness of ice is cooled in winter than can be warmed in summer, then a cold layer near the surface may survive the following summer. On glaciers where the bulk of the ice is warm this may form a thin zone of cold ice at a

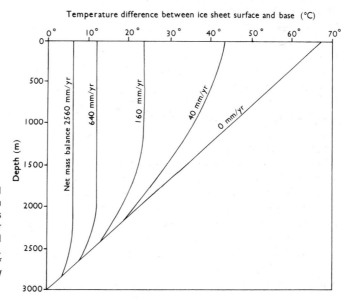

Temperature difference between ice sheet surface and base (°C)

Figure 2.6
The relationship of the vertical temperature distribution in a 3,000 m thick ice sheet to various values of accumulation (water equivalent). Negligible horizontal flow is assumed. *(Robin, J. Glaciol. 1955, by permission of the International Glaciological Society.)*

few metres depth throughout the summer, separating warm surface ice from warm ice below. Paterson (1972) suggests that this situation may be common on many glaciers previously regarded as consisting of warm ice.

Warm ice will form whenever there is sufficient heat to raise ice temperatures to the pressure melting point. Surface conditions commonly raise the firn and surface ice temperature to 0°C. Here, the most important process is the transfer of heat by surface meltwater, which penetrates the glacier and freezes if it comes into contact with ice below the freezing point (superimposed ice). Each gramme of meltwater which freezes, releases enough latent heat to raise the temperature of 160 grammes of ice by 1°C. On any glacier where firn is subjected to summer melting there is usually enough heat to bring the water saturated snow and ice to the melting point. This process is so efficient that a 13–15 m cold surface layer on Seward glacier was raised to 0°C in the first 10 days of summer melting in the years 1948 and 1949 (Sharp, 1951). Such a situation is common in the accumulation areas of glaciers where the firn easily absorbs any water produced by summer melting. In the ablation areas of some glaciers, however, where relatively impermeable ice exists at the surface for much of the melting season, meltwater may run off the glacier and not penetrate sufficiently to raise the ice temperature to 0°C. This is the reason why some subarctic glaciers, for example in Scandinavia and Svalbard, consist of warm ice in their accumulation zones and cold ice in their ablation zones (Paterson, 1972). A crucial feature of this process of warming the surface layers of a glacier is that it ceases to operate when the ice is raised to the pressure melting point. Thus it contributes to the warming of the ice but not to melting. Were this not so, warm glaciers would soon disappear!

Basal heat sources may be sufficient to raise the temperature of basal ice in an otherwise cold glacier to the pressure melting point as at Byrd Station. This is most likely to occur where some of the following conditions are met:

1   the ice is thick,
2   the surface temperature is relatively high,
3   ice velocities are high, producing frictional heat, and
4   accumulation is light or moderate.

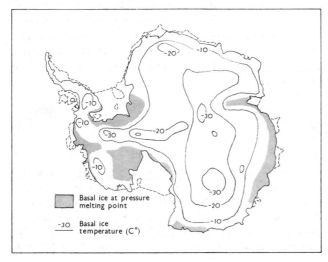

Figure 2.7
Basal ice temperatures beneath the Antarctic ice sheet. *(Budd, Jenssen and Radok, Polarforschung 1970.)* Recent discoveries suggest that this map underestimates the extent of basal melting.

These heat sources, in combination, have succeeded in raising the temperature of the basal layers of parts of the Antarctic and Greenland ice sheets to the pressure melting point. Figure 2.7 is a reconstruction of the basal temperatures of the Antarctic ice sheet. At a meeting of the International Glaciological Society in Cambridge in 1974, it was suggested that new information on ice depths and mass balance of the ice sheet implies that far greater areas are at the pressure melting point than indicated in the map. A map showing basal ice at the pressure melting point beneath western parts of the Greenland ice sheet has been compiled by D. Jenssen and is reproduced by Sugden (1974).

A point of crucial significance to glacial geomorphology is that the presence of basal ice at the pressure melting point means that water is likely to be present at the ice/rock interface. This is because, once the pressure melting point is attained, only a reduced amount of geothermal heat can be conducted away into the ice and the surplus is used to melt a thin layer of basal ice. When the temperature gradient within the lowest layers of the ice is zero or reversed because of the existence of a *layer* of warm ice, then all the geothermal heat is employed in basal melting.

Ice movement mechanisms

The reality of glacier flow is most effectively demonstrated by the use of time-lapse photography, which is unfortunately beyond the capabilities of this book. None the less, Figure 2.8 (overleaf) is an impressive demonstration of a glacier which is buckling lake ice as it advances. It is customary and helpful to separate the problem of glacier flow into two components – internal deformation within the ice mass and basal sliding (Figure 2.9, p. 25). Table 2.3 (overleaf) gives some of the few cases measured showing the relative importance of basal slip and internal deformation. Basal slip can account for 90 per cent of the total movement.

Internal deformation
A glacier flows because it deforms in response to stress (force per unit area) set up within its ice mass by the force of gravity. Any point within the glacier is subjected to a uniaxial compressive stress as a result of the weight of the overlying ice (overburden pressure). This stress can be envisaged as having two components – hydrostatic pressure and shear stresses. The hydrostatic pressure, which is the same in all directions at any one point, is related to the weight of overlying ice. Shear stresses, which tend to cause particles to slip past one another, are related to both the weight of overlying ice and the surface slope of the glacier. The shear stress of a point can be calculated from the equation $\tau = \rho gh \sin \alpha$

where τ is the shear stress
   ρ is the density of the ice
   g is the acceleration of gravity $\Big\}$ weight
   h is the thickness of the glacier
   α is the slope of the upper surface

The important principle to emerge from this is that shear stresses vary according to the thickness of the glacier and the surface slope, with high values favoured by thick ice or a steep surface slope. However the range of variation is relatively small and calculations for a wide variety of glaciers suggest that shear stresses at the base of glaciers commonly lie between 0·5 and 1·5 bars (Nye, 1952b). Thus glacier deformation is a response to shear stresses induced by a combination of the surface slope and ice thickness. Admittedly additional

Figure 2.8
The snout of the Storstrøm glacier buckling lake ice as it advances, Dronning Louise Land, Greenland. *(By permission of the Scott Polar Research Institute, British North Greenland Collection.)*

Table 2.3   The contribution of basal sliding to overall glacier flow. The measurements have been made from tunnels or boreholes. *(Paterson, 1969; Holdsworth and Bull, 1970.)*

Glacier	Country	Basal sliding (%)	Ice thickness (m)
Aletsch	Switzerland	50	137
Tuyuksu	USSR	65	52
Salmon	Canada	45	495
Athabasca	Canada	75	322
Athabasca	Canada	10	209
Blue	USA	9	26
Skautbre	Norway	9	50
Meserve*	Antarctica	0	80

* Cold-based glacier

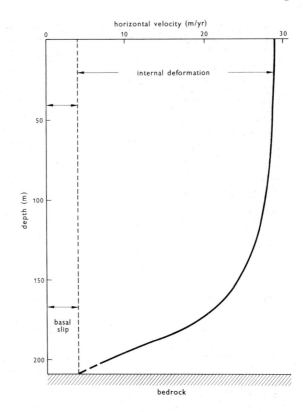

Figure 2.9
The relative contribution of internal deformation and basal slip to the forward movement of the Athabasca glacier, Canada. The glacier is warm-based. (*Savage and Paterson, J. geophys. Res. 1963.*)

stresses occur through irregularities in the flow from point to point along the long profile of a glacier, for example compression below an ice fall. However, such effects are much less important in influencing glacier flow.

Creep   The deformation of ice in response to stress is called *creep* and consists of the mutual displacement of ice crystals relative to each other. The processes involved are complex and not fully understood but include slippage within and between crystals. The rate of deformation or *strain rate* of polycrystalline ice (many crystals) has been studied in the laboratory, and the results form the basis of what is now termed Glen's law (Glen, 1955). Glen found that when a constant stress was applied to blocks of ice, the strain rate soon attained a steady value. Glen's flow law, adapted for glaciers by Nye (1957) can be written as $\dot{e} = A\tau^n$, where $\dot{e}$ represents the strain rate, A is a constant related to the temperature of the ice, τ is the effective shear stress and n is an exponent with a mean value of 3. Although the precise nature of the flow law of ice is currently the centre of much research (e.g. Colbeck and Evans, 1973; Reynaud, 1973; Thomas, 1973) it seems that Glen's law is a robust and useful model. Its importance is that it demonstrates that the strain rate is highly sensitive to the shear stress (Figure 2.10). When n = 3, it is proportional to the cube of the stress. This means for example that when the shear stress is doubled the strain rate increases 8 times. Also the strain rate is related to the temperature, though less closely. The warmer the ice the more

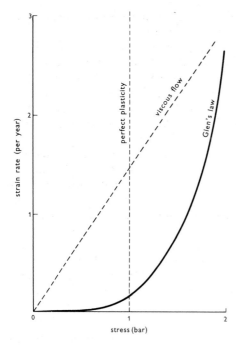

Figure 2.10
Various models of the relationship
between stress and strain rate.
Glen's law is contrasted with per-
fect plastic flow and viscous flow.
(Paterson, Pergamon, 1969.)

easily it deforms. For example, the strain rate at a temperature of $-22°C$ is one tenth of its value at $0°C$ (Paterson, 1969).

Several profound implications follow when Glen's law is applied to glaciers:

1   Most deformation within the ice mass arises at the base where stresses are highest. The movement of the surface layers is largely a result of being carried forward on the lower layers. This effect is even more marked in those glaciers where the lower layers are warmest, for example, a warm-based ice sheet.

2   On glaciers where there is no basal slip, Glen's law helps to explain how internal deformation takes place. On glaciers with some basal slip it explains an important component of the movement.

3   It helps to explain why warm glaciers flow more quickly than cold glaciers. For example, there is a marked contrast between rapid rates of creep in southern Greenland and sluggish rates of creep in the the north.

4   It helps to explain a mechanism of negative feedback whereby glaciers can regulate the discharge of ice. For example, an increase in thickness by a few per cent, such as may occur on valley glaciers in winter or on larger glaciers after a few years of climatic deterioration, will enhance basal shear stress and may cause an increase of velocity of 15–20 per cent, thus allowing the discharge of additional ice and the restoration of the original relationship. Conversely, thinning by excessive ablation decreases the ice velocity.

It is important to be aware of the difficulties involved when applying Glen's laboratory model to problems of glacier deformation. In the first place, the model can take no account of additional stresses in the glacier such as those associated with variations in longitudinal flow – for example, if the ice is being pushed from behind. Experiments suggest that such longitudinal stresses have the effect of increasing the strain rate and making the ice more easily deformed. Thus Glen (1956) found that a tunnel at the foot of an ice fall in Norway (in a zone of compression) closed more rapidly than predicted by the flow law. In the second place, the model applies to ice in which crystals are orientated randomly. However, in glaciers crystals tend to align themselves preferentially with basal planes parallel to the direction of flow and this can be expected to increase the plasticity of the ice. Gow (1970) demonstrated that this occurs at depth in the Antarctic ice sheet while Lliboutry (1970a) considered that the tendency will be accentuated in old ice, and that it contributes to the high rates of flow of some Greenland glaciers.

Glen's law is but one of several models of ice creep. It is perhaps useful to refer to another of fundamental importance which treats glacier ice as a perfect plastic. This model assumes that the ice does not deform until a critical stress or *yield stress* is attained. After this stress is reached the ice deforms indefinitely. Flow behaviour in such a model is again illustrated in Figure 2.10. In this, no deformation of the glacier is assumed until shear stresses of 1 bar are reached at the base. Then deformation near the base accounts for 100 per cent of ice movement. This model is a fair approximation to ice behaviour. Repeated calculations suggest that basal shear stresses do approximate to 1 bar. Also, as Glen's law shows, most movement is confined to the basal layers, especially in warm-based ice masses. Nye (1952a) found that the model of perfect plasticity can be used most effectively at a macro-scale, for example in predicting ice sheet profiles, ice thicknesses and surface slopes.

Fracture   Under certain conditions, ice creep cannot adjust sufficiently rapidly to the stresses within the ice and as a result the ice fractures and movement takes place along a plane. Tensional fractures, where the ice on either side of the fracture is separated, are exemplified by crevasses in the upper surface layers of glaciers. Shear fractures occur where thrusting takes place along a slip plane or fault. Such structures are most marked in thin ice near the glacier margins (Figure 2.11). At depth in a glacier where creep is more effective, clean

Figure 2.11
Shear structures in debris-carry-ing ice exposed in the margin of the Greenland ice sheet near Søndre Strømfjord. Although the bulk of the section comprises glacier ice, there may be virtually no ice in thin debris bands.

fractures are unlikely to occur. However, there is evidence of thin bands of small or highly orientated crystals reflecting localized zones of intense movement (Gow, 1970; Kamb and La Chapelle, 1964). The relative contribution of these various processes to glacier flow is not fully understood.

Basal sliding

The process of basal sliding has been discussed in a series of impressive papers over the last two decades (Weertman, 1957; 1964; Lliboutry, 1965; 1968a; 1969a; Kamb, 1970; Nye, 1973). Although there is still much to understand, three main processes have been recognized both in theory and in the field. One process, *enhanced basal creep*, occurs within the lowest layers of the glacier and can occur in ice at any temperature, although as would be expected it is more effective at higher temperatures. Two processes, *pressure melting* and *slippage over a water layer*, describe processes at the actual ice/bedrock interface. A point of particular importance is that these latter two processes can only occur when ice is at or close to the pressure melting point. This means that only warm-based glaciers slide at the ice/bedrock interface. Cold-based glaciers do not slide over bedrock for two main reasons. Firstly, the

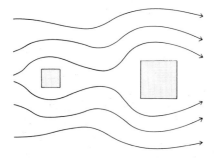

Figure 2.12
The mechanism of enhanced basal creep viewed in plan *(Weertman, J. Glaciol. 1957 by permission of the International Glaciological Society)*. The high stresses on the upstream side of obstacles cause high rates of ice deformation and allow basal ice to bypass the obstacle. The bigger the obstacle, the greater the deformation.

ice/rock bond is stronger than any bonds within the ice and thus any shear tends to take place within the ice (Tabor and Walker, 1970; Hope, Lister and Whitehouse, 1972). Secondly, there is no mechanism by which cold ice can circumvent the very small scale roughness present on any rock surface (Nye, 1970). Field observations agree with these conclusions in that tunnels excavated to the base of cold-based glaciers in Greenland and Antarctica revealed no movement at the ice/rock interface (Goldthwait, 1960; Holdsworth and Bull, 1970).

The process of *enhanced basal creep* was postulated by Weertman (1957) to explain how a glacier flows over or round large obstacles (greater than *c.* 1 m in length). If an obstacle such as a bedrock knob or a boulder lodged on the glacier bed protrudes into the bottom of a moving glacier the stresses on the upstream side increase and this increases the rate of strain of the ice. This allows the ice to flow round the obstacle (Figure 2.12). The velocity increases with the obstacle size. It is important to note that this mechanism does not necessitate sliding between ice and rock. Thus it is the main form of flow near the glacier base where the ice is below the pressure melting point. Furthermore, it determines the direction of flow of ice and any debris just above the base of all glaciers.

The *pressure melting* mechanism occurs when ice moves across a series of small bumps. The ice melts and freezes according to minor differences in pressure caused by the obstacles. Ice melts under the influence of the high pressures on the upstream side of the obstacles. The meltwater produced then flows round the bump to the downstream side where the pressure is lower and it refreezes to form regelation ice (Figure 2.13). The process operates best when the latent heat released by freezing can be transferred from the downstream side of the obstacle through the rock and assists in further melting on the upstream side. Thus it is most effective when obstacles are less than 1 m in size (Weertman, 1957). However, it is important that this figure of 1 m is not taken too literally to mean that the process does not operate on obstacles over 1 m in size. So long as the heat released by freezing can be dissipated, for example by meltwater flow in a real glacier, there is no reason to expect any upper limit. The existence of such a process of pressure melting has been confirmed by observation under active glaciers. Beneath the Blue glacier in Washington, Kamb and La Chapelle (1964) noted a layer of clear ice with few bubbles which was c. 30 mm thick over hollows and non-existent on the tops of bumps. The lowest 18 m of the 2,164 m Byrd core in the Antarctic ice sheet revealed similar ice containing dirt and few air bubbles (Gow, 1970). In both cases analysis suggested an origin by refreezing.

An important notion follows from the recognition of these two processes of sliding. Enhanced basal creep is an efficient process for large obstacles but inefficient for small ones,

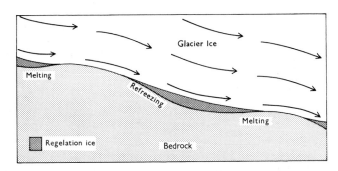

Figure 2.13
Pressure melting mechanism. Melting takes place on the upstream sides of obstacles and regelation on the downstream sides. *(Kamb, Science, 1964.)*

while pressure melting is most efficient on small obstacles. Hence there is a 'critical obstacle size' where both processes are inefficient. This size is in the range of 0·1–1 m (Nye, 1974) and offers the greatest resistance to sliding. In other words it is mainly bedrock roughness at this scale which prevents a warm-based glacier from flowing more rapidly.

The third process is *slippage over a water layer*. Weertman (1964) considered how a water layer might affect basal slip and concluded that a layer of water only a few millimetres thick could increase sliding velocity by 40–100 per cent. This is because the water lifts the ice and decreases the area of friction between rock and ice. The prominences that rise above the water layer are subjected to greater stresses and ice deforms round them more easily. Lliboutry (1965; 1968a; 1970b) stressed the role of water filled cavities between ice and bedrock. Although these cavities occur in the lee of obstacles due to ice movement, he considered that water may partly control their size and areal extent. Basal water pressure is regarded as an independent variable related to the amount of water made available by, for example, melting and rainfall. When there is sufficient water with a hydrostatic pressure greater than that of the ice, (as will be the case if the head of water extends to the glacier surface or sufficiently far upglacier), the cavities will grow (Figure 2.14, overleaf). A positive feedback mechanism may then come into play with larger, water filled cavities leading to higher sliding velocities which then lead to still larger cavities.

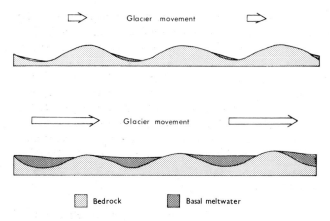

Figure 2.14
Diagram to illustrate how the growth of water-filled cavities reduces the area of contact between the glacier and its bed and thus reduces friction.

In all these various theories the rock bed is assumed to be impermeable. If it is not, then part of this water may be absorbed into the rock and this will reduce the sliding velocity of the glacier, as discussed by Weertman (1966) and Boulton (1972a).

Although the details of the mechanism of sliding over water remain to be ascertained, there is abundant field evidence to suggest that cavities and water exist at the base of glaciers near the pressure melting point and that they can affect the rate of basal slip. Cavities in the lee of obstacles on the glacier bed have often been reported (Carol, 1947; Kamb and La Chapelle, 1964; Vivian and Bocquet, 1973). That water exists in cavities in ice has been experienced by Fisher (1963), and Colbeck and Gow (1974), for example, and notably by an Alaskan subglacial miner who was unfortunately washed out of his adit (Miller, 1952). The presence of a basal water layer at the bottom of the Byrd ice core was discovered when water surged up the borehole (Gow, 1970). That basal water pressure fluctuates independently of ice pressures has been ascertained by Mathews (1964) and Vivian (1970). Furthermore, it has long been known that there is a relationship between high summer rates of glacier flow and the existence of basal meltwater, at least in the ablation areas of glaciers.

There is a growing body of evidence to suggest that basal slip takes place in a jerky stick-slip fashion. Not only is this conclusion based on detailed surface measurements of glaciers (McSaveney and Gage, 1968; Theakstone, 1967; Goldthwait, 1973), and on the interpretation of seismic events (VanWormer and Berg, 1973), but it has been observed in the glaciological laboratory set up in a cavity beneath Glacier d'Argentière in the French Alps. E. Derbyshire, who has crawled into cavities beneath some Icelandic glaciers, has described lines of regelation ice spicules which form in the immediate lee of obstacles after each 'slip' phase. They form at intervals of 20 mm and the record of past slips is visible on the glacier bottom which forms the roof of the cavity. The precise mechanisms responsible for this stick-slip motion remain to be investigated.

The role of other materials in glacier ice

Apart from ice, typical glaciers contain rock debris, air, atmospheric solid particles, and water. The latter has already been touched upon in so far as it is intimately concerned with the behaviour of ice. But there are a few points to make about the characteristics of the other components and how they affect glacier ice and glacier movement.

Rock debris is one of the most important of these and may be derived from beyond the margins of the glacier or from below. Generally volumes are small compared to the ice, and Andrews (1972), for example, estimated that rock debris accounts for only approximately 0·05 per cent of the total glacier volume in the ablation zone of the Barnes ice cap. But the proportion of rock can rise much higher (Figure 2.11). Boulton (1972b) estimated that debris carried in cold ice caps may comprise 5 per cent of the volume of the basal ice layers, whereas Holdsworth (1973) gave an estimate of c. 8 per cent for visibly dirty ice in the Barnes ice cap. In a 100–200 mm layer at the base of warm-based glaciers the volume of debris may rise to 55 per cent of the total (Boulton, 1975). Occasionally catastrophic events may make a large contribution to much wider areas of a glacier. For example, landslides triggered by the Alaskan earthquake of 1964 covered large areas of several glaciers with sheets of rock debris (Post, 1967). On Sherman glacier the debris covered an area of 8·5 km² and was c. 5 m thick (Figure 2.15, overleaf).

Rock debris can affect glacier flow in several ways. A sudden increase in glacier weight such as occurred in the rock falls mentioned can be expected to increase the rate of glacier flow locally. However, an ice/rock mixture, containing undeformable particles is usually less deformable than pure glacier ice. According to Lliboutry (1971), this is always true when rock particles are in contact with each other, but may not hold at very high stresses for ice containing scattered mineral particles. Increasing rigidity of dirty ice has long been suspected by the way in which active, debris-free ice tends to bypass and shear over debris-ridden ice, for example in overthrusting situations near glacier snouts, and in situations where a lateral moraine may be isolated to form an ice-cored debris ridge. It has been confirmed by laboratory experiments on the folding of medial moraines under ice compression (Ramberg, 1964), and in the field Dort (1967) noted that sandy layers within the ice were unable to adjust by creep to plastic deformation of the surrounding ice layers and that they fractured instead. As will be seen later, this decreasing plasticity of rock bearing ice is an important consideration when discussing subglacial and marginal deposition.

The air bubble content of a glacier has been briefly mentioned. Once compressed into bubbles the air remains under pressure and largely isolated from other adjacent air. The reality of the pressure within the bubbles is demonstrated if a piece of glacier ice (preferably from deep down) is melted, causing minor explosions reminiscent of those which occur in plates of a certain well known breakfast cereal. The snapping, crackling and popping may be heard along the shores of Greenland fjords which contain calving glaciers as brash ice fragments gradually melt.

Atmospheric debris can include both dust and salts. Dust may affect the albedo or surface reflectivity of a glacier, and thus influence firn accumulation and ablation, especially if the layer is sufficiently thick. It can be important where quantities are large, for example in the vicinity of active volcanoes as in Iceland. Salts are always present in glaciers. They can play a minor role by keeping the pressure melting point of glacier ice slightly below that of pure ice (Lliboutry, 1971; Harrison, 1972; 1975), and they may ease the intergranular movement of ice crystals in creep. As Lliboutry (1971) noted, much remains to be discovered about the role of these other component materials in influencing the behaviour of glacier ice.

Figure 2.15
Before *(above)* and after *(opposite)* photographs of a rock slide avalanche which discharged debris onto Sherman glacier, Alaska, in 1964. *(Photographs by Austin Post, US Geological Survey.)*

Further reading

ENGELHARDT, H. F., HARRISON, W. D. and KAMB, B. 1978: Basal sliding and conditions at the glacier bed as revealed by bore-hole photography. *Journal of Glaciol.* **20** (84), 469–508.

GLEN, J. W., ADIE, R. J. and JOHNSON, D. M. 1976: *International symposium on the thermal regime of glaciers and ice sheets, Burnaby, 8–11 April, 1975. J. Glaciol.* **16** (74), (316 pp). Contains 20 papers and discussion.

GOW, A. J. 1970: Preliminary results of studies of ice cores from the 2164 m deep drill-hole, Byrd Station, Antarctica. In Gow, A. J. *et al.* (eds) *International Symposium on Antarctic Glaciological Exploration (ISAGE), Int. Ass. scient. Hydrol.* **86,** 78–90.

HAMBREY, M. J. and MÜLLER, F. 1978: Structures and ice deformation in the White Glacier, Axel Heiberg Island, Northwest Territories, Canada. *J. Glaciol.*, **20,** 41–66.

IKEN, A. 1981: The effect of the subglacial water pressure on the sliding velocity of a glacier in an idealized numerical model. *J. Glaciol.* **27** (97), 407–21.

LACHAPELLE, E. R. 1973: *Field guide to snow crystals.* Univ. of Washington Press, Seattle and London (101 pp).

LLIBOUTRY, L. A. 1970: Current trends in glaciology, *Earth-Science Reviews* **6,** 141–67.

PATERSON, W. S. B. 1981: *The physics of glaciers.* Pergamon, Oxford (380 pp). Second edition.

POST, A. and LACHAPELLE, E. R. 1971: *Glacier ice.* Univ. Washington Press, Seattle (110 pp).

ROBIN, G. DE Q. 1976: Is the basal ice of a temperate glacier at the pressure melting point? *J. Glaciol.* **16** (74), 183–96.

SOUCHEZ, R. A. and LORRAIN, R. D. 1978: Origin of the basal ice layer from Alpine glaciers indicated by its chemistry. *J. Glaciol.* **20** (83), 319–28.

WEERTMAN, J. 1957: On the sliding of glaciers. *J. Glaciol.* **3** (21), 33–8.

3 Glacier systems

This chapter views glaciers as whole systems and stresses the links between the various functional components of the glacier. Beginning with some basic considerations of input and output, the chapter goes on to discuss the relationship between these and glacier movement. There then follows a discussion of the main features of movement within the glacier system and the role of the principal variables involved. If all glaciers gained and lost an equal amount of snow over the whole of their surface area and if they also flowed on plane surfaces, then there would be little need for this chapter; under such circumstances the factors considered in chapter 2 would adequately explain glacier behaviour. However, the significance of the points to be considered is a reflection of the importance of spatial variations in input/output relationships on and between glaciers, and of the irregularity of the bedrock bases on which glaciers form and flow.

Input and output

Mass balance
The mass balance of a glacier describes the input/output relationships of ice, firn and snow, and is usually measured in water equivalent (i.e. the amount of water involved if melted). Accumulation (input) includes all ways in which mass is added to a glacier; direct precipitation is the most widespread means of accumulation, but avalanching, and the growth of superimposed ice are other ways in which mass is added. Ablation (output) includes all ways in which mass is lost; surface melting is perhaps the most widespread means of ablation, but internal and basal melting, evaporation, wind deflation and calving are other means of reducing a glacier's mass which may be locally very important (Figure 3.1, overleaf). Mass balance relationships in the real world are highly complicated and for a fuller discussion the reader is directed to articles in the *Journal of Glaciology* (Anon., 1969; Mayo, Meier and Tangborn, 1972). Table 3.1 (p. 37) attempts to depict the relationships during one year. Even though it is restricted to surface conditions on a highly idealized glacier, it serves to bring out the complex changes from time to time and place to place on a glacier.

The difference between accumulation and ablation for a whole glacier over one year is called the *net balance*. Strictly this refers to a *balance year*, which is the interval between the time of minimum mass in one calendar year and the time of minimum mass in the following year (Figure 3.2, p. 37). There are several ways of calculating the net balance of a glacier. A method which is convenient in the field is to calculate a winter balance and a summer balance for the glacier as a whole. The winter balance is positive and the summer balance negative and their difference gives the net balance. A positive net balance implies that the glacier as a whole has gained snow and ice, while a negative net balance describes an overall loss of snow and ice. A zero net balance implies that the winter and summer balances are equal. Figure 3.3 (p. 38) shows data for Storglaciären in Sweden and represents the most complete series of mass balance measurements for any glacier. The net balance has been

Figure 3.1
The snout of a glacier at the head of Phillips Bay, Ellesmere Island. Most ablation is achieved by calving, but most icebergs are trapped by sea ice and are not dispersed until the summer. The occurrence of radial crevasses at the point where the glacier spreads out after leaving the valley influences iceberg size. *(Original photograph supplied by National Air Photo Library, Canadian Department of Energy, Mines and Resources.)*

positive for only five balance years in the period 1945–73, zero for one balance year and negative for the remaining balance years.

There is a wide variation in the net balance from place to place on a glacier and, as mentioned in chapter 1 it is helpful to subdivide the glacier into two subsystems. The accumulation subsystem includes all those areas where there is an excess of accumulation over ablation in a balance year, while the ablation subsystem simply describes those areas where ablation exceeds accumulation in a balance year. The boundary between the two is the equilibrium line and occurs where ablation equals accumulation in a balance year. It is important to

Table 3.1 Mass balance relationships on a medium sized idealized glacier during one year. *(Adapted from Mayo, Meier and Tangborn, 1972.)*

Season	Spatial variation	Mass balance characteristics
Autumn	Snow accumulating at higher altitudes. Ablation of ice continues at low altitudes.	Snow mass increasing Ice mass decreasing Total mass constant
Winter	Snow accumulating over whole glacier. Little ablation.	Snow mass increasing Ice mass constant Total mass increasing
Spring	Snow acumulating at higher altitudes. Ablation of winter snow at low altitudes.	Snow mass constant Ice mass constant Total mass consant
Summer	Little snow accumulation except at high altitudes. Ablation over much of glacier (snow at higher altitudes, firn and ice at lower altitudes).	Snow mass decreasing Ice mass decreasing Total mass decreasing

stress that these are highly simplified subdivisions. In reality there is a continuum of both winter accumulation and summer ablation along the length of the glacier as a whole. For example, on a valley glacier ablation may affect the whole glacier in summer but the amount involved decreases from a maximum at the snout to a minimum at the head of the glacier. Similarly winter accumulation may involve the whole glacier, the amount falling progressively from glacier head to glacier snout. It is simply the fact that accumulation and ablation balance at one position that allows the equilibrium line and thus the subsystems to be recognized.

Relationship to climate
Understanding of the mass balance of a glacier clearly demands appreciation of the relationship between the glacier and its climatic environment. The climate in the accumulation zone determines the amount of snow or ice available for net accumulation. But, important as this may be, the factors controlling ablation are just as important, because it is the preservation of the snow once fallen that is crucial to the mass balance. This can be illustrated by the observation that the world's greatest mass of ice, the Antarctic ice sheet, thrives in an area of extremely low precipitation. It is the preservation of the small amounts of snow which do fall that explains this apparent paradox.

Figure 3.2
Diagram to illustrate how curves of total accumulation and total ablation define a balance year. The winter balance is positive and the summer balance negative. If their area on the graph is equal, then the net balance of the glacier is zero.

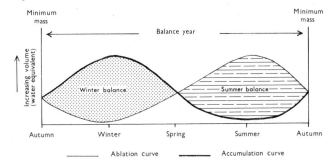

The main method by which climate influences ablation is melting. Although some ablation is achieved directly by evaporation, in general this figure is low; Paterson (1969) gave an average figure of 5 per cent for the amount of ablation due to evaporation and this figure tends to become significantly higher only where the total amounts involved are small, for example in a cold Antarctic continental climate (Bull and Carnein, 1970) and in high tropical mountains (Lliboutry, 1964). Surface melting occurs when an ice surface already at 0°C receives further heat (Loewe, 1970a). (If the ice surface is below 0°C any heat will be used simply to raise the temperature towards 0°C.) The two most important conditions for ice surface melting involve exposure to radiation and heat exchange with the air in contact

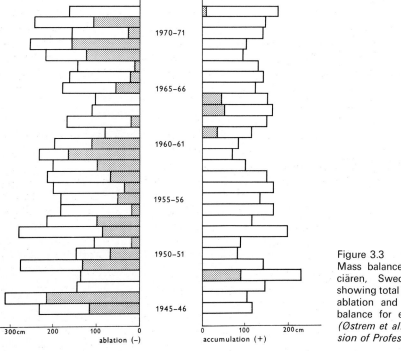

300 cm   200   100   0
ablation (−)

0   100   200 cm
accumulation (+)

1970–71

1965–66

1960–61

1955–56

1950–51

1945–46

Figure 3.3
Mass balance data for Storgla-ciären, Sweden for 1945–73 showing total accumulation, total ablation and (shaded) the net balance for each balance year. *(Østrem et al., 1973, by permission of Professor Valter Schytt.)*

with the glacier. Radiation involves short-wave solar radiation as well as long-wave radiation from water vapour and carbon dioxide in the atmosphere and their relative roles are discussed by Paterson (1969). The efficacy of solar radiation in melting is greatly influenced by the albedo of the glacier surface (Schytt, 1967). If the surface is covered with fresh snow, as is common in early summer, then the albedo is typically 0·6 to 0·9 and most solar radiation is reflected back. If the surface is glacier ice, as is common later in the season, the albedo is 0·2 to 0·4 and melting due to solar radiation becomes very significant. Heat exchange with the air at the glacier surface takes place in two main ways, by conduction of heat from the air to the ice and by condensation of water vapour on the ice surface which results in the release of latent heat. Condensation of 1 g of water vapour on the surface releases enough heat to melt about 7·5 g of ice (Paterson, 1969). Conduction and condensation are both enhanced dramatically when windy conditions cause air turbulence close to the glacier surface. They are also affected by the humidity of the air. Rain is comparatively unimportant as a cause of melting (Sharp, 1960).

The relative importance of the various processes varies from glacier to glacier. Paterson (1969) tabulated the results of 32 sets of measurements and concludes that radiation is the most important heat source, usually supplying over half the heat. In general it seems that radiation is most important in continental climates (Lang, 1968) where it may account for over 80 per cent of the total, while conduction and condensation are relatively more important in humid maritime climates (Schytt, 1967; Pytte, 1970; Gudmundsson and Sigbjarnarson, 1972). As a broad generalization, Schytt (1967) found that there was a good correlation between mean summer temperature and total ablation.

Amounts of ablation vary considerably from glacier to glacier and time to time. Fast rates of ablation are well known to many mountaineers whose tents in the ablation areas of glaciers may rest precariously on pedestals of ice after only a few days. Vivian (1970) notes mean rates of 8 mm per hour in summer in the French Alps, while about 12 m of ice are melted from the snout regions of some Norwegian, Icelandic and Alaskan glaciers in one ablation season (Embleton and King, 1975).

Effect on glacier movement
Ultimately the movement of any glacier is dependent on the nature of the input and output to the system. Two characteristics are particularly important, firstly the *magnitude* of the input and output and, secondly, its *spatial distribution* on a glacier.

The magnitude of the input and output per unit area of a glacier has an obvious effect on ice flow. To illustrate this it is helpful to assume the existence of two glaciers of identical shape both of which are in equilibrium and thus neither advancing nor retreating. Glacier A has an average input per unit area above the equilibrium line of 50 mm water equivalent, whereas Glacier B has an average input of 5000 mm water equivalent. In order that the glaciers remain in equilibrium, Glacier B must transfer 100 times as much ice into the ablation zone as Glacier A. It follows that if the glacier cross-sections are similar, Glacier B will flow at an average velocity 100 times greater than Glacier A. Clearly such differences in the total input are critical and it is worth reflecting that they do exist in reality. An index of the magnitude of this turnover of ice can be gained by extrapolating from average conditions at the equilibrium line. Where the volume of winter snowfall is low, the volume of snow melted in the following summer at the equilibrium line will also be low. Conversely, where the winter accumulation is high, so also will subsequent seasonal melting at the equilibrium line be high. Building on the work of Ahlmann (1948) and Shumskii (1950), Andrews (1972a) has used this mass loss at the equilibrium line to differentiate between the activity of glaciers. Figure 3.4 (overleaf) shows for the northern hemisphere a map of the amount of energy required to melt the accumulation at the equilibrium line altitude. The greater the amount of energy required, the greater is the mass loss at the equilibrium line. This in turn reflects the magnitude of the turnover of ice and is thus one measure of the activity of a glacier system. As the map shows, there is a tendency for the magnitude of the turnover to decrease from maritime to continental climates and from temperate to high latitudes.

The *distribution* of the total amounts of accumulation and ablation on a glacier also affects the discharge of ice. On any glacier there is a tendency for net ablation to be greatest at

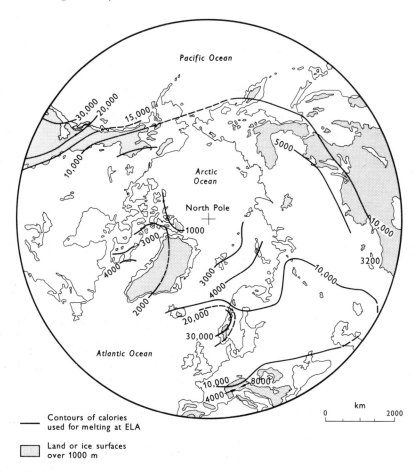

Figure 3.4
The amount of energy required to melt the accumulation at the equilibrium line altitude (ELA) in the northern hemisphere. This is an index of the magnitude of the turnover of ice and is thus one measure of the activity of a glacier system. (1 calorie=4·1868 J) *(Andrews, Z. Geomorph. 1972a.)*

the snout, falling to zero at the equilibrium line; from here as one proceeds above the equilibrium line, net accumulation will progressively rise above zero. As is shown on Figure 3.5, this tends to remove a 'wedge' of snow and ice from below the equilibrium line and add on a 'wedge' of snow and ice above it. If the surface of the glacier is to remain in equilibrium then the glacier must counteract this tendency and ice must be transferred from above to below the equilibrium line under the influence of gravity. In effect these wedges reflect the *net balance gradient* which is expressed as the increase of net balance in millimetres per metre of altitude. It is composed of the sum of the rate of increase in accumulation and the rate of decline of ablation with altitude. The higher the net balance gradient, the thicker the wedges on Figure 3.5 will be and, as a result, the more rapid the glacier flow will tend to be. It is this net balance gradient which Shumskii (1950) has termed the 'energy of glaciation'

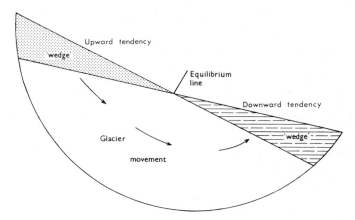

Figure 3.5
Idealized glacier with net accumulation 'wedge' above and net ablation 'wedge' below the equilibrium line.
Glacier flow is necessary in order to maintain an equilibrium surface profile.

and Meier (1961) the 'activity index'. The net balance gradient is steeper in maritime climates and decreases with increasing continentality (Table 3.2).

There is a correlation between the magnitude of total input and the net balance gradient. This can be demonstrated theoretically (Shumskii, 1950) and is observed in reality. As a generalization it can be said that both work together to make glaciers in humid areas faster flowing than in dry areas. Thus there is a latitudinal decline in activity from temperate to polar latitudes, while, superimposed on this general tendency, there is a decrease in activity from maritime to continental climates.

Glacier output is intimately and implicitly related to the input. In most land based situations, the snow and ice input has to be carried to an area with a sufficiently mild climate to dispose of it. The greater the magnitude of the input, the warmer the climate needs to be to melt the ice. This is why fast moving maritime glaciers extend their tongues into warmer climates than dry polar glaciers and may even be in dramatic contact with mature forest trees (Figure 3.6, overleaf). For the same reason, the temperature at the equilibrium line is higher in a maritime situation than in a continental situation. This may be demonstrated by the well known way in which calculated glaciation levels rise sharply with increasing distance from temperate west coasts, for example in North America and Europe (Østrem, 1966). Similar arguments explain why arid areas in dry polar latitudes such as Peary Land

Table 3.2 The tendency for the net balance gradient to decrease with increasing continentality (as expressed by magnitude of the winter balance) for three small glaciers of comparable size. *(Meier et al, 1971.)*

Glacier	Winter balance (m)	Net balance gradient (mm m^{-1})
Gulkana, interior Alaska	1·0	6
Wolverine, south central Alaska	2·5	9
South Cascade, Washington	3·1	17

Figure 3.6
The snout of an advancing glacier
in Patagonia, in contact with
mature pine forest vegetation.
(Photograph by J. H. Mercer.)

in northern Greenland can remain free of ice; although the summer warmth necessary for effective ablation is negligible, so also is the quantity of snow to be removed. These latter points are considered in more detail in chapter 5.

Movement within a glacier

Glacier movement can be envisaged as the dynamic link between the accumulation and ablation subsystems which are normally spatially distinct. Some of these movements are more or less continuous while others such as surges are periodic with brief periods of intense activity and intervening periods of quiescence. According to Miller (1973) these various types of flow may represent, in terms of the increased energy levels involved, a sequence akin to the flow stages in stream dynamics, namely laminar, streaming and shooting.

Continuous movement

Longitudinal dimension   There are important longitudinal variations in flow along the length of a glacier. Generally the discharge of ice in a glacier system is at a maximum at the equilibrium line and decreases downglacier from it. This is because the cumulative volume of ice in a glacier increases downglacier as far as the equilibrium line; beyond this there is net loss of ice and the volume must therefore decrease downglacier. It is sometimes suggested that this characteristic makes glaciers highly distinctive, when compared to rivers in humid environments where discharge tends to increase towards the mouth. However, the analogy must be treated with care. A river basin in a humid area is the equivalent of a glacier accumulation area, for the 'ablation' zone is replaced by the sea. In a comparable glacier situation, such as in the Antarctic area where the equilibrium line is near sea-level and most ablation is achieved by calving, glacier discharge also tends to increase towards the coast. A fairer river analogy is one in a zone where there is an upland source of water and where the river flows towards and terminates in a desert lowland. In such a case river discharge is at a maximum at the equivalent of the equilibrium line at the border of the mountains.

There is a significant component of vertical velocity which varies in relation to the accumulation and ablation zones in a glacier. Above the equilibrium line there is a downward velocity component as fresh accumulation builds up. Any object such as a stone or building will be buried. On the Antarctic ice sheet bases have to be periodically rebuilt on the surface as their predecessors sink to uneconomic depths and pressures. Below the equilibrium line, ablation removes the surface layers and as a result there is an upward component of movement. In this zone, for example, a stone buried beneath the surface will later emerge at the surface in spite of there being no change in the elevation of the ice surface. As a rule the greater the magnitude of the input and output of a glacier the more important the vertical component will be.

Working at a more specific level with regard to longitudinal flow along a glacier, Nye (1952b) recognized that variations must exist since the mass balance and bedrock conditions vary from place to place along the glacier. Nye showed that there are two main types of flow. Compressive flow describes a situation where the longitudinal stress is compressive throughout the depth of the glacier and, assuming a regular glacier width, is marked by a reduction in forward glacier velocity. Extending flow describes a situation where the longitudinal stress is more tensile than the overburden pressure. Under such circumstances the velocity increases downglacier. An analogy can be made with a convoy of heavy goods vehicles negotiating a gradient on a trunk road. Many car drivers will have noticed the compressive flow at the bottom of the gradient as the trucks slow down and pack closely together, and also, equally frustratingly, the acceleration and spreading which occurs as the trucks pass the brow of the hill under extending flow. Extending flow occurs where there is an addition of ice to the glacier surface, or where the bedrock slope beneath the glacier is convex. Both these conditions lead to an acceleration in glacier flow. Compressive flow occurs where there is a loss of ice at the glacier surface and where the glacier bed is concave. Such conditions favour a deceleration of glacier flow. These factors are best seen at two scales (Figure 3·7, overleaf). At the scale of a whole glacier there is a tendency for extending flow

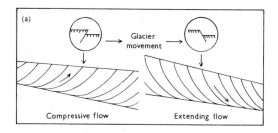

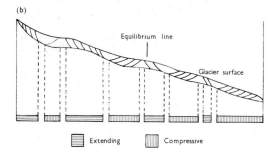

Figure 3.7
(a) Compressive and extending flow and associated slip lines. **(b)** The distribution of compressive and extending flow in an idealized glacier. *(Nye, J. Glaciol. 1952b by permission of the International Glaciological Society.)*

above the equilibrium line and compressive flow below, reflecting the accumulation/ablation pattern. Superimposed on this are local variations induced by topographic variations.

Certain types of ice movement are related to compressive and extending flow. Nye (1952b) and Nye and Martin (1968) showed 'slip lines' which define trajectories of maximum shear stresses. In extending flow these lines favour downward movement of ice, while in compressive flow they favour upward movement of ice. In both cases they are tangential to the bed but meet the surface at an angle of 45°. Figure 3.7a shows the direction of the main slip lines and also their surface expression, assuming the lines reflect actual faults within the ice. In practice movement may take place by deformation within the ice or by thrusting along pre-existing but favourably orientated lines of weakness.

Transverse dimension   Channel shape affects the amount and nature of the friction to be over-come by a moving glacier. At one extreme is *sheet flow*, where ice is unconfined in any conventional valley and is affected by friction only at the base. Such a situation is common over large areas of ice sheets and ice caps. Sheet flow may conform closely to models of ice flow such as these presented in Figure 2.11. At the other extreme is *stream flow* where a glacier flows in a confined rock valley (Figure 3.8). In these cases although overall velocity may be high there is friction not only underneath but also at the margins of the glacier. When calculating surface velocities or shear stresses in such confined channels, a shape factor can be introduced to take account of this lateral drag, and is used in conjunction with the flow law (Paterson, 1969, p. 107). On a typical valley glacier, basal shear stresses are reduced by a factor of about 0·7 (Robin and Weertman, 1973).

The main feature of flow within a channel is a zone of maximum flow in the centre at the surface and a rapid reduction in velocity towards the margins. Figure 3.9 (p. 46) shows the surface velocity profiles of glaciers of varying size. Velocity changes with depth are less well

Figure 3.8
Scott glacier, a 12 km wide outlet glacier from the east Antarctic ice sheet which cuts through the Transantarctic mountains in a rock trough. *(Copyright US Navy.)*

known. Nye (1965a), assuming ice of equal temperature, a flow law with n=3 and no basal
slip, calculated a variety of velocity gradients for channels of parabolic shape (Figure 3.10a).
The calculations explain the surface variations well. However, measurements on the warm-
based Athabasca glacier reveal markedly different basal conditions and, as a result, different
velocity profiles (Figure 3.10b). Here basal sliding is high in the centre and decreases rapidly
towards the margins. Also lines of constant velocity are semicircular in shape and significantly
different from the parabolic shape of the channel (Raymond, 1971). Raymond discussed
the two main factors which are thought to affect basal sliding – roughness of the bed and
a basal water layer – and concluded that the pattern reflects lateral variations in water
pressure underneath the glacier. As a working hypothesis, it is possible that the velocity profile
in Figure 3.10a represents a characteristic velocity distribution in cold-based ice streams,
while that in Figure 3.10b approximates to that in warm-based ice streams.

Glacier flow also involves lateral components of movement. These may be termed divergent or convergent flow, and they affect longitudinal velocities. If for some reason the ice

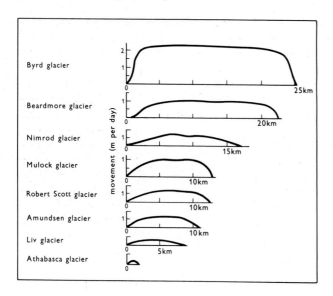

Figure 3.9
Surface velocity profiles of some
Antarctic outlet glaciers, *(Swith-
inbank, Geogr. J. 1964.)* and
Athabasca glacier. *(Paterson, J.
Glaciol. 1964 by permission of
the International Glaciological
Society.)*

is diverging then the longitudinal velocity is reduced. This conclusion may be verified visually
by the way in which, on a glacier spreading out into a piedmont lobe, regularly spaced
markers such as ogives are compressed together, and by the way in which medial moraines
are increasingly spread out and distorted towards the glacier front. Similarly, converging
ice flow will lead to higher longitudinal velocities. Thus glacier velocities increase at narrow
points of constriction in a valley glacier.

It is important to indicate the situations where converging and diverging flow are likely,

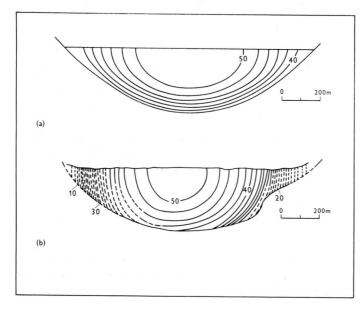

Figure 3.10
Distribution of longitudinal
velocity in a glacier channel,
(a) as calculated for a glacier
with no basal slip in a parabolic channel, *(Nye, J. Glaciol.
1965a by permission of the
International Glaciological
Society.)* and **(b)** as measured
in the Athabasca glacier,
Canada. *(Raymond, J. Glaciol.
1971 by permission of the
International Glaciological
Society.)* In **(b)** the units are
in m per year.

since the implications with regard to erosion and deposition are profound. Clearly the shape of the underlying topography is bound to divert ice to some extent. Evidence for this on a small scale comes from the interpretation of striations (Demorest, 1938) while, on a larger scale, hill massifs may play a similar role. The Cairngorm mountains in Scotland form an example where an overall ice movement towards the northeast was locally deflected round a massif rising 300–500 m above the surrounding upland (Sugden, 1968). Other less obvious situations are worth noting. Any ice cap or ice sheet has an inbuilt component of diverging flow since the ice flows radially outwards from the centre. Another noteworthy situation occurs below the equilibrium line on any glacier in a rock channel where radiation from the rock walls enhances the convexity of the surface cross-section of the glacier. Moving in response to this transverse ice gradient, there is a diverging component of ice movement. Figure 3.11, showing a cross-section of part of the Athabasca glacier, demonstrates this visually. Melting at the base and the surface causes radial movement in all transverse directions. Above the equilibrium line on a rock walled glacier, there is often an excess accumulation at the sides which produces a concave cross-profile and thus a convergent component of ice movement.

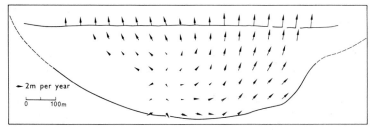

Figure 3.11
Transverse ice velocities in a cross-section of the Athabasca glacier, Canada, as measured in the ablation zone. *(Raymond, J. Glaciol. 1971 by permission of the International Glaciological Society.)*

Periodic movements
In the real world glaciers are subjected to short term variations in climate, superimposed on a variety of longer term fluctuations. Thus in order to approach an understanding of reality, it is vital to consider how glaciers respond to changes in mass balance. The subject is highly complex and is currently the centre of much research by glaciologists. This section tries to describe in qualitative terms the main issues involved.

Direct response   Variations in mass balance normally affect the whole glacier. Thus, if there is a climatic deterioration, every part of the glacier is likely to thicken immediately. This thickening will result from either an increased accumulation of firn over the whole glacier or from a reduction in ablation over the whole glacier, or both. Conversely, a climatic warming will lead to overall thinning over the whole glacier.

The response may be stable or unstable. Stable adjustment means that the glacier will thicken or thin slightly in response to change until a new equilibrium profile is reached. Suppose that a climatic deterioration means that more snow is accumulating than in the past. The glacier will quickly thicken, causing an increase in the cross-sectional area and an increase in velocity. When the adjustments are sufficient to be able to accommodate the additional snow, the surface will cease to rise and will stabilize at the new equilibrium profile.

Similar arguments apply to a reduction in snowfall. Unstable adjustment occurs when an initial change triggers off a change in the ice surface, which increases progressively with time. In this case an initial period of climatic deterioration acts as a trigger to larger changes.

Stability or instability is related to whether the ice flow is extending or compressive. In areas of predominant extending flow, such as above the equilibrium line, the response is stable whereas in areas of compressive flow below the equilibrium line it is unstable. Nye (1960) explains why this is so. In Figure 3.12, AA and BB are two cross-sections in a glacier subject to compressive flow in the ablation zone. In an equilibrium situation, the flow (volume per unit time) of ice passing BB is less than that passing AA since in the intervening space some is lost through ablation. Now suppose an extra layer of ice is artificially laid on top of the glacier. It can be shown that the increase of flow through a section following such a thickening is proportional to the original flow at the section. This is because the increase in shear stress caused by the additional layer has a greater effect on the thicker ice at AA than on the thinner ice at BB. As a result, the increase of flow through AA is greater than that through BB. It follows that ice must accumulate in the intervening section and

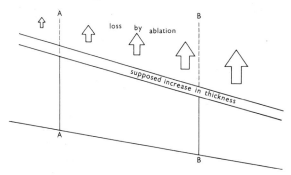

Figure 3.12
Diagram to explain why the response in an area of compressive flow is unstable. See text for explanation. *(Nye, Proc. R. Soc. 1960.)*

that the glacier must thicken. With continued flow of ice downglacier, the effect is accentuated and thus an initial change can lead to a progressive thickening downglacier. A similar argument shows that a small reduction in thickness is also unstable in a zone of compressive flow (Nye, 1960).

Kinematic waves   Kinematic waves are the means by which the effects of fluctuations in net mass balance are transmitted down the glacier (Nye, 1960; 1965b). Depending on the velocity of the waves and the length of the glacier, mass balance changes may not make themselves fully felt at the snout of the glacier until many years after the initial change. In order to understand the elusive mathematical concept of a kinematic wave, it is helpful to imagine a bulge in a glacier. The greater thickness of ice in the bulge locally affects the stress/strain relationships and increases the rate of deformation of ice in the bulge. For this reason the bulge will move downglacier faster than the thinner ice on either side. It is important to stress that it is the bulge that moves downglacier rather than the ice. Thus a morainic boulder on the surface of the ice would temporarily move at a higher velocity while it was on the top of the bulge but would then return to normal as the bulge passed it. This type of motion, classed as a kinematic wave, has been applied to river flood waves and traffic flow. Many drivers will have noticed how a car in a traffic holdup moves forward in irregular spurts. Each stoppage can be envisaged as a kinematic wave of closely packed vehicles which is passing up and down the line of traffic at a different velocity to the traffic. Another example may be seen as skeins of geese fly overhead in loose V-formation. Struggling to keep the symmetry of the line, the geese adjust their position and as a result an observer sees ripples of close or loose packing passing up and down the line. Such waves usually emanate from one end of the line and have the effect of lengthening or shortening the line. Returning to glaciers, the mechanics of ice motion mean that waves can only travel downglacier.

Kinematic waves originate in the vicinity of the equilibrium line. Figure 3.13 shows part of an idealized glacier with extending flow above and compressive flow below the equilibrium line. Suppose that, following a change of climate, the glacier has increased in thickness by an equal amount over all its surface. An increase in thickness will occur in a stable manner above the equilibrium line; below it will increase in an unstable way. As a result, a step forms on the glacier surface immediately downglacier of the equilibrium line. This step is the beginning of the kinematic wave and, once formed, it proceeds downglacier at very approximately four times the speed of the ice. The passage of the wave is analogous to a flood wave passing down a river, though clearly the velocities are very different. Since the ice velocity decreases downglacier, so also does that of the kinematic wave. Also, since the

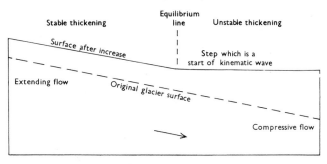

Figure 3.13
The initiation of a kinematic wave at the equilibrium line. Unstable thickening below the equilibrium line creates a step in the glacier surface which is the start of the kinematic wave.

compressive zone of a glacier responds unstably to variation in thickness there is a tendency for the height of the wave to increase downglacier. Although the wave tends to subside under its own weight by diffusion, this tendency to increase in height towards the snout is important and explains why the snout of a glacier is so sensitive to past changes in mass balance. As an example Meier and Johnson (1962) note that a wave arriving at the snout of the Nisqually glacier in Washington had the effect of increasing the ice surface level by more than 30 m. The effect of the passage of a kinematic wave is to restore a stable profile of equilibrium to the glacier (Nye, 1960; Paterson, 1969). Thus it stabilizes the initial unstable thickening of the ablation zone induced by a climatic fluctuation.

In the above discussion it has been assumed for simplicity that a kinematic wave is a zone or bulge of increased ice thickness. However, Untersteiner and Nye (1968) have stressed that kinematic waves may not be visible as surface forms. There are several reasons for this. One vital assumption is that there is a close relationship between ice thickness and ice discharge. However, thickness is also influenced by factors such as variations in valley width and the rate of diffusion of the wave, while an increase in discharge may be catered for in other ways, for example by an increase in surface slope or by a change in bed conditions causing an increase in sliding velocity. Thus it is more accurate to describe a kinematic wave as a feature carrying with it some specific property, in this case the constancy of flow (volume per unit time), which moves through the glacier at a speed different from that of the glacier itself. An additional simplifying assumption is that the climatic perturbation has been in one direction and uniform over the whole glacier. In the real world these conditions are rarely met and indeed a whole series of kinematic waves of varying amplitude and velocity will be travelling down the glacier at any one time in response to constantly changing climatic conditions. Nonetheless, in spite of these problems, surface waves have been observed to pass downglacier (Lliboutry, 1958; McSaveney and Gage, 1968), and in some cases they can be easily picked out by a zone of compression and thickening immediately in front of them and by a zone of extension accompanied by crevassing and thinning behind.

Surges   In recent years it has become increasingly clear that a number of glaciers are affected by periodic surges in which ice may temporarily be transmitted downglacier at speeds far above normal. Although most information has come from relatively small surging glaciers, Hollin (1964, 1969) and Hughes (1975) have suggested that periodic surging may also involve ice sheets on a continental or subcontinental scale. After a survey of 204 surging glaciers in western North America, Meier and Post (1969) noted that part of the glacier acts as a reservoir which fills with ice before discharging downglacier at high velocities (Figure 3.14).

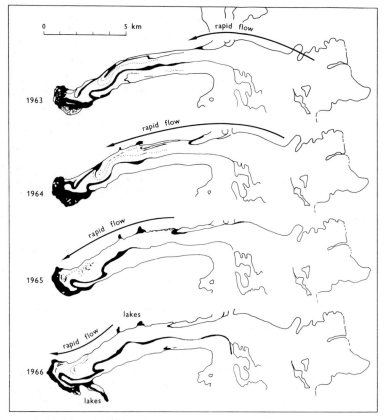

Figure 3.14
The passage of a surge down Tikke glacier, British Columbia in 1963–6, showing also the characteristic deformation of surface moraines. *(Meier and Post, Can. J. Earth Sci. 1969.)*

The surge of ice may locally increase velocities by as much as 10–100 times the normal velocity. The surge frequently consists of (a) a wave of thickening ice subjected to compressive flow, (b) a zone of high velocity ice with intensely fractured ice behind the wave crest (Figure 3.15), and (c) a zone of tension or extension where the ice is thinning and slowing down once more. Such characteristics were described vividly by Harrison (1964) in relation to the Muldrow glacier, Alaska, and in more general terms by Robin and Barnes (1969). Whereas the length of the compressive wave is comparable to the ice thickness, the high velocity section may be many kilometres in length. For example, pinnacled ice was observed over a distance of 35 km on Steele glacier in Yukon Territory, Canada (Stanley, 1969). Wave velocities as high as 350 m per day have been observed. Not surprisingly such movements are also accompanied by vibrations and rumblings audible several kilometres away (Thorarinsson, 1969).

The wave of high velocity ice may advance down the glacier into the lower reaches of the glacier and come to a standstill without extending the overall dimensions of the glacier.

Figure 3.15
Intensely fractured ice on the surface of Siðujökull, Iceland, following a surge in 1964.
(Photograph by S. Thorarinsson, 9 September 1964.)

This seems the most common type in North America. However, the wave may extend beyond the limits of the glacier and cause a catastrophic advance. A dramatic example is the 45 km front of Bruarjökull in Iceland which advanced up to 8 km in 1963–4 at speeds up to 5 m per hour (Thorarinsson, 1969). Once surged, the glacier ice involved stagnates *in situ* and may waste away or be revived in due course by a subsequent surge. It seems that this stagnating ice acts in some way as a dam and helps to induce an area of thickening compressive flow in the reservoir area, with a strong upward component of flow (Figure 3.16, overleaf). This may be of some significance in understanding the debris-transporting capability of surging glaciers.

The main effect of a surge is to shift a quantity of ice from higher to lower parts of a glacier. Horizontal displacements of ice of up to 11 km have been recorded (Meier and Post, 1969). At the same time, a surge reduces the surface slope in the upglacier reservoir and steepens it in the lower receiving area (Figure 3.17, overleaf). The effects of the passage downglacier of a surge are dramatic. Former lateral moraines are left high and dry while tributary glaciers are shorn off and hang incongruously above the main glacier. Sometimes, after a slight delay, the tributaries themselves surge and plough into the main glacier which

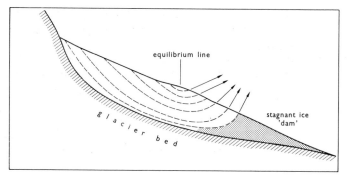

Figure 3.16
Schematic representation of flow line pattern in Rusty and Trapridge glaciers, showing the damming effect of stagnant ice after a surge. *(Collins, J. Glaciol. 1972 by permission of the International Glaciological Society.)*

by then may be quiescent. Periodic surging on a glacier may give rise to curious morainic loops which can be regarded as diagnostic of surging glaciers. If the tributary carries an arc of terminal moraine into the trunk glacier, a subsequent surge on the main glacier will carry the arc downglacier and distort it (Figure 3.14).

After their study of North American glaciers, Meier and Post (1969) concluded that most surges have a periodicity of 15–100 years. The periodicity for any one glacier seems constant. However, as Weertman (1969) pointed out, some glaciers seem to be permanently surging, at least to judge from their high velocities. For example, some outlets of the Jakobshavn Isbrae in Greenland have a velocity of 7–12 km per year (Fristrup, 1966).

At the time of writing there is much to be discovered about the causes of surges. There are two problems to be explained – the high velocities, and the trigger mechanism where it is relevant. There is some agreement that the high velocities are related to the presence of a greater than average amount of basal water (Weertman, 1969; Lliboutry, 1969; Robin and Weertman, 1973). This is confirmed by field evidence suggesting greater than average meltwater flow from a glacier during a surge (Thorarinsson, 1969). There may be a double valued law of sliding, with velocities clustering round the two peaks of 'normal' and 'surging' (Lliboutry, 1971) or there may be a continuum of velocities between normal and surging (Weertman, 1969). Since many surges seem to be periodic and frequently involve similar quantities of ice (as indicated by the tendency of waves to cover a similar area from surge

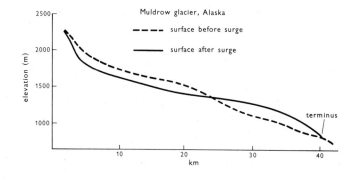

Figure 3.17
The effect of a surge on the surface profile of Muldrow glacier, Alaska. *(Post, J. Geophys. Res. 1960.)*

to surge), the mechanism is likely to be self induced, although it may be affected by external triggers such as earthquakes and precipitation changes. There is some agreement that changing conditions downstream of the reservoir cause the basal ice to cross a threshold and initiate the instability, as for example an increase in ice thickness or the amount of water and its effect on friction. The zone of instability then rapidly spreads up and down the glacier through various mechanisms of positive feedback (e.g. faster movement creates more water by frictional heat). When the ice thins after the surge, basal conditions return to a stable state. The exact mechanisms are in dispute and for a variety of explanations the reader may consult Weertman (1969), Robin and Weertman (1973), Miller (1973), and Budd (1975). The mechanisms may prove to be adequately explained in terms of kinematic wave theory (Palmer, 1972).

Main variables affecting glacier movement

Glaciers in the real world operate in response to their total environment and it is important to gain some perspective on this. Parts of this environment can be viewed as relatively *independent variables* affecting the glacier, such as the climate and the nature of the relief. Parts consist of relatively *dependent variables* such as the size and morphology of the glacier (Table 1.2). There is a complex interrelationship between all these variables which determines how any single glacier flows, and the role of the more critical variables is discussed below.

Independent variables

Geological environment   As seen in chapter 2 such variables as permeability and geothermal heat affect glacier flow. In addition the volume and type of debris contained within a glacier affect its mode of flow. In extreme situations the debris supplied to the glacier may be sufficiently abundant to hinder creep and thus flow. One example of apparently overloaded glaciers can be taken from Peru. Here with massive cliffs overlooking glaciers and an abundance of avalanches and tectonically induced rock falls, rock debris may swamp the glacier. In this case one sees what appears to be a huge ice cored moraine instead of a glacier (Clapperton, 1972). Sometimes there is a small relatively clean glacier flowing in a channel in the top of the ice cored moraine. Other examples of 'overloading' are some rock glaciers (Benedict, 1973).

Climate   In an earlier section, it was stressed that high solid precipitation totals and high ablation values favour rapid rates of flow with large volumes of ice involved. The tendency for rapid flow in such conditions is further enhanced by two other factors. Firstly, the initial snow temperature is high and the effect of warming by summer percolation of meltwater further warms the ice. As a result the ice is at or close to the pressure melting point and creep processes are rapid. Secondly, warm ice flowing fast generates a considerable amount of heat by deformation near its base and by basal sliding. As a result more meltwater may be produced and this may further facilitate basal slipping. All these factors contribute to the contrast between rapid glaciers in maritime environments on the one hand and sluggish glaciers in continental or high polar environments on the other.

The contrast is perhaps best illustrated by reference to sample glaciers. At one extreme is Meserve glacier examined by Bull and Carnein (1970). Situated on the south side of Wright

Valley, south Victoria Land, it is about 7·2 km long with an area of approximately 10 km². The climate is polar continental. Temperatures are low with a mean annual temperature of −16°C. Minimum temperatures in winter may fall below −40°C while briefly in summer daily temperatures average 0°C. Mean annual precipitation is low ($<$ 50 kg m^{-2}) and all falls as snow. Ablation occurs mainly by evaporation from a melting surface and by sublimation and, indeed, only 2–3 per cent is estimated to occur as meltwater runoff. Under such extreme continental conditions glacier flow is slow and amounts to 3–4 m per year at the equilibrium line. All movement takes place by internal deformation and none by basal slip (Holdsworth and Bull, 1970). At the other extreme is the 8 km long Franz Josef glacier in South Island, New Zealand. Originating in snowfields around Mount Cook at c. 3000 m, it flows westwards and ends only 210 m above sea-level. Its snout ends in a zone of lush forest vegetation containing tree ferns. The mean annual precipitation is not known for certain, but it is likely to exceed 10,000 mm per year in the firn area (Soons, 1971). Thus the glacier is representative of an extreme maritime climate. Ice velocities of over 300 m per year are attained and involve basal sliding (McSaveney and Gage, 1968). Clearly the difference in movement between the comparably sized Meserve and Franz Josef glaciers represents responses to two extreme climatic regimes. An intermediate response is provided by Storglaciären in northern Sweden. Here, with a mean annual precipitation of 1900 mm, the velocity at the equilibrium line is 30 m per year.

Regional relief and slope forms   The flow of glaciers is influenced by topography in many ways and it is worth mentioning some of these.

1   The steepness of the bedrock slope down which a glacier slides affects its velocity. Many localized patches of high velocity ice occur over locally steep bedrock slopes. An obvious example of this is an ice fall where ice velocities are much greater than normal. At a macro-scale, contrasts in average gradient between areas may be found to carry profound implications for glacier flow. Rutkis (1972) has calculated the relative relief for 100 km² squares covering the whole of western Europe, and his figures pose many interesting questions about the potential efficiency of glacier mass transfer from place to place.

2   Irregularities on the bedrock floor can influence glacier flow, for example by causing variations in ice thickness, by inducing divergent or convergent flow or by favouring extending or compressive flow.

3   The pattern of glacier flow is affected by whether or not the glacier ends on land or calves in water. On land a glacier thins near the snout and as a result flow decreases towards the snout. If the glacier ends at the coastline and the glacier snout is being sapped by waves with chunks of ice falling into the sea or lake, there is no thinning and no reason for ice velocity to decrease. The actual height of the cliff and the velocity will depend partly on the turnover of the ice and partly on the efficacy of wave sapping. If the snout floats the glacier loses touch with bedrock and its associated basal friction and thus accelerates. The zone of transition is often marked by a series of tension crevasses.

Dependent variables

Glacier morphology   The importance of a glacier's shape in influencing its response to its environment has been stressed above all by Ahlmann (1948). In particular he has drawn attention to the importance of the distribution of glacier area by altitude. It is important because the net mass balance at any point on the glacier surface varies from the head to the snout of the glacier and is closely related to altitude. Thus variations in altitudinal distribution affect the total amount of ice to be discharged. For example, since the amount of ice to be discharged normally increases with increasing altitude above the equilibrium line, a plateau ice cap with a high proportion of its area at a high altitude will, other variables being equal, discharge more ice than a glacier of similar size with most of its accumulation area at a relatively low altitude.

Glacier size also affects glacier activity. After analysing many glaciers in North America, Meier *et al.* (1971) noted that small cirque glaciers have anomalously high ice turnovers and net balance gradients. This probably reflects the importance of wind blown snow from a wide area collecting in sheltered basins. On medium sized glaciers the snow totals are more closely controlled by general climatic conditions.

On the other hand, at a macro-scale, ice sheets tend to be less active and have relatively low turnovers of ice due to their size. This is because at a sufficiently high elevation above the equilibrium line, precipitation may fall off owing to extreme altitude or remoteness from sources of water. Also, under these circumstances the net balance gradient also decreases with altitude.

Further reading

CLARKE, G. K. C. 1976: Thermal regulation of glacier surging. *J. Glaciol.* **16** (74), 231–50.

MEIER, M. F., TANGBORN, W. V., MAYO, L. R. and POST, A. 1971: Combined ice and water balances of Gulkana and Wolverine glaciers, Alaska, and South Gascade glacier, Washington, 1965 and 1966 hydrologic years. *U.S. geol. Surv. Prof. Pap.* **715**-A (23 pp).

NYE, J. F. 1952: The mechanics of glacier flow. *J. Glaciol.* **2** (12), 82–93.

PATERSON, W. S. B. 1981: *The physics of glaciers.* Pergamon, Oxford (380 pp). Second edition.

RAYMOND, C. F. 1980: Valley Glaciers. In S. C. COLBECK (Ed.) *Dynamics of snow and ice masses.* Academic Press, New York, 79–139.

SCHYTT, V. 1967: A study of 'ablation gradient'. *Geogr. Annlr* **49**A (2–4), 327–32.

SUGDEN, D. E. 1977: Reconstruction of the morphology, dynamics and thermal characteristics of the Laurentide ice sheet at its maximum. *Arctic and Alpine Res.* **9** (1), 21–47.

Many interesting articles on glacier surges are contained in a special theme volume: *Can. J. Earth Sci.* **6** (4), August 1969, 807–1018.

Several interesting articles may be found in *Symposium on dynamics of large ice masses, Ottawa, 21–25 August 1978. J. Glaciol.* **24** (90), (520 pp), 1979.

4 Glacier morphology

The two previous chapters have been concerned with the processes and environmental factors influencing glacier dynamics. Necessarily, when examining such processes glaciers have been treated in an idealized way, and each problem viewed with an intentionally limited perspective. This chapter aims to discuss the links between the processes and the forms and is thus concerned with glaciers as dynamic landforms on the earth's surface. The problem is viewed at two scales. The first part of the chapter examines glaciers as a whole and then there is a discussion of surface ice forms found on glaciers.

Glacier types

Table 4.1 presents a simple morphological classification of glaciers. The three main glacier types are distinguished by fundamental differences in the way their morphological expression

Table 4.1 A simple morphological classification of glaciers

Ice sheets and ice caps (unconstrained by topography)	Ice domes
	Outlet glaciers
Ice shelves	
Glaciers constrained by topography	Icefields
	Valley glaciers
	Cirque glaciers
	Other small glaciers

reflects the interactions between glacier ice and topography (Østrem, 1974b). An *ice sheet* or *ice cap* is superimposed on the underlying topography which it largely submerges; the direction of flow of the ice reflects the size and shape of the glacier rather than the shape of the ground. An *ice shelf* is essentially a floating ice sheet only loosely constrained by the shape of the coastline. A *glacier constrained by topography* is strongly influenced both in its form and its direction of flow by the shape of the ground. These three main types are then subdivided (where necessary) into component elements on the basis of morphology. Where possible the terminology used follows that suggested by Armstrong *et al.* in *The illustrated glossary of snow and ice* (1973). Where terminology differs from this usage an explanation is given.

(a) *Ice sheets and ice caps*
The difference between an ice sheet and an ice cap is normally accepted as being one of scale with the dividing line somewhere around 50,000 km² (Armstrong, *et al.*, 1973). Thus

Figure 4.1
ERTS image of Vatnajökull, Iceland. There is little detail of the ice dome but morainic patterns on outlet glaciers and the ice-dammed lake of Graenalon (centre) are clearly visible. From left to right the image covers a distance of c. 110 km. *(NASA image E—1372—12080—5, 30 July, 1973.)*

an ice mass covering Antarctica, northern America or the British Isles would be termed an ice sheet, whereas an ice mass over Wales, the Grampian mountains of Scotland or Svalbard would be termed an ice cap. Two main components can be recognized – ice domes[1] and outlet glaciers (Figure 4.1).

Ice domes An ice dome builds up so that it is situated approximately symmetrically over the land area involved. Sometimes, as in east Antarctica, the summits of ice domes may exceed 4,200 m in altitude and lie over topographic rises (Figures 4.2, 4.3). Sometimes, as in northern Greenland and in the case of the former Laurentide and Scandinavian ice sheets the high points of the domes may overlie bedrock lows. Ice thicknesses may frequently exceed 3,000 m, while a local maximum of 4,300 m is known from Antarctica. Similar though smaller scale variations occur beneath ice cap domes, but thicknesses are very much less. For example,

[1]It seems necessary to use such a term. If ice caps and ice sheets are subdivided further there is need for a term to cover those parts which are not outlet glaciers.

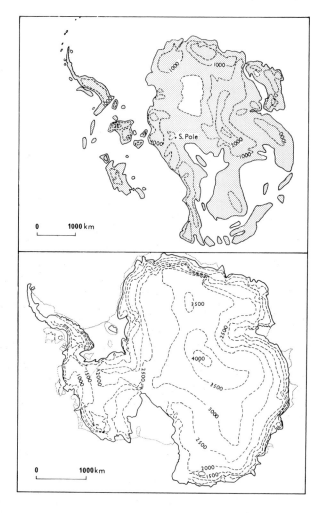

Figure 4.2
The shape of the bedrock base
(upper map) and of the ice sheet
surface (lower map), Antarctica.
Details of both maps are constantly being revised in the light
of the results of radio-echo
sounding forays. Contours are in
m above sea-level.

the central part of the Sukkertoppen ice cap in west Greenland has a thickness which varies
from 250 to 400 m (Bull, 1963) (Figure 4.3).
    The convex surface slope of an ice dome forms in response to the basic flow characteristics
of ice. Assuming that there is adequate snow, ice will build up until the shear stresses are
sufficient to deform the ice efficiently. Since shear stresses are influenced by a combination
of ice thickness and surface slope, the thinner the ice the steeper the surface slope needs to
to be to maintain flow. Conversely, the thicker the ice the less the surface slope needs to
be. For these reasons, the surface of an ice dome is gently sloping in the centre where the
ice is thickest, and steepens progressively as the ice becomes thinner towards the margin.
Many attempts have been made to describe the profile of such dome surfaces. This has usually
been achieved by assuming that the ice behaves as a perfect plastic material, and by comparing a theoretical curve with reality. In some situations a parabola fits well (Nye, 1952a),
while in other places more complicated curves are required (Weertman, 1961; Haefeli, 1961).

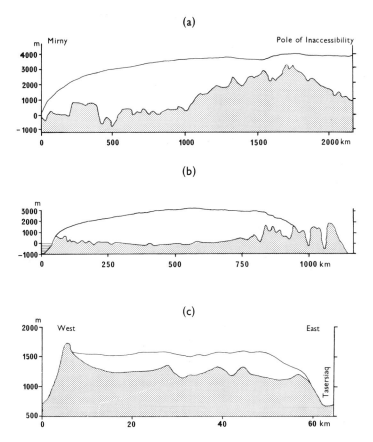

Figure 4.3
Profiles of ice domes and the underlying bedrock surface – (a) from Mirny to the Pole of Inaccessibility, east Antarctica *(Tolstikov, 1966)*, (b) from Disko Bugt to Cecilia Nunatak, Greenland *(Fristrup, Univ. Washington Press 1966)*, (c) Sukkertoppen ice cap, west Greenland *(Bull. J. Glaciol. 1963, by permission of the International Glaciological Society)*. Note the difference in scales.

Nye's solution is particularly useful as a first approximation because of its simplicity. On a horizontal bed the altitude of the ice surface at any point inland from a known margin can be found from the formula

$$h = \sqrt{2h_0 s}$$

where h = ice altitude in m, h_0 = 11 m and s = horizontal distance from the margin in metres. If the bedrock altitude is known and greater accuracy is desired, then a slightly more complex procedure may be used (Nye, 1952a). From comparison with other theoretical profiles and real profiles, it seems that Nye's parabola slightly overestimates the slope (and thus the altitude) near the centre of an ice sheet (Figure 4.4, overleaf).

The parabola can only be used as a first approximation, since a number of simplifying assumptions are made. Of these the most important are that (i) the ice dome acts as a perfect plastic material and that differences in ice temperature and variations in accumulation from place to place do not materially affect its behaviour, (ii) the underlying topography has

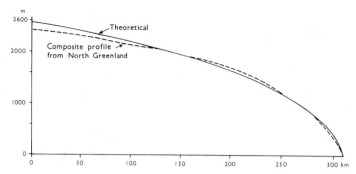

Figure 4.4
A comparison of a composite ice sheet surface profile in north Greenland with the theoretical profile of Nye *(1952a)*, assuming a horizontal base.

little effect on the surface profile, (iii) the profile follows a flow line, and (iv) the ice dome is active and in equilibrium.

Regarding the first assumption, Vialov (1958) argued that in general a cold ice sheet should be thicker than a warm one, but in practice the difference is not very great (Paterson, 1969). Also, Weertman (1973) showed that differences in accumulation are relatively unimportant in affecting surface slope, since ice velocity responds sensitively to any change in thickness. Regarding the second assumption, bedrock irregularities cause corresponding irregularities on the surface with bedrock highs overlain by locally steepened ice surface slopes, but their effects are severely damped down and small in comparison to the overall ice sheet profile

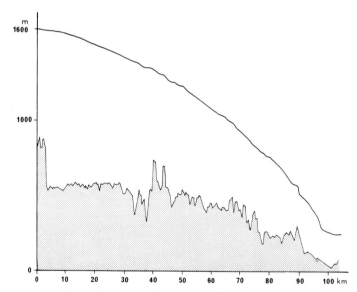

Figure 4.5
Profile of the Wilkes ice cap, Antarctica showing how bedrock highs are overlain by locally steepened ice surface. For ease of comparison, the bedrock elevation has been raised 200 m in relation to the ice surface elevation. *(Budd and Carter, J. Glaciol. 1971 by permission of the International Glaciological Society.)*

(Figure 4.5) (Robin, 1958; Budd, 1970; Budd and Carter, 1971). The third assumption is made in order to ensure that the profile is determined by the longitudinal flow paths of the ice. In this context it is fair to point out that many of the ice profiles available in the literature do not necessarily run parallel to a flow line. The fourth assumption is probably the most important in that, if it is not met, it can account for a wide divergence from the theoretical norm. It is clear that since the equilibrium profile represents a limiting condition imposed by the flow characteristics of the ice, then the ice must be actively flowing. This means that there must be an adequate supply of snow for active flow and yet no history of instability, such as might be caused by periodic surging. It is likely that failure to meet these conditions accounts for atypical profiles as illustrated in Figure 4.6. Recently it has been suggested that the west Antarctic ice sheet is out of equilibrium because of surges and accompanying disintegration which have lowered its surface (Hughes, 1972; 1975). On the

(a)

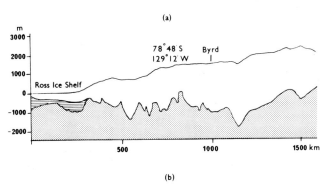

(b)

Figure 4.6
Irregular ice domes which do not attain the equilibrium profiles of active domes – (a) west Antarctic ice dome *(Tolstikov, 1966)*, (b) glaciers on Ellesmere Island. *(Hattersley-Smith et al. Can Dept of Nat. Defence, DREO Tech. Note, 1969).*

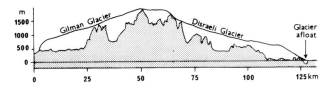

other hand, insufficient snow seems to be the reason why certain ice caps in dry continental environments fail to achieve active profiles. In such cases gradients tend to be low, and as a result ice flow is slight or virtually non-existent. Examples include the Meighen ice gap in Arctic Canada and many plateau ice caps in northern Greenland. In this situation any point on the glacier surface will respond to variations in the mass balance at that point. This latter type of glacier has been imaginatively termed a *glacier réservoir* and distinguished from the equally appropriately named normal *glacier évacateur* (Lliboutry, 1965).

Ice flow in an ice dome (excluding outlet glaciers) occurs by sheet flow. This may take the form of internal creep alone or also involve basal sliding. Budd (1969) has shown how surface ice flows outwards in a direction at right angles to the surface contours (Figure 4.7a). As it does this there is a strong vertical component as the individual particles of ice are covered by subsequent accumulation (Figure 4.7b, overleaf).

Figure 4.8a summarizes in diagrammatic form the essential dynamic characteristics of an ice dome of continental size. The contrasts between the land-bound and ocean-bound

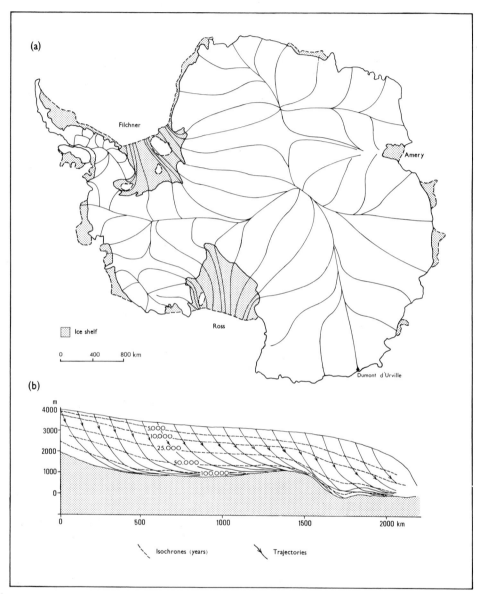

Figure 4.7
(a) Flow lines on the Antarctic ice sheet. (b) Profile along a flow line ending at Dumont d'Urville, showing particle trajectories and approximate ages of ice. *(Budd et al., Univ. Melbourne, Meteorol Dept Pub., 1971.)*

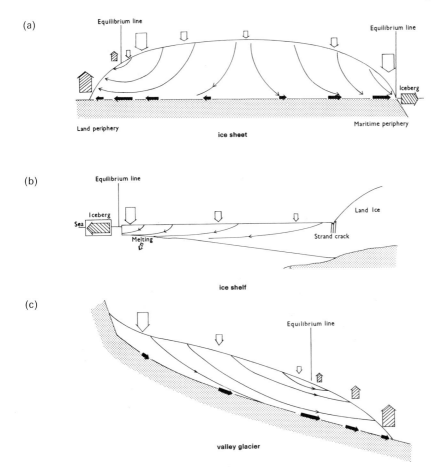

Figure 4.8
Models of (a) an ice sheet, (b) an ice shelf and (c) a valley glacier showing distribution of snow input and output and related flow characteristics. Basal slipping is assumed to occur in models (a) and (c) and is at a maximum in the vicinity of the equilibrium line.

peripheries reflect differences in the position of the equilibrium line; the zone of maximum horizontal velocity is near the edge of the ocean-bound ice dome but further in from the edge of the land-bound example. Contrasts between an ice dome of continental size and a medium sized one result from contrasts in precipitation. On a dome of 10–50 km in size, precipitation is likely to increase with increasing altitude as far as the summit. On a dome of continental size this increase of precipitation with altitude only occurs near the periphery; further inland, precipitation falls off dramatically and ice discharge is correspondingly less.

Outlet glaciers   The peripheral zone of an ice dome is often marked by a radiating pattern of outlet glaciers which extend some way beyond the dome margin. Within the ice dome they may occupy a depression and be distinguished only by a zone of rapidly moving ice bordered by crevasses. In this situation the outlet glacier is termed an *ice stream*. In the vicinity of the ice dome margin and beyond it, an outlet glacier may occupy a shallow rather irregular depression. The Antarctic boasts some of the world's finest outlet glaciers, that of the Lambert glacier being 700 km in length. An impressive series cuts through the Transantarctic mountains (Figure 3.8). Of these Beardmore glacier is 200 km long and some 23 km wide, and it flows through the mountains at a velocity of 1 m per day (Figure 3.9). Outlet glaciers flowing from the Greenland ice sheet are similar in many ways though generally they flow at much greater velocities. Outlet glaciers are also common around smaller ice caps and numerous well known examples occur around Icelandic ice caps like Vatnajökull (Figure 4.1) and Norwegian ice caps like the Jostedalsbre. Usually such glaciers leave the ice dome by means of an abrupt ice fall and then flow in a rock walled trough (Figure 4.9). Such ice falls are sometimes encountered on large ice sheet outlet glaciers. The overall gradient of an outlet glacier is considerably less than that of an ice dome. Buckley (1969) found that there was a correlation between length and gradient with longer glaciers having gentler

Figure 4.9
Outlet glaciers near Bartholins Brae, Blosseville Kyst, Greenland. *(Reproduced with the permission (A421/ 74) of the Geodetic Institute, Denmark.)*

gradients. For example, 190 km long outlet glaciers cutting through the Transantarctic mountains have overall gradients of about 1 : 95 while glaciers 60 km in length have gradients of about 1 : 40. In west Greenland where there is no comparable mountain range, gradients are gentler.

Sheet flow in the ice dome regions and stream flow within the outlet glaciers are the main means by which ice is evacuated from ice sheets and ice caps. The relative importance of each is far from known. Tentatively, it seems that the bulk of the discharge of the Greenland ice sheet is drained by outlet glaciers (Fristrup, 1966), while in the Antarctic the proportion is probably around 75 per cent (Bull, 1971). In the smallest ice caps, discharge may be entirely by sheet flow.

(b) Ice shelves

An ice shelf is a floating ice cap or part of an ice sheet which deforms under its own weight. It can be regarded as a slab of ice being squeezed between two surfaces – the atmosphere

Figure 4.10
The cliffed edge of the Ross ice shelf as sketched by R. M'Cormick in February 1842 on board *Erebus*. This was the most southerly position reached by Captain Ross's expedition. (*M'Cormick, 1884.*)

and the ocean. Ice shelves comprise unique models ideal for developing principles of ice creep, since basal friction can be largely ignored. Rates of ice movement may be between 0·8 and 2·6 km per year (Swithinbank and Zumberge, 1965). Ice shelves are most common in the Antarctic where they comprise some 7 per cent of the total ice area and 30 per cent of the length of coastline. However, small examples also occur along the northern coasts of Ellesmere Island in the Canadian Arctic (Lyons *et al.*, 1971). They persist only where they are anchored at several points, as for example in an embayment or along a coastline dotted with small islands. Also they only occur where the mean summer temperature does not rise above 0°C (Mercer, 1968a).

The main morphological characteristics of Antarctic ice shelves are described by Swithinbank and Zumberge (1965). The seaward margin generally forms a sheer cliff rising some 30 m above sea level. It was for this reason that the Ross ice shelf was originally named the Great Barrier by the first seaborne explorers to the area (Figure 4.10). Indeed, James Clark Ross noted ruefully that he might 'with equal chance of success try to sail through the cliffs of Dover' (Kirwan, 1959; p. 171). The surface, rising almost imperceptibly inland, appears flat. Undulations and crevasses occur wherever parts of the ice shelf are aground. On the inland margin there is often a slight depression between land based and floating ice and strandcracks are caused by the tidal movements of the shelf. At the seaward margin ice shelves are commonly 200 m thick. This would probably be the equilibrium thickness of any ice shelf which is free floating and able to expand in all directions, for this is the approximate thickness beneath which ice ceases to be subject to effective creep. However, since ice shelves are constrained by embayments, they cannot spread in all directions and as a result they thicken away from the seaward edge. The bottom surface may be highly irregular if inland ice streams enter the ice shelf. This is vividly brought out by Robin's map of the thickness of the Ross ice shelf compiled by radio-echo sounding (Figure 4.11). The

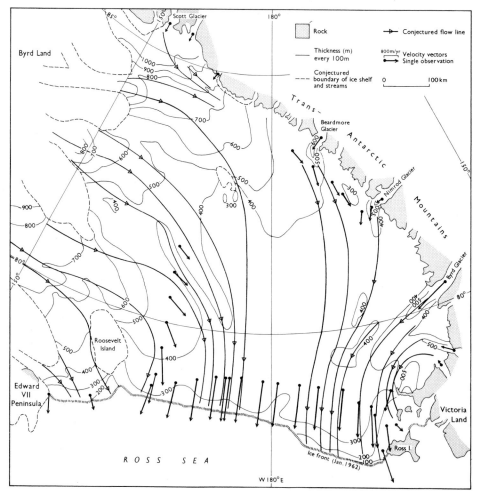

Figure 4.11
The thickness and movement of the Ross ice shelf. *(By permission of G. de Q. Robin.)*

Figure 4.12
An icefield and associated valley glaciers on Coburg Island, Arctic Canada. *(Original photograph supplied by National Air Photo Library, Canadian Department of Energy, Mines and Resources.)*

map shows great thicknesses of ice where outlets join the shelf and also how this ice thins and diffuses laterally as it moves seawards.

Ice shelves are nourished largely by snow accumulation on their flat upper surfaces, but there may be varying amounts of ice supplied from land glaciers and occasionally by bottom freezing. With the exception of small ice shelves, surface accumulation is highest near the seaward edge and decreases rapidly inland in a similar fashion to that on ice domes (Figure 4.8b). The contribution from land based glaciers varies from ice shelf to ice shelf. On the Amery ice shelf at the mouth of the large Lambert glacier, the contribution is probably relatively large, whereas in other situations it may be insignificant. Under certain circumstances bottom freezing may add ice to the bottom (Swithinbank, 1970).

Ice wastage is by calving and by bottom melting. Calving may produce massive tabular icebergs many tens of kilometres in length. Calving is periodic and, for example, the Amery ice shelf lost a massive slab of ice in 1963 after a buildup of some 40 years (Budd, 1966). Bottom melting has been observed on the Brunt ice shelf to remove 0·8 m of ice per year at distances of 70 km from the ice front, and this is thought to operate over the bulk of its lower surface (Thomas and Coslett, 1970). The latter writers feel that bottom melting is the normal situation beneath ice shelves.

It is important to note that ice shelves do not have sufficient surface gradients to rise up over landmasses. If part of an ice shelf is aground, an ice rise or small landbased ice cap builds up locally over the area of grounding (Figure 4.11). This will tend to maintain an independent radial pattern of flow regardless of the movement of the ice shelf round the obstruction.

(c) *Glaciers constrained by topography*

Icefields   An icefield[1] can be regarded as an approximately level area of ice which is distinguished from an ice cap because its surface does not achieve the characteristic domelike shape, and because flow is strongly influenced by the underlying topography (Figure 4.12). As with all classifications some features may straddle two classes and it is difficult to distinguish between a non-equilibrium ice cap and a mountain icefield. Icefields form wherever the topography is sufficently high or gentle for ice to accumulate before going on to flow in restricted valley glaciers. A large example occurs in the Pacific mountain ranges of western North America in the St Elias mountains area. The mountains, many of which exceed 4,000 m, form a series of sub-parallel chains between the Pacific and the Yukon plateau. The peaks overlook the most extensive area of ice and snow in North America. Straddling the main divide at an altitude of 2–3,000 m, is a generally level area of snow accumulation from which radiate five of the longest valley glaciers outside the polar regions. Similar though smaller icefields occur in the less dissected parts of the world's mountain chains.

Figure 4.13
Valley glaciers in southern Greenland following the dendritic pattern of pre-existing valleys. (*Reproduced with the permission (A421/74) of the Geodetic Institute, Denmark.*)

[1] The term *icefield* is not defined by Armstrong *et al.* (1973). However, usage in North America is widespread. Ommanney (1969) uses the term for features varying from near continental size to features a few hundreds of metres across. The term icefield is used in this book to describe features with areas generally greater than *c.* 5 km². Hence this excludes small features which may be little more than perennial snow patches.

Figure 4.14
ERTS image of the Malaspina piedmont lobe. The compressive flow as the glacier leaves the mountains is responsible for the intense deformation of moraines. The lobe is about 40 km across. *(NASA image E–1420–20102–5, 16 September 1973.)*

Valley glaciers   A valley glacier flows in a rock valley and is overlooked by rock cliffs. It may originate in an icefield or a cirque. Such glaciers flow in valleys radiating from the main massif on which they form, and commonly display a dendritic pattern simpler than but similar to those of river valleys (Figure 4.13). Regardless of their different sizes, tributary glaciers tend to join the main glacier with their surfaces more or less conformable to one another. Exceptionally, valley glaciers may be 120 km in length like the Hubbard glacier in north western America, but lengths of 10–30 km are more usual. As is the case with rivers, there is a tendency for the size of a glacier to reflect its importance in the hierarchy of the drainage basin. Thus a test on 40 glaciers in east Greenland revealed a close correlation between the width of valley glaciers and the number of tributaries, measured in this case as the number of cirque collecting grounds. It is common to find valley glaciers debouching from steep mountain valleys into adjacent lowlands. Freed from the constraints of the valley the glacier snouts broaden out to form piedmont lobes (Figure 4.14).

Some prominent characteristics of valley glaciers are summarized in Figure 4.8c. The altitudinal range of a valley glacier is high in relation to its size and thus the bedrock slope

Figure 4.15
Three north facing cirque glaciers in the Sarek area of northern Sweden. From the left the glaciers are Ruopsok-jekna, Åparjekna and Perikjekna. *(Photograph no. 63 Hi 270–27; copyright Statens Lantmäteriverk, Sweden.)*

is likely to be steep. Also, in most cases, net accumulation increases with altitude above the equilibrium line. For both these reasons, ice turnover is vigorous. Furthermore, this vigour is confined to a relatively narrow valley floor, especially in the vicinity of the equilibrium line which is relatively centrally situated along the length of the glacier. The valley sides are relatively ice free and serve to provide the glacier with a veneer of frost-shattered rock debris.

Cirque glaciers A cirque glacier is a small ice mass generally wide in relation to its length and characteristically occupying an armchair-shaped bedrock hollow. Sometimes the slopes overlooking the glacier are cliffed, sometimes not. The glacier may be confined to part or the whole of the hollow or may simply comprise the arcuate head of a valley glacier (Figure 4.15). In the latter case, cirques are well filled with ice and snow and the cirque part of the glacier may merge imperceptibly into the larger valley glacier, as is common in areas of intensive glacierization. Small cirque glaciers are relatively well known from the work of W. V. Lewis (1960) and associates working on Vesl-skautbreen in Jotunheimen, Norway. This glacier is less than a kilometre in width and has a mean surface slope of approximately 26°. It lies, like many other cirque glaciers, in a rock basin. Longitudinal flow characteristics as measured by McCall (1960), can be regarded as similar to those in a foreshortened valley glacier. Special features arise from the fact that much accumulation is obtained from drifting snow, which makes cirque glaciers more vigorous than adjacent larger glaciers at a similar altitude. Another point of interest is that the plan shape of an isolated cirque glacier favours strong convergent flow above the equilibrium line and divergent flow below.

Other small glaciers This category includes a wide variety of glaciers whose forms are closely controlled by the form of the underlying topography. An appreciation of the range of possibilities can be gained from Ommanney's comprehensive inventory of glaciers on Axel Heiberg Island (Ommanney, 1969). Some glaciers cling precariously to small hollows on steep valley sides. Others are thin masses of snow and ice adhering to mountain sides and are termed *ice aprons* (Figure 4.16, overleaf); they have been discussed in detail in relation to the Alps by Galibert (1965) and Röthlisberger (1974). Others are thin ice patches occupying slight depressions in gently sloping topography. Still others border a coastline, as on the western side of the Antarctic Peninsula, where they form *ice fringes*. The permutations are almost limitless and a meaningful classification of these smaller glaciers awaits more detailed study of their characteristics.

Surface ice forms

Glacier surfaces are far from being featureless white expanses of snow, although this is sometimes the case. There are many surface forms related to a series of different processes operating at the glacier surface. These processes may be associated with accumulation and ablation or with the movement of a glacier.

Surface features related to processes
of accumulation and ablation
There is a fundamental contrast between glacier surface forms affected by surface melting and those forms not so affected. In zones of no melting, surface features relate to the interplay of snow transport, erosion and deposition, while in the melting zone water creates a distinctive suite of surface forms. The relative importance of these zones varies from glacier to glacier and from season to season on a single glacier.

Figure 4.16
Ice apron on the east ridge of the Weisshorn, Switzerland. *(Photograph by H. Röthlisberger.)*

Surface forms due to wind action are common in Greenland and Antarctica. However, there is little information on the precise forms and the processes responsible (Whillans, 1975). It would seem that here is a productive field for an exchange of ideas between polar geomorphologists and those colleagues concerned with the action of wind in hot deserts. Perhaps the best known forms are sastrugi. These are dunes of hard packed snow elongated in the direction of the prevailing wind. They form a corrugated surface with individual dunes attaining heights of 2 m. The dimensions of the dunes are related partly to wind velocity (King, 1969). Sastrugi will quickly form in the lee of obstacles, for example oil drums, but they may also occur in broad zones. On the Antarctic ice sheet, where there are prevailing outward flowing katabatic winds, the sastrugi form a radial pattern. Their development is best near the steeper ice sheet periphery, where winds tend to be stronger, but at a sufficiently high altitude to escape ablation. Larger characteristic snow forms occur in association with nunataks and other obstacles rising above the smooth surface of an ice cap or ice sheet. Often there is a moatlike windscoop on the upwind side of a nunatak where wind velocities are locally high as the air circumvents the obstacle (Figure 4.17). Accidents have occurred through people mistaking this depression for a sheltered camp site only to be later exposed

Figure 4.17
Windscoop adjacent to the Pálsfjall nunatak on Vatnajökull. Ablation helps to maintain the moat. *(Photograph by Magnus Jóhannson.)*

to higher than average wind velocities during a storm. On the lee side of a nunatak a wedge shaped snowdrift may form a tail several kilometres long.

Summer melting produces a continuum of surface forms on a glacier, from bare ice near the snout, through a zone of saturated snow subject to slusher flows, to barely modified snow at higher altitudes. The zones migrate upglacier as the ablation season proceeds. Many forms, characteristic of those shaped by water, occur as soon as meltwater is sufficiently abundant to flow in surface streams. However, this change from ice to liquid marks a threshold so fundamental to glacial geomorphology, that further discussion is held over until those chapters devoted to meltwater. Surface moraines likewise are discussed later (chapter 12).

Surface features related to glacier movement
Several families of surface forms occur as a result of glacier movement. Crevasses are more
or less vertical cracks which may vary from a few millimetres to several metres in width.
They form when the surface layers of a glacier are unable to adjust to horizontal tensional
stresses by creep. According to Mellor (1964), this commonly occurs when the longitudinal
strain rate or stretching of the glacier surface exceeds 1 per cent per year. The depth of
crevasses appears to be limited to values of the order of 25–36 m, since at depths greater
than this the ice is usually sufficiently plastic to adjust to tensional stresses by creep. Differences in depth between one crevasse and another reflect such factors as the amount of
tension and the temperature of the ice.

The characteristic patterns of crevasses on valley glaciers have been discussed by Nye
(1952b) who built on ideas first suggested by Hopkins in 1862. Figure 4.18 shows these characteristic crevasse patterns – chevron, transverse and splaying. Chevron crevasses occur because
of the tensional stresses introduced by the drag of the valley walls on the moving glacier.
Crevasses may open up at right angles to the maximum tensional stress and thus are aligned
at approximately 45°. If the glacier is subjected to longitudinal compressive flow in a valley

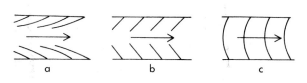

Figure 4.18
Crevasses caused by the drag of
the valley sides on a glacier: **(a)**
splaying crevasses where flow is
compressive, **(b)** chevron crevasses, and **(c)** transverse crevasses where flow is extending.
*(Nye, J. Glaciol. 1952b by
permission of the International
Glaciological Society.)*

section then the ice will tend to expand sideways. This modifies the stress pattern in such
a way that crevasses are curved upstream and meet the glacier edge at angles of less than
45°. These are splaying crevasses. If the glacier is subjected to longitudinal extending flow
in a valley section, then the maximum stresses tend to lie parallel to the line of ice flow,
at least in the centre. This means that crevasses open up at right angles to the direction
of flow and curve downstream where they are influenced by the drag of the valley walls;
these are transverse crevasses.

Once formed, a crevasse will be modified by glacier flow. As it moves downglacier the
initial crevasse is distorted by the differential flow between the glacier centre and margin.
A marginal crevasse will tend to rotate since the part nearest the glacier centre moves fastest.
Both splaying and transverse crevasses tend to be straightened. Inevitably the life of a crevasse
is limited. It will either be carried into an area of less tension where it will close or it will
be replaced by another more suitably oriented crevasse. Once closed it may be represented
by no more than a thin blue vein of ice.

In practice, most crevasses associated with extending and compressive flow reflect the influence of the underlying topography. Extensive crevassing is associated with convex long profiles; thus transverse crevasses are characteristic of ice falls and the glacier may resemble
a succession of faulted ice blocks (Gunn, 1964). Elsewhere in a valley glacier, crevassing
may be associated with extending flow where the valley widens or on the outside of a bend

(Nye, 1952b). Crevasses associated with compressive flow in a glacier are restricted to the glacier sides. Topographically induced compressive flow occurs at constriction in a valley, in basins and on the inside of bends.

Different patterns of crevasses occur on glaciers which are not constrained by valley walls. A piedmont lobe may have radial crevasses formed as a result of tensional stresses related to the spreading of the ice (Figure 3.1). On ice sheets and ice caps, tensional crevasses due to extending flow may be straight for considerable distances since they are not influenced by valley sides (Figure 4.19). Alternatively, zones of crevasses may occur due to differential ice movement as ice moves round a thinly submerged hill or mountain massif or at the mar-

Figure 4.19
Regular near-parallel crevasses on Robert Scott glacier, Antarctica. The pattern is more complicated in the middle distance *(Photograph by D. J. Drewry.)*

gins of an ice stream. A common situation where extending flow leads to crevassing occurs where a land based glacier discharges into the sea and begins to float. Relieved of bottom friction, the glacier velocity increases and the point of flotation is clearly marked by transverse tensional crevasses. A similar zone of crevasses borders the inland margin of ice shelves, and marks the point where land based ice gives way to floating ice shelf.

It is important to regard the above as a discussion of a few idealized examples. The distortions and other irregularities which occur in reality can sometimes be highly complex. None the less the models give an insight into important components of crevasse patterns and many of the features can be ascertained on the glacier shown in (Figure 4.20, overleaf).

One of the striking features of many glaciers in their ablation zones is the banding in the ice, often known as foliation. It consists of alternating layers of white bubbly ice and

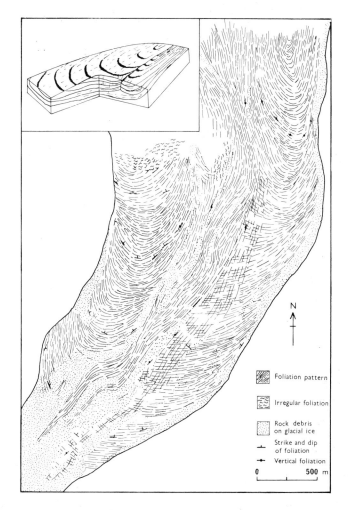

Figure 4.21
The foliation pattern on Gulkana glacier, Alaska *(Rutter, J. Glaciol. 1965 by permission of the International Glaciological Society.)* and (inset) a three dimensional simplification. *(Taylor, J. Glaciol. 1963 by permission of the International Glaciological Society.)*

N

Foliation pattern

Irregular foliation

Rock debris
on glacial ice

Strike and dip
of foliation

Vertical foliation

0   500 m

bluish ice, which may or may not have a surface morphological expression. Layers vary from several millimetres in width to several metres. The white bands are dominant and, for example, 80 per cent of the surface ice exposed at the surface of the Gulkana glacier in Alaska is white. Blue ice bands which rarely exceed 50 mm in thickness account for the rest (Rutter, 1965). The banding within the glacier is approximately spoon shaped and runs parallel to the sides of the glacier and across the glacier nearer the snout (Taylor, 1963) (Figure 4.21). It seems generally accepted that the banding is related to metamorphism as a result of glacier flow, and that it is best displayed on those glaciers which have undergone considerable deformation, as for example at the foot of an icefall (Kamb, 1964; Gunn, 1964). The structures which may form at depth are exposed at the surface further down the glacier

Figure 4.20
Crevasse patterns, ogives and surface moraines on a glacier in northern Milne Land, east Greenland. *(Reproduced with the permission (A421/74) of the Geodetic Institute, Denmark.)*

by ablation. This type of foliation is different from sedimentary layering derived from the deposition of firn.

Ogives are broad banded surface patterns which generally curve downglacier as a result of faster ice movement in mid glacier. It is becoming clear that there are several types of ogive. Some consist of alternating bands of white and dark ice. The white ice contains many air bubbles while the dark ice contains few. They are common below ice falls (Figure 4.20). They are spaced at intervals of approximately 50–200 m, which usually represents one year's flow. Some investigators relate the dark ice to that ice which descends an ice fall in summer and the white ice to that which descends an ice fall in winter. Once reconstituted at the bottom of the ice fall, the ice forms bands (King and Lewis, 1961; Fisher, 1962; Lliboutry, 1957). Other ogives may relate to avalanching in spring and winter which contributes snow and ice to the

Figure 4.22
Wave ogives on Bartley glacier, Wright Valley, Antarctica. *(Photograph by M. J. McSaveney; copyright US National Science Foundation.)*

foot of an ice fall. This gives a seasonal variation in ice texture at the foot of the ice fall, which is maintained downglacier (Kamb, 1964). Still other ogives seem to reflect ice structures related to longitudinal pressures in ice flow. For example, McSaveney (1973) described such undulating wave ogives from cold glaciers in Victoria Land (Figure 4.22). Many spectacular photographs of all these surface features may be seen in Post and LaChapelle (1971).

Further reading

ARMSTRONG, T. E., ROBERTS, B. and SWITHINBANK, C. 1973: *Illustrated glossary of snow and ice.* (2nd edn) Scott Polar Research Institute, Cambridge (60 pp).
BOULTON, G. S. and JONES, A. S. 1979: Stability of temperate ice caps and ice sheets resting on beds of deformable sediment. *J. Glaciol.* **24** (90), 29–43.
HAMBREY, M. J., MILNES, A. G. and SIEGENTHALER, H. 1980: Dynamics and structure of Griesgletscher, Switzerland. *J. Glaciol.* **25,** 215–28.
OMMANNEY, C. S. L. 1969: A study in glacier inventory. The ice masses of Axel Heiberg Island, Canadian Arctic Archipelago. *Axel Heiberg Island Res. Rept. Glaciology* **3,** McGill Univ. (105 pp).
POST, A. and LACHAPELLE, E. R. 1971: *Glacier ice.* Univ. Washington Press, Seattle (110 pp).
ROBIN, G. de Q., DREWRY, D. J. and MELDRUM, D. T. 1977: International studies of ice sheet and bedrock. *Phil. Trans. Roy. Soc. London, Series B* **279,** 185–96.
For an impression of what it really feels like to be on an ice sheet, we recommend the first recorded account of a journey across Greenland.
NANSEN, F. 1890: *The first crossing of Greenland*, Longmans, London.

Part II
Glacier distribution in space and time

5 Spatial distribution of glaciers

This chapter considers glacier distribution in the space dimension, and is thus concerned with the *glacierized* areas of the earth's surface. The next two chapters deal with glacier distribution over time, and they draw upon many more examples from previously glaciated[1] regions.

Current glacierization

Areal extent of glaciers

At present the aggregate area of the world's glaciers is about 14·9 million km² (10 per cent of the world's land area). Of this, about 12·5 million km² is accounted for by the Antarctic ice sheet, and 1·7 million km² by the Greenland ice sheet (Flint, 1971). This leaves a mere 700,000 km² (or 4 per cent of the total area) of glacier ice distributed among the other glacierized areas; this is located in many ice caps which rarely exceed 10,000 km² in extent (mostly in high latitudes), and in many thousands of small glaciers in the upland areas of the world (Table 5.1, overleaf). The figures quoted in the table are approximations only, and will be subject to modification especially for those regions (such as South America) which are imperfectly mapped. However, more and more detail is now forthcoming on the extent of glaciers within specific countries, and national and regional inventories and maps are now published at an increasing rate, partly as a result of the interest shown in glacier hydrology during the International Hydrological Decade (e.g. Lorenzo, 1959; Prest *et al.*, 1968; Meier, 1961; Ommanney, 1969; Østrem *et al.*, 1973).

Table 5.1 shows that, apart from the two ice sheets, glacierization is concentrated in the northern hemisphere, for the most part on the islands of the North Polar basin and on the uplands of the oceanic peripheries (e.g. Alaska and Scandinavia). Other highlands of the middle and low latitudes, such as the Alps, Karakoram and Himalayan ranges, have appreciable ice covered areas. The areal extent of ice on the African continent is negligible, and the only large ice caps in the southern hemisphere anywhere beyond the Antarctic ice sheet are those of the southern Andes and the Antarctic Peninsula. Thus the present day ice cover is essentially discontinuous and is in no sense balanced between the two hemispheres or between the major landmasses (Hattersley-Smith, 1974; Østrem, 1974b).

Flint (1971) has rightly stressed the importance of knowing glacier volumes if we wish to appreciate the global importance of glacierization. It is exceedingly difficult to measure the volume of ice contained within a small valley glacier, let alone a continental ice sheet. At best glaciologists can provide only estimates, but their accuracy is improving with the advent of techniques such as radio-echo sounding. Presently the ice of the Antarctic ice sheet is calculated to have a volume (water equivalent) of 21·5 million km³ as against 2·38 million km³ for the Greenland ice sheet and 180,000 km³ for all other ice. The largest of the

[1] In reality, the differentiation of 'glacierization' and 'glaciation' is of course extremely difficult. Nevertheless, the terminology is still considered useful. See Flint (1971, p. 86) and Davies (1969, p. 68).

Table 5.1   Present day glacier extent. *(Flint, 1971.)*

Region	Area (km² approximate)	Sub-totals
South Polar region		
Antarctic ice sheet (excluding shelves)	12,535,000	
Other Antarctic glaciers	50,000	
Subantarctic islands	3,000	
		12,588,000
North Polar region		
Greenland ice sheet	1,726,400	
Other Greenland glaciers	76,200	
Canadian Arctic archipelago	153,169	
Iceland	12,173	
Spitsbergen and Nordaustlandet	58,016	
Other Arctic islands	55,658	
		2,081,616
North American continent		
Alaska	51,476	
Other	25,404	
		76,880
South American cordillera		26,500
European continent		
Scandinavia	3,810	
Alps	3,600	
Caucasus	1,805	
Other	61	
		9,276
Asian continent		
Himalaya	33,200	
K'un Lun chains	16,700	
Karakoram and Ghujerab –		
Khunjerab ranges	16,000	
Other	49,121	
		115,021
African continent		12
Pacific region (including New Zealand)		1,015
Grand total		14,898,320

remaining ice caps, Vatnajökull, has a volume of only 3,100 km³ (Bauer, 1955). Again the overwhelming importance of the Antarctic ice sheet at the world scale is emphasized; if the ice sheet were to disappear, world sea-level would rise by about 59 m, as against a 6 m rise calculated for the hypothetical wastage of the Greenland ice sheet (Donn *et al.*, 1962; Guilcher, 1969; Mercer, 1968).

When the geomorphologist is viewing landscapes at a higher level of resolution (fourth or fifth order scale) he often needs to know the degree of inundation of the land surface by snow and ice. It is a simple matter to rate 'degree of glacierization' of a region according to the percentage of the land surface covered at the end of the balance year (Andrews *et al.*, 1970; Henoch, 1971). A scale of tenths can be employed on the lines of the scales used for describing pack ice or cloud cover, and the technique will be most effective if data are plotted on a grid rather than on a map of drainage basins. The geomorphologist is interested in the degree of relief inundation by ice not only as a morphological variable within a landscape system, but also as an index of the intensity or type of glacial or nival processes operating upon the bedrock base. Here he must also be concerned with detailed measurements of snow and ice thicknesses. The results now being obtained from radio-echo sounding profiles are particularly valuable in this respect (Robin *et al.*, 1969; Drewry, 1972), for they provide continuous data concerning ice surface altitude and form, ice thickness, and sub-ice bedrock relief (Figure 5.1). Until recently the technique had only proved successful for investigations of areas heavily inundated by cold ice, as in Greenland, Arctic Canada and Antarctica; but recent advances have made it possible to investigate areas subject to glacierization by warm ice (Goodman, 1975).

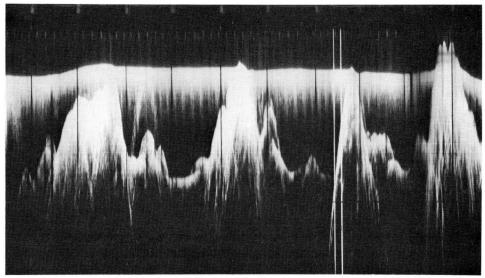

Figure 5.1
A radio-echo sounding profile across part of the Transantarctic mountains. The smooth upper surface is the ice sheet surface, which contrasts sharply with the irregular relief of the bedrock base. (*By permission of Scott Polar Research Institute, Cambridge.*)

Distribution of glacier types
It is somewhat difficult to generalize concerning the world distribution of the major types of glacier system. Certainly the two ice sheets are located in polar and sub-polar environments, but mountain glaciers are so widely distributed that one cannot allocate them to specific latitudinal or morphoclimatic zones on a world scale. In Figure 5.2 (overleaf) an attempt is made to differentiate between the glacierized zones of the northern hemisphere on the basis of glacier morphology. The glacier categories are those discussed in chapter 4, although it is difficult to decide upon the 'diagnostic types' to be used for middle and low latitude areas once intensively glaciated but now only covered by widely dispersed small ice masses.

At present it would be unwise to attempt a world map of glaciers classified according to thermal criteria. Ahlmann's (1948) simple threefold classification of glaciers, made on the basis of thermal regimes, contains inherent dangers already discussed in chapter 2. Perhaps more realistic is the fivefold classification of Avsiuk (1955) or the scheme of Court

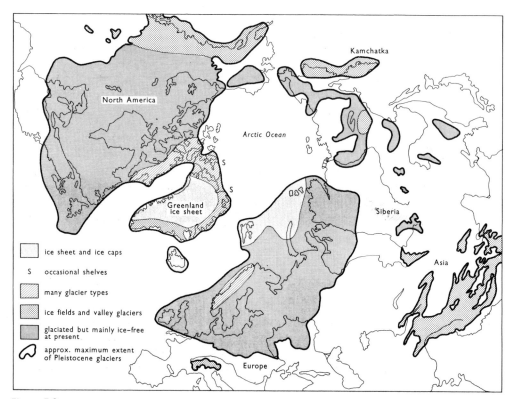

Figure 5.2
A northern hemisphere map showing the current distribution of major glacier types by morphology, within the areas affected by Pleistocene glaciation. For simplicity, the map omits periglacial areas which have never been subject to large scale glaciation, although they may contain locally significant firn fields and cirque glaciers.

(1957); but these are just as difficult to map. The recent scheme of Miller (1973), proposing four thermal glacier types (polar; sub-polar; sub-temperate; temperate) is difficult to use in view of the current widespread belief that most glaciers are composed of two or more different types of ice. Until more is known about the relationships of these ice types within specific glacier systems, no map can be even remotely reliable.

Factors influencing the current distribution of snow and ice

Permanent snow and ice is located wherever topographic and climatic factors are suitable for snow to collect and survive. Some of the variables which affect glacier distributions are considered in the paragraphs which follow. Each of them exerts a different control upon glacierization at different scales, but always the critical point is that snow should be enabled to persist and hence accumulate from one year to the next.

(a) *Precipitation*

Precipitation is a potent control over glacierization. As indicated in chapter 3, a variety of factors combine to determine the effectiveness of precipitation. High evaporation rates and low annual precipitation may combine to give a situation where there is a negative net precipitation, as in polar desert systems (Wilson, 1970). A high altitude west coast environment in the northern hemisphere is often suitable for extremely heavy precipitation (more than 3000 mm water equivalent); however, much of this may be glaciologically ineffective if it falls as rainfall during the summer season or if prolonged periods of clear skies enable solar radiation to achieve high melting rates at the snow or ice surface. It is clear from the work of Østrem *et al.* (1967), Adams (1966) and many other glaciologists that, except in high latitudes, summer precipitation in the form of rainfall contributes little to glacier mass. Most of it remains unfrozen, and with the exception of that which is converted into superimposed ice, it does not enter the glacier system store in solid form; during rapid throughput either on or beneath the glacier surface it combines with meltwater and is discharged at the snout.

The extreme aridity of parts of the ice sheet interiors, where ice thicknesses are greatest, indicates that total annual precipitation is a poor guide to the occurrence of glacier systems or to the effectiveness of regional glacierization. It is equally misleading to use maps of annual total snowfall, or the number of days on which snow falls, or 'snow intensity' (Tricart, 1969). The northern part of the Greenland ice sheet receives only 150 mm, less than in many of the world's tropical desert areas. However, almost all of this falls as snow, and the precipitation which does occur is therefore much more effective than that of marginal glacierized areas. The term 'nivometric coefficient' has been coined as an index of snowfall effectiveness (Tricart, 1969), being the ratio of snowfall (in water equivalent) to total annual precipitation. A coefficient of 1·0 implies a precipitation entirely of snow.

If one considers the nivometric coefficient and net annual precipitation together, a rough guide to the likelihood of glacierization is obtained. Hence regions with a nivometric coefficient approaching unity, and high precipitation, provide the best conditions for nourishing glaciers. Examples are the high Alps (above 3000 m) and the mountains of Svalbard and southwest Greenland. Regions with a nivometric coefficient approaching unity and low annual precipitation totals are less suitable for glacier growth but more suitable for prolonged glacier survival. The Antarctic ice sheet is an example. Areas with a medium nivometric coefficient and high precipitation (such as the higher parts of some mid latitude maritime mountain massifs) are more marginal from the point of view of glacierization, but may nevertheless contain extensive ice fields and vigorous valley glaciers.

Helpful as the above concept may be, it should be remembered that it concentrates entirely upon the accumulation side of the balance equation and that it is therefore of limited glaciological value. Any glacier seen as a system depends for its existence upon the interaction of a large number of variables.

(b) *Temperature*

As noted in chapter 3, a glacier surface receives heat from a number of different sources. Of these, solar radiation received at the ground or ice surface probably exerts the most important influence upon ablation. Thus there is a close relationship between regional temperature characteristics and glacierization. From Figure 5.3 (overleaf) it can be seen that present day glaciers exist in areas which receive widely differing amounts of solar energy, from about 251 kJm⁻² per year in Svalbard to over 670 kJm⁻² in the South American cordillera.

There have been many attempts to generalize concerning the thermal requisites of glaciers, but few of them have been successful. For example, Peltier (1950) rather unwisely defined one characteristic of his glacial morphogenetic regions as those experiencing a range of average annual temperatures between −17·8°C and −6·7°C. However, many glaciers exist in environmental conditions outside these limits, and Paterson (1969) stressed that glaciers can exist in unlikely situations as a result of local peculiarities of climate. As mentioned in

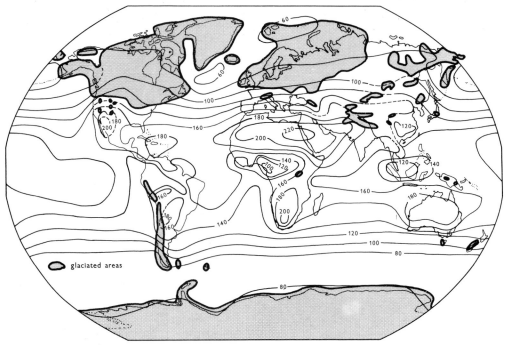

Fig. 5.3
Map of average annual solar radiation on a horizontal surface at ground level (measured in kilolangleys per year). Also shown on the map are the areas intermittently subjected to glaciation. *(In part after Sellers, Univ. Chicago Press 1965.)* One langley equals 41·868 kJm⁻².

chapter 3, glaciers generally exist in localities where mean annual air temperatures are sub-zero; in parts of Antarctica mean annual air temperatures may be below −50°C (Loewe, 1970), whereas the mean is −19°C on parts of Axel Heiberg Island and −23°C at Camp Century, Greenland. However, few glaciologists consider mean annual air temperature as a significant climatic parameter as far as glaciers are concerned; as long as air temperatures are below zero snow can accumulate, and as long as air temperature is above zero ablation will occur. Even the annual amplitude of temperature appears insignificant, although fluctuations around freezing point can exert some control over glacier regimes (Loewe, 1970a).

Of much greater importance from a glaciological point of view is the *mean summer temperature*, preferably measured for the ablation season as recognized in the field (Orvig, 1951).

In many studies, degree days above 0°C are calculated for the ablation area, and there seems to be a linear relationship between number of degree days and rate of ablation on warm glaciers (Orheim, 1970; Outcalt and McPhial, 1965). For instance, on Supphellebreen in 1965 a total of 1,104 degree days centigrade was equated with 7·19 m of mean ablation, whereas in 1966 a total of 1,161 degree days was equated with 7·12 m of mean ablation. The calculated constants for the two years varied by only 6 per cent. However, Orheim's sample of only 2 years is too small for comfort, even if Liestøl (1967) and Schytt (1964) have found similar relationships on other glaciers (Figure 5.4). Glaciers can exist in an approximately steady state where the number of degree days is well above 1000, but only if accumulation rates are high. To conceptualize, the lower the solar radiation, the greater are the chances of glacier survival, with the overriding proviso that solid precipitation must be adequate for glacierization.

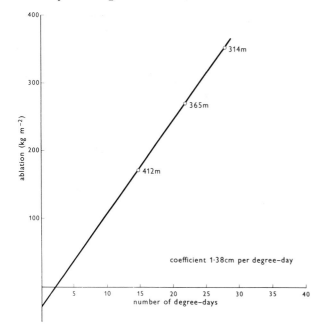

Figure 5.4
The straight-line relationship between accumulated degree days and ablation on Vestfonna, Svalbard. *(Schytt, Geogr. Annlr 1964.)*

(c) *Latitude*
It has often been considered that glaciers are by preference polar in their habits (Figure 1.1). According to Tricart (1969), 'glaciation is zonal and the ice sheets occupy some of the coldest regions of the globe'. Also, Davies (1969) referred to glacierization within the context of 'climatically controlled systems', and on latitudinal grounds distinguished the ice cover zone, the frost rubble zone and the tundra zone as the three major cold climate morphogenetic zones of the northern hemisphere at the present day. In this he followed the style of Büdel (1948), Peltier (1950) and French geomorphologists such as Tricart and Cailleux (1965) and Birot (1968a). While such an approach, stressing the control of latitude, is undoubtedly valuable in some respects, it tends to oversimplify the relationship which exists between glacierization and the standard climatic parameters, and to ignore such critical factors as winter precipitation totals.

Nevertheless, latitude is an independent variable at a world scale, with air temperatures and, to a lesser extent, precipitation dependent upon it. The world's high latitude areas are favourable for the existence of glaciers because of the fundamental control exerted on solar radiation (Hattersley-Smith, 1974). These areas receive relatively low amounts of annual radiation (Figure 5.3) and experience prolonged winters with more or less unbroken sub-zero temperatures. It is no coincidence that the North Polar ocean basin is covered with pack ice and the South Polar landmass is buried beneath a continental ice sheet. Both of these qualify as large-scale world features of the second order (Table 1.1). As early as 1912 Paschinger showed how the regional snowline (i.e. the line at which annual accumulation equals annual ablation) is influenced by latitude at a world scale (Figure 5.5). Also, at a continental scale we might expect a good correlation between latitude and glacier altitude. Meier (1960) has demonstrated the relationship between mean glacier altitudes and latitude along the North American cordillera (Figure 5.6) and Hastenrath (1971) has shown how

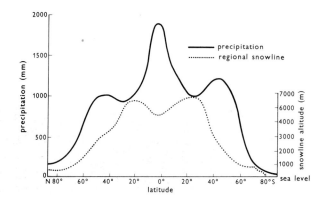

Figure 5.5
Generalized curves showing the variations of the regional snow-line with latitude, together with latitudinal precipitation averages. *(Modified after Paschinger, Peterm. Geogr, Mitt. 1912, and Sellers, Univ. Chicago Press 1965.)*

the modern snowline gradually falls by about 2300 m along the crest of the South American cordillera between latitudes 24°S and 33°S. However, at a regional scale it is less easy to demonstrate the effect of latitude on snowline or glacier distribution, and one has to search for the influence of other variables in combination.

(d) *Altitude*
Altitude can be considered an independent variable at the regional or local scale, and again it exerts fundamental control over climatic parameters and hence on glacier distribution (Østrem, 1974b). Flint (1971, p. 22) summarized the local influence of altitude thus: 'No one who examines the present day distribution of glaciers can fail to realize that glaciers are related to highlands. Without high and extensive mountains some of them situated in the paths of moist winds, extensive glacierization cannot occur'. Drewry (1972) illustrated this point by reference to Antarctica, the glacierization of which was probably initiated by a phase of upland glaciation in the Transantarctic mountains.

Evans (1969) referred to the altitudinal zonation for mountainous areas as glacial/nival, sub-nival, alpine and sub-alpine. This is very much a middle and low latitude viewpoint, and it must not be forgotten that glacierization occurs down to sea-level in high latitudes, but

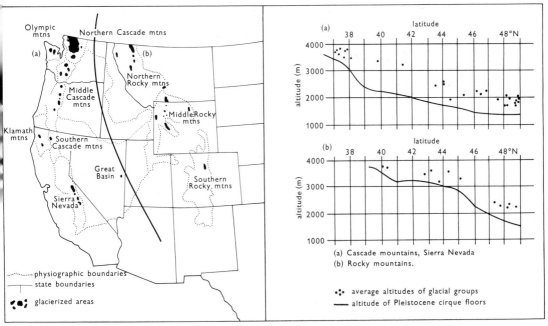

Figure 5.6
Glacierized areas and the control of latitude over glacier altitudes, western USA. The map on the left shows the distribution of glacier groups in two main regions, separated by the heavy line. The graphs on the right show a steeper gradient for the inland region **(b)** than for the continental periphery **(a)**. *(Meier, Int. Ass. scient. Hydrol. 1960b.)*

only at high altitudes in low latitudes. For example, the only three glacierized mountain peaks of Mexico have summit altitudes above 5,000 m (Lorenzo, 1969). Overall, as pointed out by Klute (1921) and many others since, the regional snowline falls irregularly from the tropics towards the poles, and this irregularity is again due to the interaction of many environmental variables. However, glacierization cannot occur unless there are upland areas above certain critical altitudes, these altitudes varying partly in relation to latitude (Figure 5.7, overleaf).

(e) *Relief*
An important morphological variable is surface relief. This interacts with the variables already described to exert its own specific control over glacierization at a number of different scales. At a continental scale, for example, an expanding Antarctic ice sheet would have its dimensions limited simply because beyond the edge of the continental slope the bedrock base is too far beneath sea-level for the ice to remain in contact (Hollin, 1962). At the scale of individual mountain massifs, surface form again exerts a powerful influence over the types of glaciers which inhabit them; hence in east Greenland the Staunings Alper, with many steep peaks rising precipitously to 3,000 m have a less complete ice cover than the adjacent plateau areas of Renland and Milneland (Figure 5.8, p. 91).

At more local scales, individual plateau glaciers and mountain valley glaciers have obvious links with topography which are indicated by the descriptive names used. Manley (1955)

Fig. 5.7
Part of the western Alps and Haute Savoie, with Lake Geneva at the top of the image. Here the snow-covered Alps stand out clearly, as does the pattern of glacial troughs. The glaciers in this area are for the most part more than 2000 m above sea-level. *(ERTS image No. E—1060—09554—5 01; reproduced by permission of NASA.)*

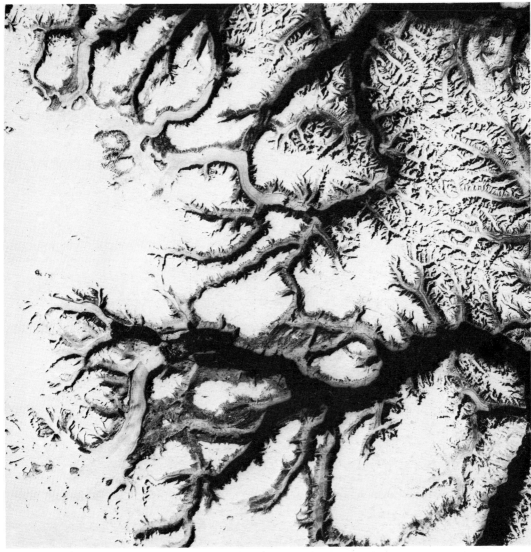

Figure 5.8
The contrast in ice cover between the Staunings Alper (upper right) and the adjacent plateau areas of east Greenland. The long fjord running from left to right in the lower part of the picture is Nordvestfjord. *(ERTS image No. E–1368–13255–7 01; reproduced by permission of NASA.)*

has shown how the breadth of an individual summit determines whether or not a glacier can be supported, and this concept has also been used by a number of authors in evolving the idea of the 'glaciation level'. Østrem (1964, 1966) followed Ahlmann (1937) in using the Partsch–Bruckner method of defining this limit, recognizing the part played by relief in determining glacier distribution (Fig. 5.9, overleaf). In short, glaciers must have suitable topographic situations in which to exist.

In highly dissected alpine uplands, glaciers are restricted to the valleys; nourished largely by avalanches, such glaciers often reach lower altitudes than usual. Steep slopes often remain ice free, even well above the glacierization limit. Such conditions are typical of glaciers in the Pamirs and other mountain areas of central Asia and steep islands such as South Georgia. In similar vein, Andrews *et al.* (1970) have drawn attention to the 'snow fence' effect of jagged mountain scenery in Baffin Island in trapping snow and allowing cirque glaciers to develop at lower than normal altitudes.

The positions of individual glacier snouts are often influenced by the surrounding topography. Whereas mass balance conditions will determine the overall position of the snout, the precise location is strongly influenced by the morphology of the channel. Mercer (1961) has stressed the importance of this in Alaskan fjords where calving snouts tend to stabilize at points of local widening, such as at tributary junctions. Funder (1972) has confirmed

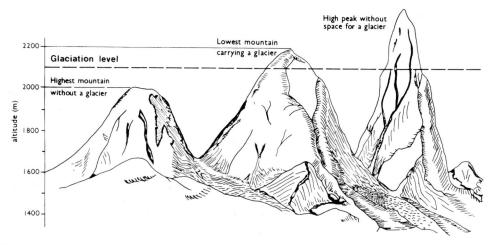

Figure 5.9
The Partsch-Bruckner method of determining the 'glaciation level', previously termed the 'glaciation limit'. The regional glaciation level lies between the summits of the lowest suitable mountain carrying a glacier and the highest suitable mountain without a glacier. (*Østrem, Geogr. Annlr 1966.*)

this tendency in the east Greenland fjord region where moraines reflecting stillstands of the fjord glaciers during deglaciation are concentrated at local points of widening. In the South Shetland Islands, glacier margins are often located at the position of the contemporary shoreline. Any moraines formed at such sites are the result of the lie of the topography, and have little climatological significance.

At the local or small scale, the control of relief is exerted in circumstances where microclimatic factors can induce the growth of perennial snowfields or ice masses. For example, topographic irregularities (particularly hollows) on hill slopes may induce drifting snow to lodge and accumulate. In northwest Iceland the broad plateau which once held the extensive Glamujökull is now ice free, but there are small cirque glaciers at lower altitudes on the lee side of the plateau edge (John, 1974). Andrews *et al.* (1970) have devised a 'shape ratio' which can be used as a morphological variable in the statistical analysis of glacier distribution. The shape ratio is defined as a ratio of elevation to area, where the area refers, for example, to land over 600 m. The ratio increases as a mountain becomes more and more peaked. In the Home Bay and Okoa Bay areas of Baffin Island, the presence or absence of glacierization on particular mountains is considered to be a function of land area over 600 m and the 'shape ratio'. Plateau surfaces above the glacierization level may also remain ice free if prevailing winds are powerful enough to sweep them clear of snow, whereas they may support numerous snowfields and glaciers if there are sufficient surface undulations. At the local scale, Young (1972) has shown how snow accumulation within a river basin catchment varies according to the geometric properties of the surface; he has defined a 'roughness index' which is of great potential value at larger scales also.

Highly dissected topography with precipitous slopes can sometimes inhibit glacierization. However, if snowfall is ample and effective, and particularly if conditions are favourable for the accumulation of rime ice, glaciers can still maintain themselves apparently without difficulty. As examples one may cite the steep ice covered ridge of Mount Friesland, Livingston Island (in the South Shetland group) and at a larger scale the Graham Land peninsula itself (Linton, 1964; Holtedahl, 1929). Also, tropical glaciers can exist on slopes of more than 40° if a high proportion of total precipitation is in the form of hail or rime ice (Tricart, 1969).

(f) Aspect

The orientation of the ground surface with respect to incoming solar radiation is particularly important at the local scale. Chorlton and Lister (1971) in their studies of Norwegian glacierization, found that aspect exerted relatively little control over glacierization at the regional and larger scales. However, as mentioned in chapter 3, slope orientation or aspect exerts a profound control particularly in marginal upland situations where the regional snowline is not far below the mountain crests (Evans, 1969).

In such areas the snowline may vary through a vertical range of several hundred metres according to slope aspect. This is because aspect affects the surface receipt of both solar radiation and precipitation, particularly in middle and high latitudes. Steep north facing slopes receive the least direct radiation in the northern hemisphere, but slopes facing east of north are the coolest. This, combined with the fact that northeast facing slopes are the leeside slopes in areas of prevailing southwesterly winds, explains why the cirque glaciers and firn fields of many upland areas are preferentially orientated towards the northeast (Figure 5.10, overleaf). In some situations which are apparently marginal for glacierization, upland asymmetry has contributed towards the growth of perennial ice masses; an example is the polar Urals, where Dolgushin (1961) has described cirque glaciers in an area whose regional snowline is supposedly above the highest summits.

(g) Distance from nearest ocean

The importance of this factor has been stressed by Chorlton and Lister (1968; 1971). From their series of regression analyses they found that the pattern of snow accumulation over Antarctica was influenced by one geographical parameter above all, namely distance to the nearest ocean which is ice free in summer. Similarly, from a study of glaciers in Norway they suggested that the only real independent variable influencing the glaciation level (and hence the actual distribution of glaciers also) is distance from the nearest ocean, defined in their study by the position of the 200 m submarine contour (Figure 5.11, p. 95). This of course is a measure of continentality, with an influence upon both total precipitation and annual temperature regimes.

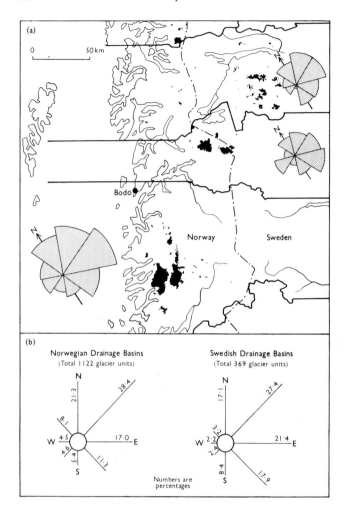

Figure 5.10
Present day glacier orientation in Scandinavia. *(Østrem et al., Atlas over Breer i Nord-Skandinavia 1973.)* In **(a)** the diagrams show the proportions of glacier catchments in eight compass sectors, for three arbitrary glacierized zones in northern Scandinavia. In **(b)** the diagrams present data for the whole of northern Scandinavia.

(h) *Summary of interrelationships*

In reality the glacierization of a continent, or a highland region, or a cirque glacier catchment, is controlled by a complex interaction of all the variables mentioned above (Figure 5.12). At every scale there are probably additional factors which should be considered. For example, at a large scale one could discern a relationship between ocean currents and contemporary glacierization. One could also discuss the width of landmasses as an effective control on ice sheet development. Both of these ideas could be used to explain the great imbalance of glacierization between northern and southern hemisphere landmasses. Thus Africa and Australia are seen as tropical and subtropical continents which have overall altitudes too low for the existence of anything more than minute ice masses. Also they are inadequately located with respect to planetary wind systems and ocean currents. Fundamentally, however, north of Antarctica the southern hemisphere has too little land for large scale glacierization.

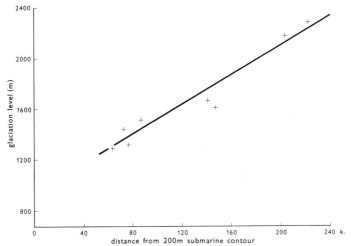

Figure 5.11
Graph for Norway showing the relationship between the altitude of the glaciation level and 'ocean distance', defined as distance from the 200 m submarine contour. *(Chorlton and Lister, Norsk geogr. Tidsskr. 1971.)*

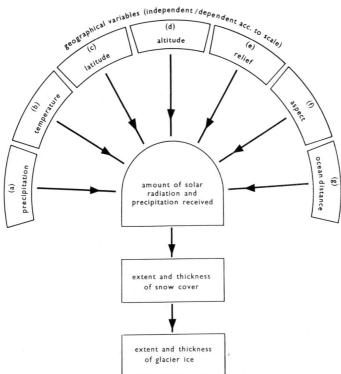

Figure 5.12
A simplified diagram showing how geographical variables exert a control, though a chain of linkages, over the extent and thickness of glacier ice. In reality there are also many linkages between the variables themselves.

From the above consideration, it seems that the significance of specific morphological and dynamic variables as controls over glacierization varies according to the level of resolution being employed (Table 1.1). Hence aspect and relief are seen to be relatively unimportant for glacierization at a large scale. Patterns of atmospheric circulation and seasonal variations in storm tracks exert a profound influence upon precipitation, temperature, and wind direction and force (Barry and Hare, 1974); undoubtedly they affect the glacierization of second order and all smaller scale landscapes (Flint, 1971). Fourth and fifth order landscape systems exhibit strong positive correlations between the degree of glacierization and local climate, influenced by such factors as relief and aspect; the relationship is well illustrated by the glaciers of New Zealand (Porter, 1975). Sixth order glacial landforms may well be related to geological controls also, including the character of the ground and local site conditions (Tricart, 1969). For example, areas of dark rock have a low albedo and hence they absorb heat relatively rapidly; in areas of marginal glacierization these areas will be less favourable for the growth of snow-patches and névé fields than areas of light coloured rock.

In an attempt to summarize the points made above, the significance of variables at different scales is suggested in Table 5.2.

Table 5.2   A tentative indication of the factors which control glacier distribution at different scales.

Variables	Larger scale	⟶	Smaller scale
	(a) Ice sheets and ice caps	(b) Ice shelves	(c) Glaciers constrained by topography
(a) High latitude	Yes	Yes	Yes
(b) High altitude	Yes	No	Yes
(c) Low temperature	Yes	Yes	Yes
(d) High precipitation	Yes	Yes	Yes
(e) High relief	No	No	Yes
(f) Poleward aspect	No	No	Yes
(g) Proximity to ocean	Yes	Yes	Yes

Glacier 'inertia'

Studies of many glaciers (particularly ice caps and ice sheets) have commonly revealed that they are out of equilibrium with their climatic environment and even devoid of any clear relationship with such morphological parameters as altitude, aspect and relief (Andrews *et al.*, 1970). Clearly, glaciers have considerable resilience or inertia which enables them to survive for prolonged periods in areas which are no longer environmentally favourable. Glaciers can achieve self perpetuation and even advance during periods of slightly negative balance, creating problems for regression analyses of geographical parameters and contemporary glacier distribution. The most important factor here is glacier response time. As mentioned in chapter 1, this can delay glacier reaction to a change in environmental conditions. Thus historical factors have to be considered, even in this discussion of present day glacierization on the surface of the globe. Glacier reaction to environmental change may be measured in periods of a year or in millennia, depending on the size of the ice mass; exactly how the glacier responds, and why the response may be delayed or more or less instantaneous, depends upon such matters as ice velocity, ice thickness and the transmission of kinematic waves. These matters have already been referred to in chapter 3, and their implications will be discussed in more detail in chapter 6.

Another important point is that all glaciers, particularly ice caps and ice sheets, exert some control over their own local climate, so that glacial climatic conditions may be maintained even in unfavourable regional climatic situations. Small glaciers are largely the progeny of regional and local climate but, as glacier dimensions increase, an important feedback mechanism comes into play whereby glaciers increasingly affect their own climates. This is particularly important on ice sheets whose large size and high altitude may be major factors responsible for the cold climate necessary for the continued preservation of snow. It is often argued, for example, that if the ice were suddenly removed from Greenland and Antarctica, the climate would perhaps be too mild to nourish new ice sheets. The implications of this argument, at least on the continental scale, are examined in more detail in chapter 7. However, at this stage it is important to stress glacier-climate relations at much smaller scales also. There is evidence from Norway and Iceland, often in the form of local folklore, that the climate around ice caps which are often no more than a few tens of km² in area, is both sunnier and drier than would be expected if the ice cap were not present. Indeed, one farmer in northwest Iceland claims that the success of his glasshouse close to Drangajökull is partially due to higher than average sunshine hours.

Unglacierized enclaves and nunataks

In areas subject to a high degree of glacierization it is not uncommon to find extensive land surface areas devoid of perennial snow or ice. Nunataks, for example, are common on the peripheries of both the Antarctic and Greenland ice sheets, where mountain summits project above the ice sheet surface (chapter 4). If peaks are precipitous, snowfall may not lodge on mountain summits or flanks, and will simply avalanche down on to the ice sheet or glacier surface beneath.

Another critical factor is the effect of bedrock upon the regional albedo. As noted earlier, a dark bedrock surface with a low albedo will readily absorb heat from solar radiation, and consequent snow melting will further assist in the maintenance of an ice free condition. Where a nunatak projects starkly above an ice surface it will promote increased ablation – on the flanks of the nunatak (Figure 4.17). At a larger scale a considerable problem concerns the continued existence of some non-glacierized enclaves or 'oases' of Antarctica which are not strictly projecting mountain summits or upland massifs at all (Hatherton, 1965; Wilson, 1972; Calkin and Nichols, 1972). Some of these areas do not project far above the ice sheet

Figure 5.13
An oblique air photograph of part of the Schirmacher Ponds oasis in Dronning Maud Land, east Antarctica. *(Copyright Norsk Polarinstitutt.)*

surface and they are not too steep to support glaciers (Figure 5.13). Indeed, the Victoria Land dry valley system (Figure 5.14) and Bunger's Oasis have extensive areas below the level of the adjacent ice sheet surface. However, these occur in particularly arid areas, like Peary Land in north Greenland, and it seems that their bedrock surfaces, once exposed, have been able to remain ice free as a result of their increased rates of surface heat absorption. Bunger's Oasis, for example, is heated so effectively that its mean annual air temperature is only about $-10°C$ (Tricart and Cailleux, 1965). Under such circumstances precipitation is largely ineffective; most available moisture is removed each summer by sublimation even at sub-zero temperatures (Wilson, 1970).

The occurrence of different glacier systems

To conclude this chapter, it is useful to look again at the three main types of glacier system referred to in chapter 4, concentrating this time on the overall conditions necessary for their initiation and survival. Some of these conditions are summarized in Table 5.2.

(a) *Ice sheets and ice caps*
These form where precipitation input into a landscape system is too great to be dealt with by glacial discharge through troughs, so that output cannot compensate. In other words,

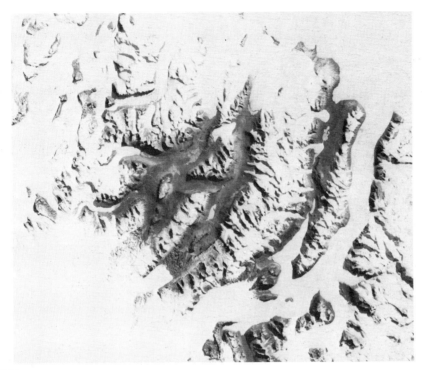

Figure 5.14
ERTS image of the ice-free valley system of southern Victoria Land, Antarctica. The main troughs comprising the oasis are (from left to right) Barwick Valley, Victoria Valley, McKelvey Valley, Wright Valley and Taylor Valley. *(ERTS No. E–1174–19433–7 69; reproduced by permission of NASA.)*

Figure 5.15
Oblique aerial photograph of Nupsfonn in southern Norway – an example of a decaying ice cap. *(Norsk Polarinstitutt photograph.)*

precipitation is highly effective once the threshold has been crossed and the glacier is growing. The glacier system, once initiated, overflows existing topographic irregularities and acts as a greater and greater storage reservoir. The reason for the inadequacy of the system to achieve rapid throughput may be insufficient land surface gradient or insufficient discharge capacity in existing valleys. If the glacier has a sufficiently broad base (such as an extensive plateau or undulating upland region), an ice cap or ice sheet equilibrium profile can be built up and maintained. As mentioned earlier, the two current ice sheets now maintain an approximate state of balance even where precipitation input is relatively low; this is because of the climatic positive feedback, in which ice masses influence the regional climate extensively enough to reduce rates of summer melting to a level where there is no drastic deficit in glacier budget. At present the two ice sheets are in a somewhat healthier state than most ice caps. For example, the Hardangerjökull and Barnes ice cap are in decline, and there are a number of small ice caps, such as the Drangajökull in Iceland and the Nupsfonn in Norway (Figure 5.15), which perhaps no longer enjoy the support of local climate. Indeed, they may only have a short life expectancy (Ostrem et al., 1969; Thorarinsson, 1943). Already glaciologists mourn the passing, less than a century ago, of the youthful Glamujökull in northwest Iceland, which was only 500 years old but which had been in a state of poor health for some time (Thorarinsson, 1943). On the other hand some ice caps in alpine uplands (such as the Himalayan or the North American cordillera) are adequately maintained by a high and sustained input of snowfall.

(b) *Ice shelves*
These are specifically polar in their distribution. They occur widely around the peripheries of Antarctica, but in the northern hemisphere they are restricted to one or two favourable high latitude areas on the northern fringes of the Canadian Arctic archipelago. They require mean annual air temperatures below zero, and relatively high precipitation. As indicated in Table 5.2, aspect is unimportant, for while the northern hemisphere ice shelves face polewards their Antarctic equivalents face equatorwards. Relief is, however, a significant control, for ice shelves require either broad coastal embayments or groups of islands in order to remain safely anchored. Their growth may be inhibited in coastal areas where there are powerful ocean currents or where there is a large tidal range. They are also related to sea-level in a sensitive way; a drop of sea-level would convert many ice shelves into simple extensions of the ice sheet, whereas a rise of sea-level would cause many ice shelves to become unstable and break up. Indeed, sea-level control and the sensitivity of ice shelves to air and sea temperatures has been stressed by Mercer (1968a) and Hughes (1975) in explaining the erratic glacial history of west Antarctica.

(c) *Glaciers constrained by topography*
These tend to form where precipitation is effective, as long as other variables also favour their growth. However, precipitation must not be excessive, for it must be dealt with within the limits of efficiency of valley glacier systems with their characteristic patterns of energy expenditure and geometry (Miller, 1973). Often there appears to be a nice state of adjustment to prevailing climatic conditions, and many valley glaciers today have mass balances in approximate equilibrium, being either slightly positive or (more frequently) slightly negative. In some situations, valley glaciers form where ice sheets and ice caps cannot build up, perhaps because land surface gradients are too steep for an ice sheet profile to be maintained.

Examples are high alpine uplands, steep continental edges, and local massifs which are sufficiently elevated to rise above an ice sheet surface.

Further reading

CHORLTON, J. C. and LISTER, H. 1971: Geographical control of glacier budget gradients in Norway. *Norsk geogr. Tidsskr.* **25,** 159–64.

DERBYSHIRE, E. and EVANS, I. S. 1976: The climatic factor in cirque variation. In Derbyshire, E. (ed) *Geomorphology and Climate*, Wiley, London.

IVES, J. D. and BARRY, R. G. (eds) 1974: *Arctic and alpine environments*. Methuen, London (1,024 pp).

MEIER, M. 1960b: Distribution and variations of glaciers in the United States exclusive of Alaska. *Int. Ass. scient. Hydrol.* **54,** *Gen. Assembly of Helsinki, 1960*, 420–29.

MEIER, M. 1965: Glaciers and climate. In Wright, H. E. and Frey, D. G. (eds) *The Quaternary of the United States*, Princeton Univ. Press, NJ, 795–805.

ØSTREM, G. 1966: The height of the glacial limit in southern British Columbia and Alberta. *Geogr. Annlr* **48A** (3), 126–38.

PORTER, S. C. 1977: Present and past glaciation threshold in the Cascade Range, Washington, U.S.A. Topographic and climatic controls and palaeoclimatic implications. *Journal of Glaciol.* **18,** 101–16.

6 Short and medium term glacier fluctuations

The relationship between glaciers and climate has a long tradition of careful study by glaciologists, for it is an essential part of mass balance research. The topic is also now the subject of intense debate among the general public, partly because of the somewhat catastrophist tone of Calder's (1974) book entitled *The Weather Machine and the Threat of Ice*. It has long been appreciated that glaciers respond sensitively to short term fluctuations of climate, and hundreds of papers have been published over the past twenty years on this topic. Figure 6.1 shows the chain of linkages between climate and glacier snout behaviour. Smaller valley glacier systems and cirque glaciers may respond recognizably to changes of input in 100 years or less (Table 6.1), and for this reason we are largely concerned in this chapter with these rather than with ice caps or ice sheets. Larger ice masses, and their response to low frequency and high magnitude climatic events, are considered more fully in chapter 7.

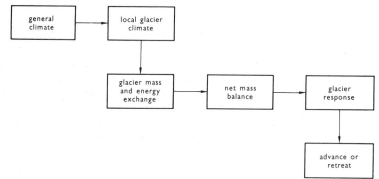

Figure 6.1
The chain of linkages which exist between general climate and glacier snout behaviour. *(Meier, Princeton Univ. Press 1965.)*

The relationship between glacier behaviour and climatic events is less simple than is commonly supposed (Paterson, 1969; Andrews *et al.* 1970), and glaciological research is now revealing a great deal of information on the different ways in which ice masses can respond to meteorological and climatic oscillations in the short and medium term (Bradley, 1973b). This chapter attempts to discover just how glacier systems respond over time to climatic and other changes on a time scale of less than 10,000 years. It takes examples some of the glacier adjustments which have occurred during the Holocene, and particularly during that period covered by reasonably accurate historical records.

From the dark ages and middle ages there are scattered historical records of glacier variations, although such records are not frequent enough for reliable palaeoclimatic reconstructions until after about AD 1700. Within the present century there is a wealth of detail concerning glacier snout fluctuations, derived from direct measurements, maps, published papers and photographs. However, comprehensive mass balance data have only become available since

Table 6.1   Relationship of different time scales of climatic change and the response level of glaciers. *(Modified after Andrews, Barry and Draper, 1970.)*

Wave length order of climatic change	Approximate time scale (years)	Glacier type affected
1 Long term fluctuations (low frequency high magnitude events)	100,000–10,000	Ice sheets
2 Medium term fluctuations	10,000–1,000	Ice caps, ice fields and larger valley glaciers
3 Medium term fluctuations	1,000–100	Smaller valley glaciers
4 Short term fluctuations	100–10	Cirque glaciers
5 High frequency low magnitude oscillations	10–1	Small névé fields and snow beds

about 1920, when Ahlmann began his important researches around the North Atlantic coasts with various collaborators. Over the last decade continuous mass balance data have become available for many glaciers in the mid latitudes; but there are few high latitude polar or sub-polar glaciers which have been continuously studied, and there is still no glacier anywhere which has continuous mass balance data for a period of 40 years. There are derived mass balance data of uncertain reliability for the Aletsch Gletscher back to 1922 (Kasser, 1967) but of considerably greater importance is the work on Storglaciären, Sweden (direct records since 1946) undertaken by Schytt *et al.*, and on White glacier on Axel Heiberg Island (direct records since 1959) undertaken by Müller *et al.* Hence, reliable assessments of the relationships between glaciers and climate can only be made thus far for high frequency events of low magnitude, such as the minor oscillations of the last few decades.

Factors controlling glacier response over time

In the last few centuries, glacier snout fluctuations of many kilometres have frequently been recorded, and mid latitude upland glaciers seem particularly sensitive to climatic change. Glacier response over time has to be examined in the context of glaciological principles, and again it is worth searching for some of the laws which control glacier behaviour.

As mentioned in chapter 1, most glacier snout fluctuations can be seen as adjustments to changes of input into the glacier system. They indicate attempted self regulation by glaciers, in which they try to achieve a new state of dynamic equilibrium with their environment (Miller, 1973). The time interval between the change of input and the achievement of a new equilibrium by the glacier has been termed the *relaxation time* (Chorley and Kennedy, 1971) or *response time* (Paterson, 1969). Probably glaciers never achieve true equilibrium with their environments, for climatic oscillations of various amplitudes and return periods ensure that glacier environments are anything but constant. One factor to be considered in this context is the 'amplification factor', by which a small change in mass balance may initiate a large change at the glacier terminus. This has already been considered in chapter 3. Another factor is the extent to which a glacier can influence its own climate, and Paterson (1969) and Dugdale (1972) have suggested that glaciers tend to acquire specific mass balance characteristics which may 'damp down' minor climatic oscillations. The means whereby slight increases of accumulation (for example) are accommodated by the glacier can be treated as a mild case of negative feedback (Figure 6.2). As mentioned in the previous

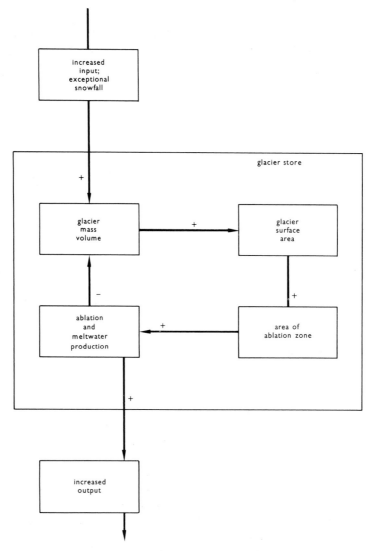

Figure 6.2
A negative feedback loop representing a chain of events which occurs due to an increase in precipitation. The glacier 'damps down' the effects of the increased input in its attempt to attain a new equilibrium.

chapter, however, the scale of the glacier is all important in this respect, for each glacier will only respond to climatic changes of appropriate magnitude and return period. The theoretical *steady-state* situation (in which glacier net balance remains at zero for many years and in which glacier dimensions remain constant) is probably rarely attained, although it may be approached by some Antarctic alpine glaciers. The Antarctic ice sheet may have the appearance of a steady-state glacier, but this may be because its response time is so enormous that it cannot adequately be measured.

Glacier morphology
The morphological parameters of glacier systems play an important part in determining how they will respond to climate. Variations in the altitudinal distribution curves of glaciers

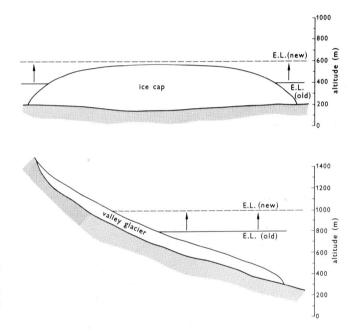

Figure 6.3
Illustration of the effects of a 200 m rise in the altitude of the equilibrium line for a small plateau ice cap and a valley glacier. Note the greater sensitivity of the ice cap.

influence their sensitivity to changes in input. Generally speaking glaciers with a large proportion of their area close to the equilibrium line are more susceptible to change than those with a more evenly distributed area/altitude graph. This is illustrated in Figure 6.3 which shows two hypothetical glaciers with the equilibrium line drawn so as to divide the glacier into accumulation and ablation zones of approximately equal vertical extent. The ice cap has an equilibrium profile and the bulk of its surface area at higher altitudes. The typical valley glacier has its area distributed more equally throughout the vertical range of altitude. Imagine a climatic warming of approximately 2·5°C, which raises the equilibrium line vertically by *c*. 200 m. The change has a proportionately greater effect on the ice cap because of the limited altitudinal range of its collecting ground; indeed, if the ice cap were about 35 km across, the change would be sufficient for the equilibrium line to clear the ice cap completely and thus reduce the accumulation area to zero. On a valley glacier the reduction is relatively less important. For example, a fluctuation of this size even on a small glacier like South Cascade glacier (3 km long) would still leave it with one quarter of its accummulation area. On a mountain glacier with a larger surface area, then of course the effect would be much less.

The shape of the glacier in a strictly plan view affects the response time of a glacier if only because of variations in the distance between the accumulation area and the snout. Given a climatic fluctuation of a certain size, a narrow valley glacier will, other variables being equal, take longer to respond than a more compact ice cap of similar size. This is simply a reflection of the greater distance over which the change in glacier mass must be transferred (Miller, 1973).

Behaviour in response to mass balance changes is different for a calving glacier than for a glacier which ends on land. On land a glacier can react to changes of input by extending or withdrawing its snout. An extension of the snout means that an increased surface area is exposed to ablation since the glacier advances into warmer areas at a lower altitude. A withdrawal has the opposite effect and less ice is subjected to ablation. A glacier in a deep fjord of constant width cannot achieve a new equilibrium so easily. The ablation area of the glacier is limited to the lower reaches of the glacier and, more importantly, to the amount of ice that can be melted or otherwise discharged from the cross-sectional area of the snout. If there is a shift to a positive net mass balance the glacier begins to advance. Since it cannot extend to lower altitudes to enhance ablation, it will continue to advance until it can spread out and increase the cross-sectional area exposed to melting and calving (Figure 3.1; Figure 6.4). This will only occur in a less constricted place in the fjord. This means that fjord glaciers are particularly sensitive to changes in mass balance and that a relatively minor climatic fluctuation can cause spectacular variations in the position of the snout. Mercer (1961) described some dramatic examples from Alaska. He also makes the point that any land based glacier in a constricted trough with little altitudinal variation is likely to behave in a similar fashion. Funder (1972) used this thesis effectively in his analysis of east Greenland fjord glacier moraines, and there are profound implications for many other areas also.

Mass balance changes
There are many illustrations in the literature of delicate year-to-year glacier snout fluctuations which relate to climatic events just a few seasons previously; and in some cases glacier snouts seem to respond to exceptional climatic events within a single season. Responses which are so rapid should be termed 'direct' according to Nye's (1965) terminology. As pointed out in chapter 3, the minor oscillations of glacier snouts are the result of direct response to annual climatic oscillations, whereas major advances or retreats are the result of indirect, or lagged, response to more significant longer term changes in input to the glacier system.

The direct response of the snout of a small glacier depends partly on the character of short term mass balance change. A negative mass balance change may be reflected in one season, whereas a positive change may not be apparent for several seasons. In general, it is known from studies of small, relatively active valley glaciers that climatically induced snout retreats occur more rapidly than climatically induced snout advances. This is because increased ablation tends to affect the snout region immediately, whereas increased accumulation generally affects the snout only after a time lag. This is a point too often disregarded in response time studies. On the other hand small valley glaciers and outlet glaciers (such as those around Drangajökull, northwest Iceland), whose snouts may be only a few hundred metres below

Figure 6.4
Oblique aerial photograph of two outlet glaciers which reach the sea on either side of Kap Alexander, Ingefield Land, west Greenland. Note that their calving snouts are stabilized at the points of trough widening. *(Reproduced with the permission (A421/74) of the Geodetic Institute, Denmark.)*

their collecting grounds, may occasionally remain covered by fresh snow throughout a whole ablation season. In this situation, one or two snowy seasons with colder than average summers can convert a snout retreat into a stillstand or even an advance. Hence if a valley glacier has a velocity of 5 m per year at the snout, it only requires two ablation seasons during which the snout remains snow covered to create a 10 m glacier snout advance. Even on a sluggish glacier like Storglaciären, a few years of unusually high accumulation and unusually low ablation (as occurred in 1961–5) may be sufficient to interrupt a long term retreat with a short term stillstand or even ice edge advance (Karlén, 1973).

The indirect response of many glaciers can be related to the behaviour of kinematic waves and surges. Kinematic waves must be considered because they are perhaps the means by which most mass balance changes are transmitted to the glacier snout. The mechanics of such waves have already been discussed in chapter 3, but it is important here to remember that it is often extremely difficult to pick them out on glacier surfaces. Glacier systems are continuously adjusting themselves to changes of input, and kinematic waves are associated with a type of spontaneous energy release (Miller, 1973). On a single glacier system there may be a number of waves moving towards the snout at the same time; some of these will move rapidly and others slowly, and nearly all will be unrecognizable as a result of diffusion and the effects of relief-induced irregularities of the glacier profile.

Glacier activity is important in influencing the velocity of kinematic waves. For example, very active glaciers such as those on the western side of New Zealand respond rapidly to climatic oscillations, whereas there is a much greater time lag in the response of the more sluggish eastern glaciers (Suggate, 1950). The north Greenland outlet glaciers are so sluggish that their response may lag several centuries behind that of the glaciers of southwest Greenland (Weidick, 1968).

Generally there seems to be good agreement between the theoretical behaviour of waves and observations made in the field. As a result of the mass balance studies of recent decades and from theoretical work, it is possible to calculate how a particular glacier may respond to a particular climatic event. For this work, data on some (or probably all) of the morphological parameters of a glacier system are required; for example, Untersteiner and Nye (1969) used data on glacier length, width, height of glacier surface above mean sea-level, and slope of surface, together with mass balance data, in order to compute the response time of the Berendon glacier in British Columbia. This work has been brought up to date with improved mass balance data by Fisher and Jones (1971).

Using specific mass balance histories, Nye (1965) calculated a specific response time for South Cascade glacier as 26 years and for Storglaciären as 55 years. Young (1972) also suggested that glacier margin response to climatic fluctuations in the Venedigergruppe, Austria, could be measured in decades. Nevertheless, indirect glacier response varies within certain limits according to the precise nature of the stimulus and the length of time over which it is applied. Glacier size is also important. The larger the glacier, the more difficult it is to calculate its response time; for example, Nye (1961) suggested that the response time of the Antarctic ice sheet is 'roughly 5,000 years', whereas Paterson (1969) calculated a response time of 2,500 years. Clearly such a difference indicates that there is still much uncertainty in this whole field.

It is worth bearing in mind that the work referred to above relies on the employment of simplified mathematical models. Although such models explain some observed features satisfactorily, one can expect many refinements as work on actual glaciers reveals more detail about how glacier response varies through time. Also a number of important assumptions are embodied in the mathematical models and these call for care before indiscriminate application. Apart from technical details concerning the manipulation of the equations, several assumptions are commonly made:

1   Climatic fluctuations are small.
2   Any effects caused by changes in the quantity of meltwater at the base can be ignored.
3   Changes in the temperature of the ice (and thus the relation of stress to strain rate) can be ignored.

If these assumptions are unjustified, then the calculated response times will clearly be unreliable.

To summarize, it seems that much of the response of glaciers to mass balance change is measurable and probably predictable, being essentially linked with the behaviour of kinematic waves. Glacier response time, like time lag, varies according to many factors, and each glacier has its own particular characteristics which will affect the type and speed of each response.

Surge behaviour
There is no doubt that many spectacular snout advances are related to surges (Schytt, 1969; Thorarinsson, 1969). One such advance is illustrated in Figure 6.5. As suggested in chapter 3, there is a strong possibility that the surging behaviour of many glaciers is periodic, and that the return periods for surges are often measurable in decades. Meier and Post (1969) proposed that surging glaciers have a cycle of activity. However, it is fundamental that the cyclic behaviour is not necessarily climatically induced (Budd, 1975). Surging glaciers appear to be unstable to the extent that positive changes of input are not continuously transmitted downglacier. Instead there is prolonged storage of surplus mass until a critical stage of instability or threshold is reached. The attainment of this critical stage may be predictable.

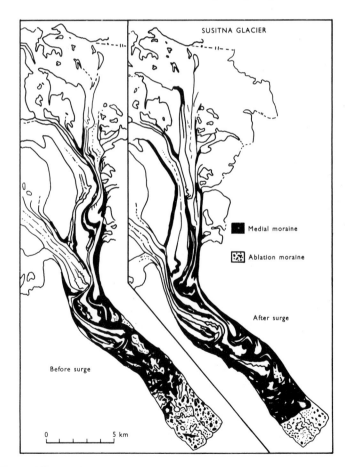

Figure 6.5
Maps of the Susitna glacier in the central Alaska range, showing the ice displacements resulting from the 1952 surge. Note that much of the stagnant debris-covered snout was 'reactivated' during the surge, resulting in a distinct advance. *(Meier and Post, Can. J. Earth Sci., 1969.)*

Table 6.2 Characteristics of some surging glaciers. *(Meier and Post, 1969.)*

Glacier	Total area (km^2)	Total length (km)	Length of surging part (km)	Average slope, surging part	Cycling period (yrs)	Duration of active phase	Maximum annual velocity (km/yr)	Maximum displacement
Bering, Alaska	5800	200	153	0.7°	30+15	3 yrs	—	9·7 km
Klutlan, Yukon	1072	55	40	1·3°	30+10	3 yrs	3·2	6·5 km
Walsh, Alaska	830	89	86	1·0°	50+10	4 yrs	5·6	11·5 km
Muldrow, Alaska	393	63	46	2·2°	50+10	2 yrs	6·6	6·6 km
Variegated, Alaska	49	20	19	4·2°	20+0·5	2 yrs	5·0	5·0 km
Tikke, British Columbia	75	19	18	3·6°	20+1	3 yrs	1·0	2·0 km

As with other types of glacier it seems possible to refer to the response time of surging glaciers, as long as it is realized that climatic oscilltions (particularly changes in precipitation) are capable of speeding up or slowing down the surging process. As is the case with other forms of glacier response, the wavelength and the amplitude of the surge cycle should be much shorter for small valley glaciers than for large outlet glaciers or ice cap margins. However, the illustrations on Table 6.2 serve to dispute this point; and it may well be that climatic environment and glacier characteristics, particularly at the glacier bed, are of much greater importance in the surging process than glacier morphology and dimensions (Miller, 1973).

Fluctuations of less than a year

At the smallest scale, there are many illustrations in the literature of the manner in which glacier velocity reacts to hourly, daily, weekly or seasonal fluctuations in weather patterns. There is no doubt that these fluctuations can cause changes in glacier dimensions. Generally the meteorological oscillations are low magnitude events of high frequency; some (such as the day/night melting cycle or the accumulation season/ablation season cycle) are predictable and have recognizable return periods, while others (such as exceptional sunny spells or periods of continuous rainfall) may be termed random (Figure 6.6).

There is a widespread impression that glaciers move slowly and smoothly, but in reality this is seldom the case. Glen and Lewis (1961) found rapid and irregular alterations in the

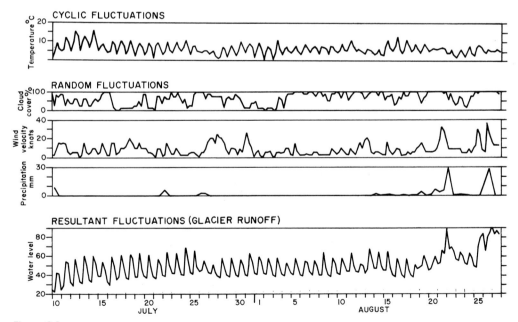

Figure 6.6
High frequency low amplitude oscillations in the environment of Tungnaarjökull, Iceland *(Guðmundsson and Sigbjarnarson, J. Glaciol. 1972 by permission of the International Glaciological Society.)* The temperature fluctuations are predictable, but cloud cover, wind velocity and precipitation oscillations are random. The cycle of water-level fluctuations is clearly affected by the rainy spells in late August.

speed of ice movement on Austerdalsbreen, and correlated these with periods of heavy rainfall and increased air temperatures within the day-to-day weather pattern. Meier (1960) suggested a correlation between the fluctuations in velocity of Saskatchewan glacier and daily meteorological conditions, as did Ives and King (1955) for Morsarjökull in Iceland. Hourly rates of ice movement may vary by 100 per cent or more, and variations in the rate of ice movement are seen to be even greater when the interval of measurement is decreased (Goldthwait, 1973). This suggests an overall ice movement by small jerks. These jerks may not be synchronous on different parts of a glacier, and they represent localized releases of accumulated strain. The analogy with the situation along an active terrestrial fault is obvious.

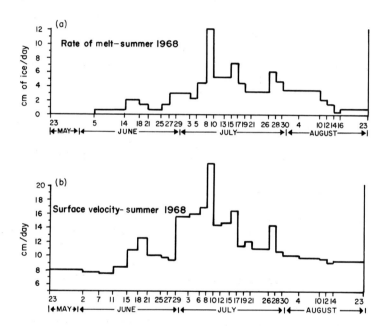

Figure 6.7
The relationship between (a) rate of surface melting and (b) ice surface velocity fluctuations, White glacier, Axel Heiberg Island. *(Müller and Iken, Int. Ass. scient. Hydrol. 1973.)*

As with a crustal fault, these releases of strain may be more rapid if there is adequate meltwater on the surface of contact, as suggested in the theory of basal sliding (Weertman, 1957; Lliboutry, 1968); hence it is not surprising that spells of very rapid ablation, or short rainy spells, can markedly affect the rate of movement of warm glaciers after a lag of a few hours or a few days (Paterson, 1964).

Weekly, monthly and seasonal variations in ice velocity are also well documented. Variations over days and weeks can often be related to ablation rates and meltwater discharge, even on some polar glaciers. Müller and Iken (1973), working on the sub-polar White glacier (Axel Heiberg Island), have shown that it reacts very sensitively (sometimes almost instantaneously) to changes in ablation and hence to meltwater lubrication (Figure 6.7). Velocity fluctuations for days, weeks and months could all be correlated with the rate of meltwater production on the glacier, suggesting that basal meltwater plays a significant part in glacier movement during the ablation season.

There is much published data for other glaciers also, particularly concerning seasonal fluctuations in ice movement. The Casement glacier in Alaska had an average slip rate (1966–1967) of 2·3 cm per day in winter and 2·9 cm per day in summer. During 1845–6 the surface velocity of Unteraargletscher was measured by Agassiz and found to vary between a maximum of over 36 cm per day in early summer and a minimum of about 23 cm per day in winter (Haefeli, 1970). Part of the seasonal variations in glacial movement (up to 15 or 20 per cent of the total) may be explained by the 'loading' of the accumulation zone by winter and spring snowfall, but this factor is clearly not adequate to explain the more seasonal velocity fluctuations recorded on many glaciers. For many temperate glaciers, the average summer and winter rates of movement may vary by 10–20 per cent or more (Paterson, 1969), and again meltwater lubrication must be invoked as a mechanism. From the work of Stenborg (1969; 1970) and Elliston (1973), it is apparent what variations there are in seasonal meltwater discharge rates from temperate glaciers. Many authors have related rapid ice movement in late spring and early summer to the time of peak meltwater storage (Hodge, 1974). Often there is a decrease in ice velocity in late summer relating to the evacuation of most meltwater from the glacier bed, representing reduced water storage and water pressure in the glacier. Minimum rates of temperate glacier movement are thus recorded during the winter months, when the presence of meltwater on the glacier bed is negligible (Hodge, 1974).

Glacier fluctuations over years and decades

In some cases where glacier margins have experienced shortlived advances interrupting an overall period of retreat, it is possible to explain them as rapid responses to minor climatic oscillations. Hence three of the glacier outlets of Drangajökull, northwest Iceland, advanced sharply during the decade 1933–42 (Figure 6.8). These advances were slightly out of phase; the Leirufjörður glacier, which was the last one to start moving forward, advanced about 1000 m in a three year period (Eythorsson, 1949). It is reasonable to relate this phase of glacier advance to a period of exceptionally heavy snowfall, particularly in the years 1931, 1934, 1936 and 1938. Similarly the Vatnajökull outlet glaciers appear to have slowed down

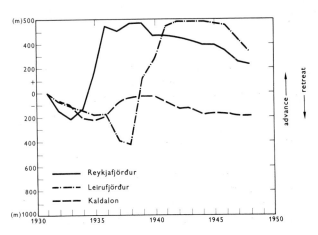

Figure 6.8
The advances of the Drangajökull outlet glaciers in the decade 1933–42. (Eythorsson, J. Glaciol. 1949 by permission of the International Glaciological Society.)

their rate of retreat (with some of them even advancing slightly) during the period 1950–1953, following a phase of falling air temperatures and increased precipitation in the late 1940s (Björnsson, 1970). In Arctic Canada a small ice cap near the glaciation limit in northeastern Ellesmere Island has begun to expand during the past decade, showing a very rapid response to a few years of cooler summers with increased snowfall (Hattersley-Smith and Serson, 1973). Similarly snowbanks have grown and small ice bodies have begun to advance in the mountainous Cumberland peninsula region of Baffin Island as a result of a recent summer cooling trend and increased precipitation during the last decade (Bradley and Miller, 1972; Bradley, 1973a). In all of these cases, glacier reactions to climatic oscillations can be measured in less than a decade, and there are other examples of indirect responses of twenty or thirty years.

At present it is not possible to correlate glacier fluctuations on a world scale over the past two or three decades, although regional trends have been recognized by a number of authors. As mentioned above, there seemed to be a trend towards cooler summers and heavier precipitation during the 1960s in Arctic Canada, which had a near instantaneous effect upon the extent of perennial snow cover. During the same decade the Alps began to experience heavier winter snowfalls, colder and more prolonged springs, and lower mean summer temperatures (Young, 1972). Vanni (1968) and Tonnini and Rossi (1968) related this trend to an increased number of stationary and advancing snouts, particularly in the Italian Alps. In the Caucasus, the percentage of advancing glaciers was increasing, and in Iceland the trend towards cooler summers was beginning to affect glacier snout fluctuations (Björnsson, 1970). A much greater body of data concerning recent high latitude climatic trends was summarized by Lamb (1972) and Namias (1969), again suggesting a hemispheric cooling trend. However, it is premature to talk about the onset of a 'snowblitz' or the establishment of a climatic 'downward spiral' (Calder, 1974); and there is as yet nothing to justify any concern about the imminent coming of a new Ice Age (Kukla *et al.*, 1972). In Svalbard there were signs that the early 1970s marked the beginning of a new warming trend, while western Europe experienced a sequence of unusually mild winters. For every paper which talks of imminent glacier buildup, there is another which suggests that the overall trend of glacier recession will continue.

Glacier fluctuations over centuries: the example of the 'Little Ice Age'

During the last few centuries many parts of the world have experienced a Little Ice Age (Figure 6.9). This period, dated in a general way to AD 1500–1920[1], represents a climatic cooling during which the majority of glaciers advanced, often on more than one occasion (Lamb, 1963; 1972; Manley, 1971). Many spectacular moraines were constructed, particularly by small ice cap outlet glaciers and valley glaciers in the uplands of the middle latitudes (Figure 6.10). Glacier snout advances from the inhabited parts of the northern hemisphere are well documented, and the impact of increased glacierization upon human communities was considerable.

---

[1]From a purely climatic point of view, there seems to be justification for limiting the Little Ice Age to a shorter period than this (see Lamb, 1972); while from glaciological data, Denton and Karlén (1973) argued that the phase lasted from the fourteenth century at least until 1920. In the present book, as a compromise conforming to most common usage, the Little Ice Age is defined as covering approximately four centuries, AD 1500–1920.

Figure 6.9
The Little Ice Age in the Alps. The snout of the Rhonegletscher sketched by F. Meyer about the year 1720.
It is in a very advanced position and is almost confluent with the Muttgletscher, seen on the right. *(Repro-
duced by permission of E. Le Roy Ladurie.)*

Evidence for the climatic fluctuations of the Little Ice Age has been collected from a vast
variety of sources, and it is now becoming possible to reconstruct some of its characteristic
weather patterns (Lamb, 1974). Some of the historical records of glacier fluctuations in Norway and the events (sometimes locally of catastrophic proportions) associated with them
were described by Hoel and Werenskiold (1962). Ladurie (1971) made great use of the records
of vineyards to trace Alpine climatic fluctuations, and Lamb (1959) referred to British fruit
growing, settlement history, cereal growing, the ease of ocean travel and even the aspect
of individual dated settlement sites as indicators of prevailing climatic conditions. Vegetation

Figure 6.10
The Alpine Glacier de la Brenva, 1767 and 1966. The sketch (above) by Jalabert is remarkably accurate, showing an advanced glacier position in 1767. The photograph (below) from 1966 shows the massive Little Ice Age moraine. *(Reproduced by permission E. Le Roy Ladurie.)*

studies (for example pollen analysis, lichenometry, and studies of changing tree-line altitudes) have provided valuable information on climatic oscillations in the last millennium (Pennington, 1969; Fritts, 1965; Miller, 1969; Davis and Wright, 1975). Elsewhere archaeological and pedological investigations have provided information on recent climatic fluctuations close to existing ice masses (Norlund, 1924; Alexander and Worsley, 1973; Birkeland, 1974). Of fundamental importance to many of these studies has been the technique of radiocarbon dating. A great deal of data on these and other lines of evidence is collected and correlated in the publications by Lamb (1966), Ladurie (1971) and Denton and Karlén (1973).

Iceland provides a particularly useful example of the history of climatic fluctuations over the past few centuries, and settlement in this sensitive environment has often been threatened by climatic as well as volcanic events. The historical evidence for settlement advance and retreat since the Landnam (or colonization) phase has been recounted in fascinating detail

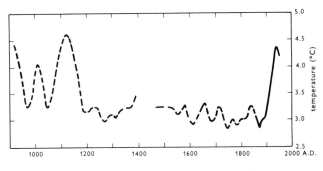

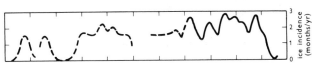

Figure 6.11
Graphs for air temperature (30 year running means) and sea ice incidence off the coasts of Iceland. *(Bergthorsson, Jökull 1969.)*

by Thorarinsson (e.g. 1943; 1944; 1961), Fredriksson (1969) and Eythorsson and Sigtryggsson (1971). A useful indicator of the severity of the Icelandic climate is the occurrence of sea ice off the Icelandic coasts. There are long records of sea ice occurrence (Koch, 1945), but the historical data are very incomplete and cannot be used for reasonable quantitative assessments of the frequency of sea ice until the late nineteenth century. Nevertheless, the old annals do give some indications of the most severe ice years in the foregoing centuries, and Bergthorsson (1969) has attempted to plot ice incidence in months per year for the whole period of Icelandic settlement. On his graphs (Figure 6.11) there seems to be a very clear inverse relationship between the 30 year running means of estimated temperature and the estimated incidence of ice at the coast; and there is little doubt that the years of most abundant ice and strongest cooling (for example, the period around 1300 and the period 1600–1890) are prominent in the written records as periods of poor grass growth, frequent animal deaths and indeed human starvation. Conversely the periods AD 900–1200 and 1350–1600 seem to have been times of relative ease, when settlement and economic development proceeded quietly without any major climatic disruption (Eythorsson and Sigtryggsson, 1971).

An analysis of Icelandic glacier fluctuations also reveals a close correspondence with the climatic data. There is scanty evidence that some Icelandic glaciers advanced as a result of the thirteenth-century climatic deterioration; but there is little doubt concerning the significance of the later cool period. After 1350 the climate cooled intermittently, with the most drastic deterioration at the beginning of the seventeenth century. Glaciers began to advance (Tomasson and Vilmundardottir, 1967), and the following two centuries were the coldest in Icelandic recorded history. Sea ice was extensive, farming conditions extremely poor, the national state of health precarious and deaths from famine frequent. Cirque glaciers developed on previously snow free sites. Tungnaarjökull advanced about 10 km between 1600 and 1890, and many of the Vatnajökull glaciers expanded greatly. In Vestfirðir there were spectacular glacier advances in the Drangajökull outlet glacier troughs, culminating in the formation of moraines about 1756 and 1840 (Eythorsson, 1935). The Glama plateau acquired a small plateau ice cap where previously there had been none (Thorarinsson, 1943), and further south the mountain of Ok was reoccupied by ice. The expansion of glaciers proceeded

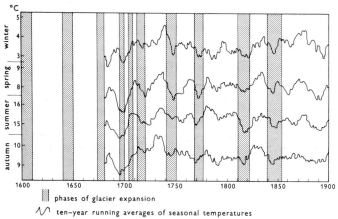

Figure 6.12
A plot of some periods of marked glacier advance for the Alps, Norway and Iceland during the period 1600–1900. The diagram shows the relationship between these advance phases and reconstructed seasonal temperature trends. *(Manley, R. Met. Soc. 1966).*

intermittently for 250–300 years, with local maxima grouped around 1750 or 1850–94. At each of these maxima farms were destroyed by ice and great hardship was experienced.

It is sometimes assumed that these twin peaks of Little Ice Age glacier expansion can be recognized in records of glacier history throughout the northern hemisphere. Moraines loosely labelled as 'the 1750 moraine' and 'the 1850 moraine' are known at the margins of Icelandic and Norwegian glaciers, and also in the Alps (Lamb, 1964; Ladurie, 1971). Weidick (1972) referred to moraines of these ages in Greenland; and in the United States the North Cascades range has several glaciers which constructed readvance moraines around 1850 (Miller, 1969). However, although there is no doubt that the Little Ice Age (and indeed some of the climatic oscillations within the Little Ice Age) can be recognized on a world scale, glacier advances in specific regions were never properly in phase. For example, Figure 6.12 shows Manley's (1966) plot of periods of marked glacier advance in Europe 1600–1900; and he has certainly not plotted them all. Indeed, it would be surprising if regional ice advances had been in phase, given the considerations discussed earlier.

The Little Ice Age maxima mentioned in the literature show just how large the discrepan-

cies are. For the Alps, Ladurie (1971) has recorded a great number of different glacier oscillations for the period 1600–1860. In South America and the subantarctic there are particular
problems of correlation (Mercer, 1968; Sugden and John, 1973) even though there are well
marked moraines dating from glacier readvances during the past few centuries. In the South
Shetland islands, the maximum Little Ice Age glacier advance seems to have occurred anomalously early, between AD 1200 and 1450. Other early advance phases are recorded by
Denton and Karlén (1973), including one in parts of the United States at about AD 700–
900 and another in Sweden at about AD 1500. The Chickamin glacier attained its maximum
Neoglacial extent as early as the twelfth century (Miller, 1969). In some areas the Little
Ice Age maximum seems to have occurred inside the present century, Karlén (1973) has
shown that in Swedish Lapland, glaciers generally attained advanced positions around 1916
and in Norway there was a strong glacier advance in 1902–6. Theakstone (1965) has shown

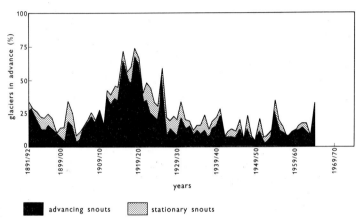

Figure 6.13
Glacier snout advances in the Swiss Alps for the period 1891–1965. Note that in 1915–20 more than 60
per cent of the measured glaciers were advancing. *(Data after Kasser, Int. Ass. scient. Hydrol. 1967.)*

that the ice margin of Engabreen was in an advanced position until about 1930, and the
Store Supphelle outlet of Jostedalsbre did not retreat from a prominent terminal moraine
until 1916. Other glacial advances are known from Iceland around 1910–20 and in the Alps
around 1915–20 (Figure 6.13). In north Greenland the maximum post-glacial advance of
the inland ice margin appears to have occurred in 1890–1900 (Weidick, 1972), and in South
Georgia, Smith (1960) noted a glacier readvance as late as 1926–39, which was a period
of general ice retreat in most other parts of the world.

It needs to be stressed that within the last few centuries the precise details of depression
tracks and precipitation totals have generally been of much greater significance than is often
recognized in determining the details of glacier mass balance and hence regional ice advance
(Lamb and Johnson, 1959; Namias, 1969; Marcus and Ragle, 1970). In northwest Iceland,
the recent advances of the three Drangajökull glaciers referred to on p. 113 may well have
been due to surges or kinematic waves triggered off not so much by a shortlived temperature
fluctuation as by a series of years with anomalously high precipitation totals (Thorarinsson,
1969). Further north, the glaciers of Jan Mayen advanced strongly after 1954 following a

supposed doubling of the precipitation since 1920 (Lamb, Probert-Jones and Sheard, 1962). In Patagonia, the glacier advances of the period 1930–50 are attributed by Lliboutry (1953) to increased precipitation and cloud cover, in spite of higher temperatures. And on Baffin Island, the recent growth of ice masses is similarly related to increasingly vigorous atmospheric circulation and rising precipitation totals (Andrews et al., 1972). Similar oscillations of depression tracks must have had marked effects upon the course of regional and local glaciation through the whole of the Little Ice Age. Considering worldwide glacier oscillations, Orheim (1973) has even suggested from his glaciological studies on Deception Island that there is an anti-phase relationship between the northern hemisphere and the southern hemisphere, but this is by no means proved.

Glacier fluctuations over millennia

It is now becoming clear that the Little Ice Age was not the culmination of the Holocene climatic cooling at all, but simply one of several major glacial phases which have punctuated the period since the last main glaciation (Figure 6.14). These phases probably differed somewhat in strength and persistence, but they are becoming easier to define as a result of recent advances in radiocarbon dating, palynology, dendrochronology and other techniques. Collectively, some of these phases are referred to as the Neoglacial (Porter and Denton, 1967; Denton and Porter, 1970). This term seems reasonable because it signifies the rebirth of glaciers in many areas following the 'Hypsithermal Interval' which culminated about 6,000 years ago. However, it is now realized that some Holocene phases of glacier expansion occurred before and even during the Hypsithermal Interval, and it would be unwise to define the Neoglacial by reference to any specific dates.

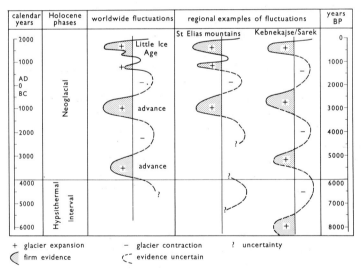

Figure 6.14
A reconstruction of worldwide Holocene glacier fluctuations after 8,000 BP, together with regional examples from the USA and northern Europe. The time scale on the right is based upon corrected C¹⁴ age determinations. (Denton and Karlén, Quat. Res. 1973.)

In many areas it is now being recognized that glacier advances previously assumed to belong to the Little Ice Age are in reality datable to the period 3,300–2,400 BP. In Lapland, many glaciers are fronted by moraines dated by C^{14} and lichenometric techniques to 2,370–2,720 BP, showing that they were up to 25 per cent longer than at present and even longer than during their Little Ice Age maxima (Karlén, 1973). In Iceland several of the glaciers of Vatnajökull reached their Holocene maximum positions before 2,000 BP (Thorarinsson, 1956), and there is evidence from Iceland and Norway that various other Little Ice Age moraines are in reality around 2,000 years old (Worsley and Alexander, 1973). Mercer (1968; 1970) recorded a marked glacier advance in Patagonia at about 2,700–2,200 BP, and Goldthwait (1966) gave evidence for extensive Alaskan glacier advances which are more or less synchronous. In the Alps also, an advance at this time was probably widespread although many traces of it were obliterated during the Little Ice Age advances (Heuberger and Beschel, 1958; Mayr, 1964).

A glacier advance dated to approximately 5,000 BP by Denton and Karlén (1973) is thought to have been less extensive than later advances, and it is thus difficult to discern in many areas. Nevertheless, there are scattered records of it from the Northern American cordillera and from Swedish Lapland, and a marked climatic deterioration at about this time is also attested by dendrochronological and lichenometric studies and by a reconstruction of tree-line altitudes. There was a significant readvance on the west and east coasts of Baffin Island around 4,800 BP (Andrews, 1972; 1974). In the southern hemisphere there is also good evidence for this phase of glacier expansion. Hence Heusser (1960) argued that morainic evidence indicated a strong ice advance in southern Chile 4,000–5,000 BP, and this has since been confirmed by the studies of Mercer (1970). The latter author concluded in a 1973 paper that there was evidence for a phase of 'maximum Neoglacial cold' around 4,500 BP in South America. Wardle (1973) presents evidence for glacier expansion at about the same time in the Southern Alps of New Zealand. On a wider front, Mercer (1967) gives worldwide evidence for other correlative readvances.

Another cold phase, this time pre-dating the Hypsithermal Interval, is dated to c. 7,000–8,000 BP. It is dominated by events associated with the wastage of the Wisconsin/Weichselian ice sheets, and in North America it took the form of a powerful readvance. The evidence for this readvance, termed the Cockburn readvance, is presented by Andrews and Ives (1972). There are correlative readvances in Greenland (Weidick, 1968), and in Alaska the Brooks range experienced an increase in glacierization around 8,200 BP (Porter, 1964); but elsewhere evidence for this phase is extremely rare. Benedict (1973) tentatively suggests that certain small cirque moraines in the Colorado Front range, USA, may have formed at about 8,000 BP. Mercer (1973) found no trace of glacier expansion in South America at this time, concluding that the whole period between 11,000 BP and 6,000 BP was one of 'hypsithermal' conditions. On the other hand, there are moraines slightly younger than 8,500 BP in New Zealand (Mcgregor, 1967).

From an extensive review of the literature and from an analysis of their own carefully collected field and laboratory data, Denton and Karlén argued for a series of distinct post-glacial cold phases in the periods mentioned above. These are summarized in Table 6.3, overleaf. The glacial phases themselves were thought to culminate approximately every 2,500 years, and they break down into approximate 600–900 year expansions and then contractions lasting up to 1,750 years.

A good case can be made for similar oscillations extending back into the late-glacial period.

Table 6.3   *A proposed sequence of Neoglaciation in the north-
                   ern hemisphere. (After Denton and Karlén, 1973.)*

Glacier expansion phase	Coldness peaks	Duration of cold interval
Little Ice Age (fourth phase of Neoglaciation)	200 BP (1750 AD)	450–30 years BP (1500–1920 AD)
Third phase of Neoglaciation	2,800 BP	2,400–3,300 years BP
Second phase of Neoglaciation (following Hypsithermal Interval)	5,300 BP	4,900–5,800 years BP
First phase of Neoglaciation (during Hypsithermal Interval)	7,800 BP	7,000–8,200 years BP

In northern Europe there seem to have been a number of distinct cool phases interrupting the overall retreat of the Scandinavian ice sheet margins. The best known of these are now dated by a variety of studies, and are termed the Zone I (Older Dryas) and Zone III (Younger Dryas) phases. These were separated by the well known Allerød interstadial, which may or may not be something of a European anomaly (Mercer, 1969). Recent studies by Mangerud (1970b) from western Norway have revealed two late-glacial readvances of the ice margin, as indicated on Figure 6.15. In the Oslo area Sörensen (1974) has also shown that there were occasional ice front readvances at the same time. In the British Isles Coope *et al.* (1971) have demonstrated some of the problems in correlating the ice margin oscillations

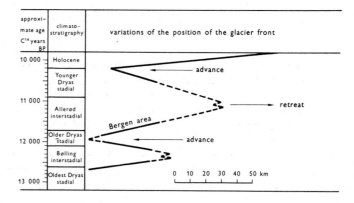

Figure 6.15
Variations in ice margin positions in the Bergen area during the period 13,000–10,000 BP. *(Mangerud, Norsk geogr. Tidsskr. 1970b.)*

of Zone I with the evidence of pollen and fossil coleoptera (Figure 6.16), but the events of Zone III are becoming reasonably well understood. For example, in Wales there was a short phase of intense periglacial activity at this time in the lowlands (John, 1973) and this coincided with the growth of cirque glaciers and the construction of moraines in the uplands (Seddon, 1957).

In Scotland, although there is continuing debate about the status and timing of a Zone I readvance, most authors agree that there was a Zone III (Loch Lomond) readvance, dated to 10,500–10,000 BP (Sissons, 1975). This correlates very closely with the age of the Fenno-scandian stadial of the Scandinavian ice sheet in Sweden, dated by varve chronology to 10,850–10,050 years ago (Tauber, 1970). Within the Late Wisconsin of North America there seem to have been similar readvances. The Port Huron readvance, along a wide front of

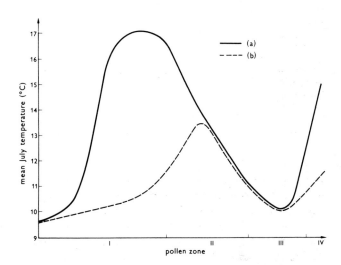

Figure 6.16
Variation of mean July temperatures for lowland Britain during the Late Weichselian, inferred from **(a)** fossil Coleoptera and **(b)** conventional pollen data. The insect fauna responds to a change in climate much more rapidly than the local flora. (*Coope, Rev. Géogr. phys. Géol. dyn. 1970.*)

the Laurentide ice sheet, culminated about 13,000–12,700 BP, and the later Valders re-advance (if it was not a surge) could possibly be an early equivalent of the European Zone III readvances (Manley, 1971). From a survey of southern hemisphere late-glacial pollen analytical data, van Zinderen Bakker (1969) argued for an overall synchroneity of the main phases with those of the northern hemisphere. Hence he claimed that the evidence from Fuego-Patagonia, southern Chile, Colombia and parts of Africa correlated well with the late-glacial (Older Dryas-Allerød-Younger Dryas) sequence referred to above. These phases may also have their equivalents in New Zealand (Suggate, 1965).

Earlier still, the Cary readvance of North America, dated to 15,000–14,000 BP, may be a correlative of one of the cold phases grouped within the European Zone I or Older Dryas. Perhaps it was the equivalent of the 'Oldest Dryas' stadial of Mangerud (1970b) in western Norway and the cold phase which preceded the Bölling interstadial of the northern European mainland (Hammen et al., 1967). In Wales, there seems to have been a readvance by the Welsh ice cap and also an expansion of the Irish Sea ice across parts of north Wales between 16,830 BP, and 14,468 BP (Coope et al., 1971; Foster, 1970); perhaps this was another correlative, although it must be explained that there is some doubt both about the reliability of the quoted radiocarbon age determinations and the interpreted drift stratigraphy.

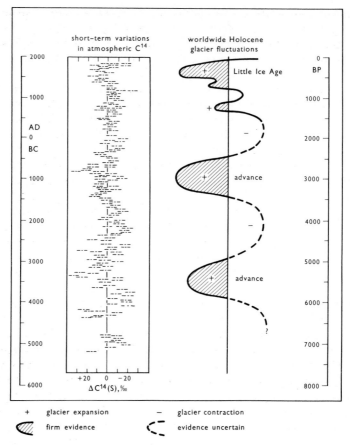

Figure 6.17
A plot of short term C^{14} variations *(Suess, 1970)* and worldwide Holocene glacier fluctuations. *(Denton and Karlén, Quat. Res. 1973.)*

The climatic oscillations responsible for these glacier expansions and contractions are in step with oscillations in the concentration of atmospheric C^{14} (Figure 6.17), which in turn depend partly upon variations in the emission of solar radiation. As pointed out by Suess (1970), atmospheric C^{14} levels have fluctuated on a number of different wavelengths during the Holocene. Denton and Karlén argue that the major intervals of high atmospheric C^{14} activity coincide with intervals of glacier expansion, while the intervals of relatively lower C^{14} activity coincide with intervals of glacier contraction.

The problem of correlation

While it is tempting to assume a causal relationship between atmospheric C^{14} variations and glacier behaviour on a medium scale, the hypothesis still needs careful testing. Also, it would be unwise to disregard the great difficulties involved in the demonstration and acceptance of worldwide periodic glacier behaviour on a wavelength of about 2,500 years. After all, there are many well documented advances of glaciers throughout the world which lie well outside the limits of the glacial phases mentioned above. Also, there are a number of problems involved in worldwide correlation at this scale:

1   Among the thousands of radiocarbon age determinations referred to in the literature, many will be unreliable. Many so called readvances are erroneously dated, and the problem is compounded by the fact that some authors calculate their ages in years before Christ (BC) while others use unadjusted 'BP' dates and others (e.g. Denton and Karlén, 1973) use adjusted or corrected dates.
2   At least some of the readvances recorded within the Holocene cold phases (particularly the 'anomalous' readvances) may be the result of surges. Therefore they can only be used with extreme caution for palaeoclimatic reconstructions.
3   Climatic oscillations of the amplitude and wavelength considered above are certain to affect a very wide range of glacier types. Variable glacier response times ensure that individual glacier termini, reacting to the same climatic deterioration, will not reach their maximal positions at anything like the same time. Individual readvances may be separated by hundreds of years.

Of much greater importance is the fundamental problem of whether we should expect approximately coincident glacier readvances at times of climatic cooling. Much of the literature prior to 1970 (see, for example, Sawyer, 1966) was characterized by a search for correlations on a worldwide scale, and the same may be said of the foregoing paragraphs! However, since 1970 there has been a heightened appreciation of the manner in which 'normal' shifts in the atmospheric circulation can lead to cooling in one climatic zone at the same time as warming in another (Lamb, 1974). This is demonstrated by the difficulties involved in establishing regional, continental and intercontinental pollen-zone conventions. Hence it may be more reliable to expect the details of regional glacial sequences to be out of phase with one another, except where exceptional phases of global cooling are involved. Already Andrews and Ives (1972) have pointed out that the eastern margin of the Laurentide ice sheet was advancing in eastern Baffin Island at about 8,000 BP, just at a time of rapid ice sheet disintegration on its other margins. This apparent anomaly is explained by the removal of the Laurentide ice sheet barrier to the south, with rising temperatures leading to a reduction of sea ice in Baffin Bay and then to locally increased precipitation on its adjacent landmasses. Perhaps the anomalous readvances on the peripheries of the Greenland ice sheet can be explained in a similar fashion (Andrews et al., 1974). Partly these apparent anomalies are the result of glacier size differences and variable response times, as pointed out above. But climatic variability over space must also be an important factor. Andrews (1973) concluded thus: 'Worldwide chronocorrelation of glacial response to mass balance changes (climatic and others) need not be the rule, and, indeed, the glacial response to mass balance changes of a worldwide climatic shift will be *time transgressive*.'

Cycles of glacier behaviour

The ideas referred to on the foregoing pages, concerning medium term 'pulses' of glacier expansion and retreat, can also be applied at a variety of other scales. At the shortest time scales, certain oscillations are predictable and regular, such as the day–night oscillation (Figure 6.6) and the seasonal oscillations which affect glaciers each year. But even these oscillations are subject to variations of amplitude, being influenced by the complex inter-actions of many meteorological factors. The random element in the oscillatory behaviour of the environment becomes more and more marked as the time scale increases, and it appears increasingly dangerous to attempt a correlation of climatic events or ice margin fluctuations over decades or centuries.

Nevertheless, even for the time scales considered in this chapter, there is good evidence for a number of cyclic or quasi-periodic pulsations in the environment. The 11 year cycle of sunspot activity has been long accepted, and in addition there are a number of other longer period solar cycles which might have had an influence upon the events of the Neo-glacial (Bray, 1967; 1968; Schove, 1955). The relationship between measurable solar and astronomical events and recorded climatic fluctuations has exercised meteorologists and physicists for many years (cf. Lamb, 1972), and it now seems probable that many astronomical periodicities (for example, cycles with approximate wavelengths of 800 years, 170 years and 19 years) have had a recognizable influence upon recent world climate. All of these, together with any of the multitude of other oscillations proposed by many authors, must have in-fluenced, in some way, the events of the Holocene.

One may hypothesize that all externally induced glacier fluctuations (whether due to direct or lagged response) are related to a series of periodic climatic oscillations of varying amplitude and wavelength. As shown above, Denton and Karlén (1973) have made a con-vincing correlation of a 2,500 year cycle of glacier margin fluctuations with Suess's (1970) data for medium term variations in atmospheric C^{14} concentrations. On the other hand there are greater problems in the correlation of events at the scale of centuries. Suess showed that there is a smaller scale cycle of atmospheric C^{14} oscillation with peaks separated by intervals of about 200 years. An approximately similar wavelength can also be recognized in the fluctuations of oxygen – 18 values from the Camp Century ice core (Johnsen et al., 1970). However, it can hardly be recognized for the main phases of glacier advance during the Little Ice Age, or for the individual glacier advances which occurred during the other late-glacial and Neoglacial cold phases. This is partly due to the fact that many other short term climatic and meteorological fluctuations appear to be random, and partly due to the diffi-culties involved in comparing the O^{18} and C^{14} time scales with the calendar years of the historical record. There is also the influence of events related to volcanic eruptions, severe dust storms, and minor shifts in atmospheric circulation. Each of these affects sea ice cover, land surface albedo and ocean surface temperatures. Again, since the early 1800s there is the effect of the introduction of industrial waste (and then radioactive materials) into the atmosphere by man (Matthews et al., 1971). Arising from all this, it seems that the real climatic oscillations which are liable to have affected glaciers in the short and medium term are composite or quasi-periodic, showing evidence of both periodic and random fluctuations.

Prediction of glacier behaviour

Glacier behaviour over time is manifestly extremely complicated in the short and medium term. From the foregoing paragraphs it seems that while some glaciers expand, others remain relatively stable, experiencing only a minor redistribution of mass within their old limits. The reasons for these differences in behaviour are in part explained by climatic factors, in part by the thermal characteristics of ice, and in part by morphological factors. Among surging glaciers, some appear particularly stable and sluggish while others react vigorously and almost instantaneously to changes of input. As stressed by Lamb (1972), every glacier behaves uniquely because its precise environmental circumstances and its own internal physical properties are unique. Nowadays, with surging glaciers being recognized in all climatic environments, it seems foolish to attempt any detailed correlation of current glacier behaviour, let alone any prediction of future advances or retreats. It may be that all glaciers are capable of surge behaviour if certain critical conditions can be met.

But climatic and glacial oscillations can be recognized for the Holocene, and as glaciology advances so will the knowledge of cyclic patterns of climatic behaviour and glacier response. Just as flood prediction on major river systems has become an important task for hydrology, so the prediction of future glacier behaviour becomes an increasingly realistic task for glaciology. At the present time glaciologists have a fine opportunity to communicate reliable ideas to the public, when there is intense interest and speculation about the coming of 'the next Ice Age'. Thus a record of climatic variability and glacier response times, together with the prediction of surges and kinematic waves (and their behaviour once initiated) becomes one of the subjects' most important applications. At the largest scale there is increasing concern about the future behaviour of the Antarctic ice sheet in view of the discovery that parts of its periphery are wet-based and in view of its known control over world sea-level; and at the smallest scale the investigation of the behaviour of Berendon glacier was encouraged by a mining company whose installations were rather too close to the glacier snout for comfort.

There are other manifestations of the current interest in climate and glacier behaviour in the short and medium term. There is little doubt that the national glacier inventories encouraged for the International Hydrological Decade 1965-74 will have lasting value, for the morphometric data which they contain can be of direct use once the science of glacier response prediction has been further refined. Indeed, the importance of this work has been recognized by the setting up of a permanent secretariat in Zurich, charged with the continuation of work towards a full world inventory of glaciers. Of related interest are the numerical modelling experiments of the CLIMAP project (Imbrie, 1974), the NCAR project (Williams et al., 1974), and other experiments designed to reconstruct the factors which control the workings of the world air-sea-ice system.

Remote sensing techniques are now being applied to the study of glacier oscillations. As pointed out by Østrem (1974a) and Krimmel and Meier (1975), the potential of ERTS and other satellite imagery for the monitoring of changes in transient snowlines and glacier firn lines is enormous. The satellite data can be related to recorded field data on runoff variations and snout oscillations, with clear practical and economic implications.

Finally, improved reconstructions can now be made of the thermal characteristics of Holocene and Weichselian/Wisconsin glaciers. Some of the models proposed have already been referred to in chapter 3; through their use it should soon be possible to predict the occurrence of different types of glacier during past glacial phases. For example, it may be possible to predict that cirque glaciers should have been present in a specific upland area during Zone III, and that they should have been active enough to create substantial moraines. Such a possibility is exciting not only to the glaciologist but also to the glacial geologist and the student of Pleistocene landforms.

Further reading

ANDREWS, J. T., DAVIS, P. T. and WRIGHT, C. 1976: Little Ice Age permanent snowcover in the eastern Canadian Arctic. *Geogr. Annlr. Stockh.* **58** (A), 71–81.

DENTON, G. H. and KARLEN, W. 1973: Holocene climatic variations – their pattern and possible cause. *Quat. Res.* **3**, 155–205.

EYTHORSSON, J. and SIGTRYGGSSON, H. 1971: The climate and weather of Iceland. *The zoology of Iceland* **1** (3), 1–62.

GROVE, J. 1979: The glacial history of the Holocene. *Progress in Physical Geography* **3** (1), 1–54.

GORDON, J. 1980: Recent climatic trends and local glacier margin fluctuations in West Greenland. *Nature* **284**, 157–159.

LADURIE, E. le Roy 1971: *Times of feast, times of famine.* Doubleday, New York (428 pp).

LAMB, H. H. 1972: *Climate: present, past and future* **1**, *fundamentals and climate now.* Methuen, London (613 pp).

MERCER, J. H. 1978: West Antarctic ice sheet and CO_2 greenhouse effect: a threat of disaster. *Nature* **271**, 321–5.

THORARINSSON, S. 1943: Oscillations of Icelandic glaciers in the last 250 years. *Geogr. Annlr. Stockh.* **25**, 1–54.

UNTERSTEINER, N. and NYE, J. F. 1968: Computations of the possible future behaviour of Berendon glacier, Canada. *J. Glaciol.* **7** (50), 205–13.

7 Long term glacier fluctuations

This chapter comprises an analysis of glacier fluctuations on a large scale; it is concerned mostly with the expansions and contractions of ice sheets and ice caps rather than of small upland glaciers, and with a time scale of over 10,000 years. The relevant climatic and glaciological adjustments are low frequency high amplitude events (Tables 1.1 and 6.1). Most of the field evidence relates to the Pleistocene period and particularly to the last glaciation (Wisconsin or Weichselian), and inevitably in a synthesis of this type one must use much data from the last 120,000 years or so. Nevertheless, these data can be used in the search for generalizations or laws concerning earlier ice sheet and ice cap fluctuations in space and time.

There is clear evidence, as mentioned in the last chapter, for the cyclic oscillations of glaciers, although the random element must still be borne in mind. The characteristics of the oscillatory waves appear to vary according to the dimensions and location of the ice mass. It has long been agreed that mid latitude ice sheets waste rapidly if not catastrophically (Flint, 1971; Bryson *et al.*, 1969); but it is becoming apparent that they can build up rapidly also (Mathews, 1972). This is because mid latitude areas are those where climatic instability is most marked. Here radiation values, precipitation totals and seasonal distributions, and the routes followed by cyclonic disturbances, may be altered significantly during glacial ages as a result of the sensitivity of the middle latitudes to some quite small change in climate (Lamb, 1971; Namias, 1970; Lamb, Lewis and Woodroffe, 1966). The middle latitudes are particularly vulnerable to the effects of 'climatic positive feedback' (Fairbridge, 1970). However, this certainly oversimplifies the relationship between glacierization and climate.

The mechanisms involved in the build-up or decay of continental ice masses are gradually being established (Kukla, 1972; Shaw and Donn, 1971; Barry, 1973); and computer simulations of shifts in atmospheric circulation patterns are yielding promising data on the possible links in the feedback chain (Gates, 1974). In much of the work done so far the Arctic basin pack ice cover plays a critical part (Fletcher, 1966), as does the snow cover and albedo of high and middle latitude land surfaces in North America and Eurasia (Hare, 1968; Adam, 1969; Dickson and Posey, 1967). Many authors (e.g. Kukla and Kukla, 1972) have suggested that the associated reflection of radiation and circulation effects brings the world dangerously close to the threshold of large scale glacierization. In the mid latitudes the strength of seasonal circulation shifts and the receipt of summer heat generally ensure that glacierization is avoided; but in high latitudes the situation is much more sensitive, and it may be that a few years of abnormally persistent summer snow cover can set in motion a chain of events which may culminate in local, then regional, then continental glacierization (Budyko, 1969; Bradley and Miller, 1972; Ives *et al.*, 1975). This is the 'snowblitz', expressed in somewhat alarmist terms by Calder (1974) and the subject of some recent controversy. Calder wrote:

> In the snowblitz the ice sheet comes out of the sky and grows, not sideways, but from the bottom upwards. Like airborne troops, invading snowflakes seize whole counties in a single

winter. The fact that they have come to stay does not become apparent, though, until the following summer. Then the snow that piled up on the meadows fails to melt completely. Instead it lies through the summer and autumn, reflecting the sunshine. It chills the air and guarantees more snow next winter. Thereafter, as fast as the snow can fall, the ice sheet gradually grows thicker over a huge area (p. 118).

The cold comes instantly, but then the snow piles up for 5000 years at perhaps 18 inches a year. 'Instantly' may mean a hundred years, or a single bad summer. So ice ages can start very suddenly – that is the implication of this research and of the snowblitz theory (p. 121).

The research referred to is a paper by Lamb and Woodroffe (1970), in which they suggested that a series of snowy winters in high latitude areas which are close to the threshold of glacierization could lead to a positive feedback mechanism. A critical area might be Arctic Canada, where an increased snowcover would raise the albedo, leading to a reduced absorption of solar radiation and perhaps overall cooling. In turn this might affect northern hemisphere circulation patterns, as suggested by Kukla and Kukla (1972) and Mathews (1972). Attractive as this theory is, it is based upon very simplified assumptions about hemispheric energy exchange, as indicated by numerical modelling such as that currently in progress with the CLIMAP project. The possibility of violent environmental positive feedback as proposed in the snowblitz theory is far from established even for the marginal parts of the Canadian Arctic archipelago, let alone for the lush pastures of England.

Arising from this discussion, it needs to be stressed that the birth and death of ice sheets must not be related simply to the crossing of environmental thresholds. All glaciers, and particularly large ones, can exist for prolonged periods where little real equilibrium with climatic factors exists. Many glaciers of the present day are probably relics which demonstrate some past, rather than present, relationship with environment. Hence one must consider glacier response time and the time lag which occurs between climatic change and glacier generation; and also the manner in which glaciers are affected by climatic oscillations of various wavelengths. While mid latitude ice sheets may grow and waste rapidly (Figure 7.1b), polar ice sheets are clearly more difficult to get rid of. The very fact that the Antarctic and Greenland ice sheets are still with us argues that their fluctuations in time and space are more regular, as suggested on Figure 7.1a. Possibly small upland glaciers (such as most of those present in the world today) are characterized by a third type of oscillatory behaviour over time; since such glaciers have short response times, there may be rapid glacier buildup following a climatic coding, and then a much slower glacier diminution. Figure 7.1c attempts to portray this oscillatory behaviour in an idealized way.

In general, the alternations of glacial and interglacial stages can be seen as major adjustments to major changes in energy input (Larsen and Barry, 1974; Adam, 1975). Probably a strong positive feedback loop operates during the buildup of a Pleistocene ice sheet, and once it has achieved approximately its maximum mass it is capable of strong control over its own climate and thus exercises self regulation. Nevertheless, ice sheets are no more capable than valley glaciers of remaining in a steady state. Because they adjust slowly to climatic change, their response time is difficult to appreciate within the confines of human experience (Catchpole, 1972). Often the understanding of ice sheet behaviour is handicapped by very limited geochronological data; hence the geomorphologist must consider hypotheses such as those presented here. In the rest of this chapter some of the collected evidence for long term glacier fluctuations is examined before we return to hypotheses of glacier behaviour

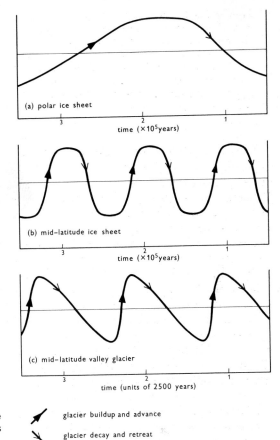

Fig. 7.1
Highly idealized portrayal of the
oscillatory behaviour of ice sheets
and upland glaciers over time.

over time. The causes of world glacierization are not examined in detail, since there is already
a vast literature (see Fairbridge, 1967; Flint, 1971; Kukla, 1972).

Glacier fluctuations over tens of thousands of years

The Weichselian/Wisconsin glacial stage
The period between approximately 110,000 BP and 10,000 BP is now generally thought to
have encompassed the whole of the last glacial stage (papers in Olsson, 1970; papers in Turekian, 1971; Wright, 1972; Shackleton and Opdyke, 1973). There is a formidable amount
of evidence from many different lines of research to show that the full interglacial climate
of 125,000 BP gradually gave way during a global climatic cooling phase. Ocean water
characteristics were changed, and as more and more moisture became locked into the growing ice sheets, world sea-level gradually fell. About 70,000 years ago, sea-level dropped
rapidly to about −80 m, and full glacial conditions were experienced in many mid latitude
and high latitude areas. World temperatures were lowered by about 6°C (Flint, 1971); climatic belts were compressed, patterns of atmospheric circulation were modified (Lamb,

1972), and increased glacierization was experienced even in the equatorial belt. For example, on Mount Kenya the snow line was depressed by about 900 m (Baker, 1967).

The rapid cooling at about 70,000 BP is attested by a variety of studies, described for example by Sancetta *et al.* (1972), Shackleton and Opdyke (1973) and Fairbridge (1972). Among the most frequently quoted evidence is that of the O^{18}/O^{16} ratios in ice from the Camp Century ice core in Greenland (Dansgaard *et al.*, 1971). From the data plotted on Figure 7.2, it seems that after 70,000 BP there was a gradual but oscillating deterioration of climate until the Late Weichselian maximum was reached after 20,000 BP. The Camp Century core suggests that northern hemisphere glacierization should have been most intense and widespread during the Late Weichselian, and this seems to be supported by the work of Shackleton and Opdyke (1973) showing that sea-level reached its lowest point of -120 m only about 17,000

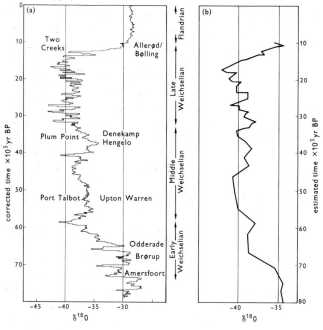

Figure 7.2
Fluctuations in O^{18} values for Camp Century (Greenland) and Byrd Station (Antarctica) for the past 80,000 years. The Byrd Station graph **(b)** is probably rather unreliable, since it is largely interpolated from widely spaced sample points, particularly in the lower part of the core. *(Dansgaard et al., Yale Univ. Press 1971; Epstein et al., Science 1970.)*

years ago. The oxygen isotope data from the Byrd Station ice core in Antarctica (Epstein *et al.*, 1970) shows certain close similarities to that of Camp Century, but it should be pointed out that the interpretation of the O^{18} data from ice sheet cores is a risky occupation which involves many precarious assumptions (Johnsen *et al.*, 1972), especially in the 'corrected' section of the time scale for material older than *c.* 65,000 BP. Not least is the problem of extrapolating climatic data from the high interiors of ice sheets to the lowlands of middle latitudes; a component of the cooling during the Early Weichselian, for instance, may be simply the result of the increasing altitude (and hence stronger cooling) of the ice sheet surface as it grew (Raynaud and Lorius, 1973).

Correlation of stadials and interstadials
Details of the Weichselian chronology worked out for specific regions suggest that one should be wary of too easy extrapolations from the sequence portrayed on Figure 7.2. It now seems

that parts of the Antarctic ice sheet might have been most extensive in the Early Weichselian. Andrews *et al.* (1972) have shown that in parts of the east Canadian Arctic the Early Wisconsin ice advance was stronger than anything subsequently, and that the stadials of the Early Wisconsin in eastern Baffin Island occurred before the onset of the Wisconsin as normally understood. Malaurie *et al.* (1972) have even questioned the wisdom of using the Camp Century data for reconstructing the climatic history of the nearby coast of northwest Greenland.

In most areas, however, it does seem that the Late Weichselian was the time of maximum ice advance. In the British Isles the Irish Sea glacier (the only large Pleistocene glacier which discharged ice from the British western highlands towards the south) reached the English midlands after 25,000 BP (Coope, Morgan and Osborne, 1971), and its southern margin in south Wales almost certainly after *c.* 18,000 BP (John, 1972b). In Pembrokeshire the ice was ineffective; it was incapable of removing many of the periglacial materials which had accumulated during the earlier part of the Weichselian. There is now some evidence that the Weichselian glacier maxima on the eastern and western coasts of Britain were not quite synchronous; the ice margin in the west extended much further southwards, and the Irish Sea glacier was probably more active than its ill defined equivalent in the North Sea.

Similar difficulties have been encountered in dating the maximum extent of the Weichselian Scandinavian ice sheet. In north Germany, the Brandenburg moraine marks the southernmost extent of the ice, but it is overlapped both to the east and west by the younger Frankfurt moraine, and in the northernmost part of Schleswig-Holstein the even younger Pomeranian moraine marks the maximum over a short distance (Woldstedt, 1969). Hence there are a number of overridden terminal moraines which help to confuse the chronostratigraphic picture.

In the United States, the Wisconsin Laurentide ice sheet behaved even more erratically. In the east the maximum advance occurred between 17,000 and 18,000 BP; in Illinois, Iowa and Wisconsin, Frye and Willman (1973) note that the ice advanced in a series of broad lobes to its maximum position somewhat earlier, before 19,000 BP (Figure 7.3, overleaf); on the northwest margin the maximum was possibly not attained until around 13,000 BP.

From the above, and from what has been said in chapter 6, it seems clear that even the largest Weichselian glaciers behaved in a pulsatory manner during both their waxing and waning phases. Relatively little is known of the fluctuations of the Scandinavian and Laurentide ice sheets during the long period 70,000–18,000 BP, for the simple reason that the culminating ice advance removed or modified most of the earlier deposits. However, there are various clues from both North America and Europe. There is no doubt that the Laurentide ice sheet was well developed during the Early Wisconsin, for a number of drift sheets and interstadial deposits have been recognized for the period prior to 50,000 BP (Figure 7.4, overleaf). In Europe at least three climatic ameliorations (the Amersfoort, Brorup and Odderade interstadials) punctuated the overall cooling phase between 70,000 and 55,000 BP (Hammen *et al.*, 1971), and the problematical 'Chelford interstadial' in Britain may be the equivalent of one of these (Coope *et al.*, 1971).

The most significant of the global Wisconsin/Weichselian fluctuations is the interstadial which occupied the period *c.* 50,000–25,000 BP, during which most ice edges retreated several hundred kilometres before their final powerful advance. This interstadial is discernible in the climatic record for Camp Century (Figure 7.2) but the reconstructed data from other areas differ enormously in detail. In northern Europe, dated interstadial deposits are known

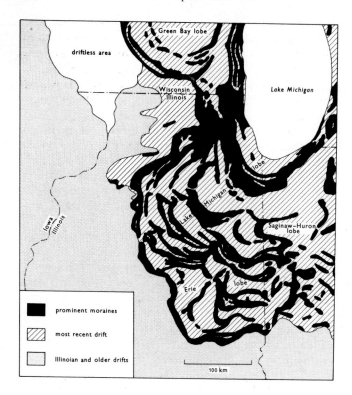

Figure 7.3
A map showing the complexity of the morainic loops of the Wisconsin 'Lake Michigan lobe'. These demonstrate many ice edge pulsations around the local Wisconsin maximum between 22,000 and 13,500 BP. *(Frye and Willman, Mem. geol. Soc. Am. 1973.)*

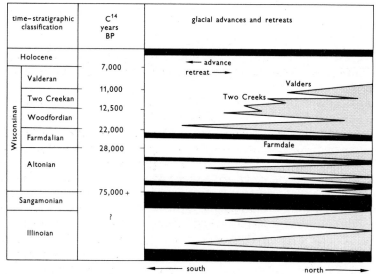

Figure 7.4
Generalized plot of Wisconsin and late Illinoian ice edge fluctuations in the central interior United States. The relevant time-stratigraphic classification is given on the left of the diagram. The dotted areas represent glacial events and the black bars represent non-glacial intervals. *(Frye, Quat. Res. 1973.)*

from Norway, Sweden and Finland (Mangerud, 1970b; Lundqvist, 1967; Korpela, 1969). Occasionally, till is found beneath these deposits (Bergersen and Follestad, 1971) but dating so far has allowed only the recognition of one 'Middle Weichselian interstadial'. In the Netherlands, Denmark and north Germany there are signs that the interstadial comprised several fluctuations including the Moershoofd, Hengelo and Denekamp phases (Hammen *et al.*, 1971; Mörner, 1972), whereas Coope *et al.* (1971) suggested that in the British Isles the Upton Warren interstadial was probably simpler with the warmest section peaking around 40,000 BP. In the midlands there is conflict between the evidence of fossil coleoptera and pollen for the reconstruction of interstadial temperatures (Morgan, 1973). Some difficulty is therefore encountered by the Pleistocene enthusiast in search of easy correlations.

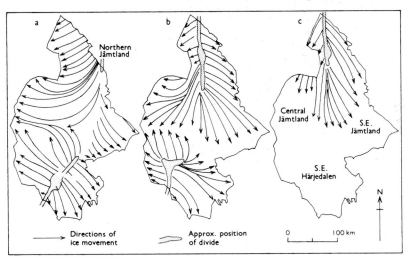

Figure 7.5
Fluctuations of the ice divides and directions of ice movement during the disintegration phase of the Weichselian ice sheet in the uplands of Jämtland, Sweden. *(Lundqvist, J., Sveriges geol. Unders. 1973.)*

Recent studies have revealed an increasingly complex sequence of climatic and ice margin oscillations on the peripheries of the wasting Wisconsin/Weichselian ice sheets. In Scandinavia there were some ice edge readvances, but generally the interstadials of Norway (including the Friesland interstadial) were times of more rapid retreat compared with the slow retreats or stillstands of the stadials. The same is generally true of Sweden and Finland, although there has been great discussion on the significance of the Ra, central Swedish and Salpausselkä stages (Tauber, 1970). The isochrones of ice margin retreat in Sweden and Finland show variable rates of recession, between 2–6 m per year during the Younger Dryas and 35 m per year during the Bölling (Nilsson, 1968). The most 'catastrophic' rates of ice disintegration occurred during the Bölling, and then in the period after 10,000 BP. As the ice retreated into its final resting place in the central Swedish mountains there were extremely complex changes in the position of ice divides and ice movement directions (Figure 7.5).

When one attempts a correlation between the deglaciation history of northern Europe and elsewhere, certain similarities emerge, but there are also fundamental differences. For instance, Mercer (1969) and Manley (1971) supported the idea that the Allerod was a peculiarly European phenomenon, since it cannot be reliably correlated with any late-glacial phase from other parts of the world. However, the Camp Century data reveal a clear correlative of the Allerød phase in the midst of a complicated sequence of climatic fluctuations during the period 14,000–10,000 BP; and this sort of complexity is also demonstrated by work on fjord glacier moraines on the peripheries of the Greenland ice sheet (Weidick, 1968; 1972; Funder and Hjort, 1973). The isochrone maps for the disintegration of the Laurentide ice sheet (Figure 7.6) also mask many shortlived marginal stillstands and other fluctuations in different sectors (Andrews, 1973; Bryson *et al.*, 1969). As in Scandinavia, rates of ice edge recession varied during different phases of deglaciation and from one region to another.

Factors controlling ice margin fluctuations
There seems to be a broad global synchroneity of the Weichselian events reviewed above. But time and again detailed correlations fall down because the main stages and sub-stages in different regions are separated by up to several thousand years. No matter which time scales are used (radiocarbon years, pollen zones or the varve sequence), no amount of correction or adjustment will produce the ideal fit of events which many workers seem to crave.

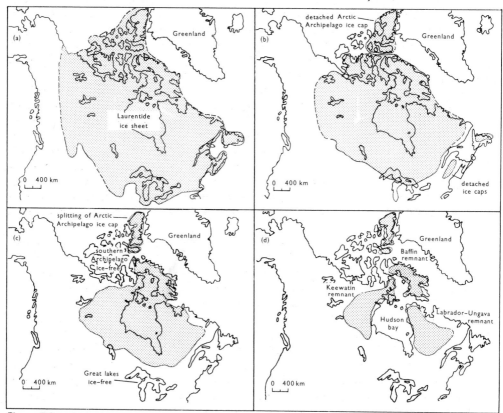

Figure 7.6
Maps showing the extent of Laurentide ice at various stages of deglacierization. (a) 13,000 C^{14} years BP; (b) 10,000 C^{14} years BP; (c) 8,000 C^{14} years BP; and (d) 7,000 C^{14} years BP. *(Bryson et al., Arctic and Alpine Res. 1969.)*

With Andrews (1973), it should be emphasized that the normal inter-hemispheric or inter-regional correlation of events at this time scale should be a time-transgressive one. The main factors contributing towards this lack of precise correlation are as follows:

1   The length of a glacier or ice sheet slope, and its gradient, between snout and main accumulation area will strongly influence its speed of reaction to climatic fluctuations. Hence Mangerud (1970b) has shown that the glaciers of the maritime west coast of Norway, being active, short and steep, reacted violently to the Zone I, II, III climatic oscillations whereas the ice front in Sweden and Norway reacted hardly at all. In Scotland, also, there was a similar contrast; the Zone III Loch Lomond readvance moraines are more spectacular in the west than in the east (Clapperton et al., 1975).

2   The influence of water bodies upon local glacier behaviour can be great. Ice free seas act on the one hand to 'damp down' climatic change (thus slowing the response of glaciers on oceanic islands, for example); and on the other hand to provide moisture which may raise glacier accumulation rates considerably. The exact nature of the land-sea feedback is still imperfectly understood, but Andrews et al. (1972) have suggested that the Cockburn readvance occurred at a time of climatic warming when more open water in Baffin Bay and Davis Strait provided the conditions for increased local precipitation. Similarly the events of the southern hemisphere must have been profoundly influenced by the amount of sea ice around Antarctica and by the shifting position of the Antarctic convergence (van Zinderen Bakker, 1969; Lamb, 1969; Mercer, 1973).

3   Differing climatic circumstances between high latitude and mid latitude areas must have resulted in vastly different response times for various ice masses. One can speculate, for example, that the Alpine and Himalayan glaciers will have responded rapidly to Weichselian global climatic oscillations, whereas the Laurentide and Scandinavian ice sheets as a whole will have responded more slowly. Even on the peripheries of the Laurentide ice sheet, various segments of its component ice domes had different thermal characteristics and different response times. The pulsations of the ice margin in the Great Lakes region indicate a rapid and sensitive response to climatic change, but it will be surprising if these pulsations have any equivalents on the northern ice sheet margin adjacent to the Arctic Ocean.

4   The effects of Weichselian ice sheet and glacier surges are as yet largely unknown. Mercer (1973) has built upon the ideas of Wilson (1964) and Hollin (1965) to propose a Holocene Antarctic surge, and Prest (1969) has considered the possibility that several of the Laurentide second-order readvances (including the substantial Valders readvance) were the result of localized surging and hence of no climatological significance. Elsewhere (such as in Svalbard and parts of east Greenland) the possibility should be considered that many glaciers have behaved in the past as they do now, by habitually surging.

5   As mentioned in the last chapter, the factor of 'normal' shifts in depression tracks and increased temperature and pressure gradients in the mid latitudes must now be considered seriously by all those interested in Pleistocene glaciology (Lamb, 1969; 1972; Mitchell, 1972). The maps of the NCAR and CLIMAP projects, giving hemispheric circulation patterns at different stages of the Weichselian and Holocene, show that the ice sheet areas cannot have experienced synchronous precipitation increases or synchronous phases of surface cooling and warming. As input–output relations have varied, so must the response of the large Weichselian ice masses.

6   The control exerted by sea-level on ice sheet growth and decay has already been noted in chapter 5. Here it needs to be stressed again. The maximum Weichselian extent of the Antarctic ice sheet was probably attained at a time when lowered sea-level allowed the grounding of ice shelves (Denton *et al.*, 1971). In Arctic Canada the coalescence of individual ice caps over the archipelago was favoured by a fall in sea-level, just as during deglaciation the rise of sea-level contributed to rapid ice disintegration (Bryson *et al.*, 1969; Andrews *et al.*, 1972). Between 8,000 and 7,500 BP the central portion of the Laurentide ice sheet disintegrated in just a few hundred years, as sea water flooded into Hudson Bay (Figure 7.6). Andrews (1973) argued that the sea and temporary pro-glacial lakes enabled calving to operate as a crucial ablation mechanism, thus contributing to the huge rates of ice margin retreat (up to 600 m per year) which cannot otherwise be explained by calculations of total energy receipt. In many areas, individual moraine systems are related not to climatic oscillations but simply to the grounding position of coastal glacier fronts. This will come as no surprise to Scandinavian workers, who have long realized the close relationship which existed between the eastern and southern margins of the Scandinavian ice sheet and the dimensions and relative altitudes of a succession of water bodies (Fredén, 1967; Lundqvist, 1967; Okko, 1965). Almost all of the traceable ice limits and retreat stages are related to the grounding position of the ice edge.

7   Finally, the local control of topography upon ice margin positions must be mentioned. The influence of fjord systems and flat-floored troughs upon glacier dimensions has already been mentioned; but it must not be forgotten that many of the other factors referred to in chapter 5 (such as surface roughness, aspect, and gradient) will have controlled glacier behaviour on parts of ice sheets and around their peripheries.

Now that a little more is known about glacier behaviour and the fluctuations of climate over space as well as time, it may be reasonable to expect the major events of the Weichselian and other glaciations to be *time-transgressive* and considerably variable in character.

Glacier fluctuations over hundreds of thousands of years

Earlier Cenozoic glaciations
We can assume that the mechanisms involved in the continental glaciations which occurred earlier than 100,000 BP were basically similar to those referred to in the last section. We shall not, therefore, consider these glaciations in any detail – a task which would in any case be difficult because of the scarcity of supporting evidence. The problems arising from these early glaciations become increasingly the concern of the geologist and the oceanographer as one extends back in time (Turekian, 1971). Nevertheless, some instructive comparisons with the Weichselian can be made.

From areas which have been inundated by thick ice during the last glaciation, there is generally only scanty evidence of earlier glacial phases. Nevertheless, a frequent impression is gained that the Weichselian glaciers were geomorphologically not very effective. The occurrence of interglacial deposits in Scandinavia (Lundqvist, 1973; Mangerud, 1970a) and Arctic Canada (Skinner, 1973), all in locations close to centres of Weichselian ice outflow, demonstrates that the ice has not affected much lowering of the bedrock base even if it may have been locally effective in streaming situations. In the South Shetland Islands even unconsolidated raised beach deposits were not completely destroyed during the last glaciation (John and Sugden, 1972), and elsewhere there is evidence that older erosional features have been only slightly modified during the Weichselian (Ahlmann, 1919; Sugden, 1969). It may be that the Weichselian glacial stage was either too short or glaciologically unsuitable for the scale of landscape modification achieved by its predecessors. Often, however, it is difficult in upland areas to differentiate these early glaciations. Sets of erosional landforms (such as streamlining features or components of glacier trough systems) which indicate variable directions of ice flow may be the end-products of a number of Pleistocene glaciations or simply of a single glaciation composed of a number of different phases. Hence in Scotland there are no reliable grounds for recognizing more than two glaciations – one early (perhaps Saalian) and one late (Weichselian). The same is true on the peripheries of Greenland, where deposits of the last interglacial are now known, and in the South Shetland Islands. In this island group the bulk of glacial erosion, including the cutting of a series of major transverse troughs has been assigned to one 'early glaciation', for there is no reason as yet for it to be subdivided.

On the other hand the marginal zones of the continental ice sheets provide abundant evidence for multiple Pleistocene glaciations. On the peripheries of the Alps the pioneer studies of Penck and Bruckner (1909) showed that the Alpine glaciers had expanded on no less than four occasions, although there has always been some doubt about whether these glaciations were separated by real interglacials (Henberger, 1974). The Alpine terminology was used for many years in the search for correlatives in northern Europe and elsewhere, and there now seems little doubt that approximate correlatives do occur. In north Germany and Denmark there are glacial deposits (separated by interglacial sediments in places) from three glaciations, generally referred to as Weichselian, Saalian and Elsterian, which are thought to correlate with the Würm, Riss and Mindel glaciations of the Alps (Hammen *et al.*, 1971). However, the correlation of the alpine Günz and Donau glaciations with phenomena in northern Europe and the Mediterranean is problematic, for no pre-Elsterian glacial deposits are known. As far as one can tell from current evidence, the Elster and Saale glaciations of the Scandinavian ice sheet were of approximately equal extent; however, in several sectors the Saalian drifts extend well to the south of the Elster ice margin.

A similar situation is encountered in North America, where the main drift sheets of the north central United States have been assigned to four glaciations – Wisconsin, Illinoian, Kansan and Nebraskan. Here again, the outer limits of these glaciations were approximately coincident, with the Kansan drift sheet on average the most extensive (Flint, 1971; Wright and Frey, 1965; Frye, 1973). As in northern Europe, the evidence for full interglacials separating these glacial stages is convincing. Occasionally there is evidence for multiple Pleistocene glaciation from areas which are still intensively glacierized. One such area is the McMurdo Sound region of Antarctica (Denton *et al.*, 1971), and another is Patagonia and Tierra del Fuego (Fidalgo and Riggi, 1965; Mercer, 1969). Other convincing stratigraphic evidence comes from Tjörnes in Iceland, where Einarsson, Hopkins and Doell (1967) have shown that there are at least ten tillites, each one representing a major Icelandic glaciation and each separated by a lengthy non-glacial interval.

Beyond the limits of the Pleistocene ice sheets there is abundant evidence for a complicated succession of cold and warm phases throughout Cenozoic time (Cooke, 1973). From the excellent record in the Netherlands, Hammen *et al.* (1971) have recognized at least eight cold stages. It is not known whether each of these was severe or prolonged enough for large scale continental glaciation, but it is quite likely. Several of the north European cold stages can be correlated with the British sequence (Sparks and West, 1972), but in England there

Table 7.1 Main Pleistocene stages in the British Isles (*Sparks and West, 1972.*)

Stage	Climate	Type site or area	Correlation of glacial stages with northwest Europe
Flandrian (postglacial)	temperate	postglacial peats and lake deposits of Britain	
Devensian (last glaciation)	cold, glacial	glacial deposits of Cheshire Plain	Weichselian
Ipswichian (last interglacial)	temperate	interglacial lake deposits at Bobbitshole, Ipswich	
Wolstonian	cold, glacial	glacial deposits at Wolston, Warwickshire	Saalian
Hoxnian	temperate	interglacial lake deposits at Hoxne, Suffolk	
Anglian	cold, glacial	glacial deposits on the coast at Corton, Suffolk	Elsterian
Cromerian	temperate	lake deposits at West Runton, Norfolk	
Beestonian	cold	silts and fluviatile gravels at Beeston, Norfolk	
Pastonian	temperate	tidal deposits at Paston, Norfolk	
Baventian	cold	marine sands and silts on the coast at Easton Bavents, Suffolk	
Antian / Thurnian / Ludhamian	temperate / cold / temperate	marine deposits in borehole at Ludham, Norfolk	
Waltonian		crag on coast at Walton-on-the-Naze, Essex	

is good evidence for only three glacial phases (Table 7.1). Apart from the pollen analytical evidence now building up from many countries, other techniques have contributed much to the reconstruction of Pleistocene climates. Among these are the measurement of oxygen isotope fluctuations in oceanic deep sea cores by Emiliani (1955; 1966) and Shackleton and Opdyke (1973); the pioneer geochemical analyses of Arrhenius (1952) on Pacific deep sea cores; the studies of radiolarian assemblages by Sachs (1973) and of planktonic foraminifera by Imbrie *et al.* (1973); and other work on the faunal content of marine sediments by Wollin *et al.* (1971), Kennett (1970) and Hays and Opdyke (1967). Some of the work published to date appears to be reliably fixed by absolute dating techniques, and most of the palaeotemperature graphs demonstrate neat quasi-periodic fluctuations which must be the result of real temperature oscillations during the Cenozoic. Hence the latest 'coolings' on the graphs shown by Chappell (1973) and others must correlate with the Pleistocene glaciations referred to above. However, the details of the various graphs show considerable discrepancies, and the papers in Turekian (1971) show a wide divergence of opinion on the absolute dating of the 'cool' and 'warm' periods. In general, it may be safest to conclude that there have been clear oscillations of world climate during the Cenozoic, and that at least 10 'cold' stages are recognizable. The last four of these cold stages were severe enough to induce continental glaciation in the middle latitudes of the northern hemisphere.

The polar regions have probably experienced more or less continuous glacierization through the Pleistocene. Emiliani (1969) has suggested that the Greenland ice sheet may have experienced 'significant melting' during the last interglacial, and that it has grown to its present dimensions in Weichselian and Holocene time. Mercer (1969) disputed this, but he has allowed the possibility that the west Antarctic ice sheet may have disappeared entirely during the same interglacial. On the other hand there is little doubt that the main Antarctic ice sheet has been present throughout the Pleistocene and the Pliocene. Rutford *et al.* (1972) have shown that parts of Antarctica were glacierized about 7 million years ago, and Denton *et al.* (1971) argued that the east Antarctic ice sheet had assumed its full bodied form before 4 million BP. It has probably never disappeared since. Mercer (1973) argued that the west Antarctic ice sheet may have formed at about the same time, although it may have disintegrated and re-formed several times since. On the other hand, a number of studies of cores from the Southern Ocean have confirmed that no full interglacial stage has interrupted the pattern of Antarctic continental glaciation since about 4 million years ago (Goodell *et al.*, 1968). Recent discoveries arising from the Antarctic drilling programme of the *Glomar Challenger* indicate that there was an Antarctic ice sheet very much earlier than previously suspected – about 20 million years ago. The ice sheet seems to have attained its maximum extent some 5 million years ago, since which time it has stabilized at approximately its present dimensions (Kennett, 1973).

Glacier fluctuations over millions of years

Ice Ages through geological time
There is now a wealth of evidence from the stratigraphic record to suggest that Ice Ages have occurred at intervals through geological time (Harland and Herod, 1975). In explaining them we are no longer concerned simply with the assumed fluctuations in solar radiation and their associated climatic and glaciological feedback relations; we have to consider the

Figure 7.7
Grooved bedrock surfaces from the Ordovician glaciation of the central Sahara (above) *(photograph by kind permission of Dr Anders Rapp)*, and a Cambrian glaciation at Simrislund, southern Sweden (below) *(photograph by Sven Stridsberg)*.

endogenetic processes of tectonics and continental drift in any attempt to provide a logical picture of world glaciation over 3,000 million years or more.

According to Steiner and Grillmair (1973) there are seven major glacial episodes which are well documented and dated in the geological record. The earliest recorded Ice Ages occurred in the Pre-Cambrian, for tillites are known from Pre-Cambrian rock sequences in many of the shield areas of the world. The best documented of the older Pre-Cambrian glaciations is the Gowganda glaciation of Canada and the United States, dated to 2,288 ± 87 million years (Fairbairn *et al.*, 1969). There are others which are probably older. For the later Pre-Cambrian (1,000–600 million years ago) there is evidence of glaciation from all the continents except Antarctica, including tillites, glaciated pavements and glacio-marine deposits (Dunn *et al.*, 1971; Reading and Walker, 1966; Perry and Roberts, 1968; Grabert, 1967). It may be that there were more than twelve Ice Ages during the immensely long period between 2,500 million and 600 million years ago, but at present it is difficult to generalize concerning their occurrence either in space or time on the surface of the globe.

The next important Ice Age appears to have occurred during the Silurian and Ordovician. There is evidence from Argentina, Brazil, the Yukon and Spain, but much of this is inconclusive. On the other hand, studies by French workers over the last decade or so have revealed widespread areas of striated pavements and Ordovician tillites in the central Sahara (Figure 7.7) (Bannacef *et al.*, 1971). Fairbridge (1970; 1971) reviews some of the evidence, referring also to the occurrence of eskers, faceted boulders, crescentic gouges, and even subglacial meltwater channels at various localities over a distance of over 4,000 km on the Saharan shield. There seems little doubt that the Ordovician glaciation was by a true polar ice sheet, for the central Sahara coincides closely with the position of the Ordovician south pole. Generally, the ice movement across this region as from south towards north, and there are signs of considerable ice edge fluctuations possibly associated with the periodic development and break up of ice shelves. The Ordovician ice sheet centre was undoubtedly located at one time to the south of the Sahara, and ice action on its other margins may account for the tillites and other traces of glaciation reported from South Africa and South America.

Evidence for the Gondwanaland ice sheet of Permo-Carboniferous times is even more solid; for example, the Dwyka tillite at the base of the Karroo System in South Africa rests

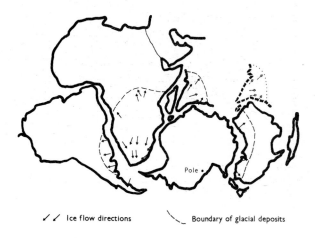

Figure 7.8
A reconstruction of Upper Carboniferous Gondwanaland, showing the extent of the Gondwanaland ice sheet and ice-directions inferred from tillites and other traces in the geological record. *(Tarling and Tarling, Doubleday, 1971.)*

↙↙ Ice flow directions   Boundary of glacial deposits

Pole

on a striated pavement in places, and indeed several hundred metres of glacial debris (including varved clays and fluvio-glacial materials) can be found on all the Gondwanaland continents (Hamilton and Krinsley, 1967; Frakes and Crowell, 1969; 1970; Frakes *et al.*, 1971). In Brazil there are up to 100 m of glacial deposits of this age, distributed over an area of about 4 million km². When the deposits of this glaciation are plotted on a map of the Gondwanaland landmass/ocean pattern, they form exactly the grouping that might be expected from a large polar ice sheet (Figure 7.8). The area covered by this ice sheet may have been more than twice that of the contemporary Antarctic ice sheet.

It now seems that fluctuations in the Gondwanaland ice sheet axis and other spatial variations can be discerned from the stratigraphic record (Adie, 1975). It is known, for example, that not all of the tillites were laid down by a lowland continental ice sheet; some of the tillites of New South Wales have been related to local Permo-Carboniferous mountain glaciation. Also the glaciation of the different parts of Gondwanaland was not synchronous. The approximate order of glacial initiation in the different areas was as follows:

1 Earliest – South America and Africa
2 Later – Madagascar and India
3 Latest – Antarctica and Australia

This sequence of glaciation can be related quite simply to the slow migration of Gondwanaland through the southern polar regions broadly from southeast towards northwest. The final stages of glaciation in Australia were contemporaneous with the appearance of hot deserts in eastern South America and northern Africa which were by then only some 20° from the Equator. Also, there are certain anomalies among the reconstructed directions of ice movement from the various segments of Gondwanaland; some of these may be explained by multiple glaciation over time, for it would be surprising if the glaciated regions did not experience violent ice margin fluctuations (such as those of the Pleistocene) especially when they were located near the peripheries of the ice sheet. Again, it is probable that the Gondwanaland ice sheet was in reality made up of several domes just like the Wisconsin Laurentide ice sheet and the contemporary Antarctic ice sheet. These domes acted as more or less independent centres of accumulation and ice dispersal, and they will have had their own particular response characteristics. At times they will have coalesced, and at other times they may have become detached from one another. Hollin (1969) has even extended current glaciological theory to the Gondwanaland situation, and has suggested that parts of the ice edge which were calving in the sea were subject to periodic surges. Almost certainly, as studies of the Gondwanaland glaciation advance, it will be possible to apply further glaciological and climatological principles to its pattern of behaviour; conversely, it should be possible to learn much from the history of Gondwanaland glaciation about the future behaviour of the Antarctic ice sheet.

Many authors have attempted to relate the sequence of global Ice Ages to theories of oscillating climate, but it is difficult to discern any reliable glacial periodicity in spite of the work of Steiner and Grillmair (1973) and many other authors. As far as can be seen, the major Ice Ages occurred about 2,300, 900, 750, 600, 450, 300 million years ago, and within the last 6 million years. If only someone would find a convenient continental glaciation from the middle of the Mesozoic era we should be able to propose an approximate periodicity of 150 million years for most global Ice Ages; but so far nothing has been forthcoming. However, this may partly be due to the fact that the late Jurassic and early Cretaceous poles were inconveniently situated with respect to the landmasses of the time (Wellman et al., 1969; Mercer, 1973). In reality the world has probably experienced continuous glaciation since the Pre-Cambrian, with oscillations between the continental Ice Ages and intervening periods when glaciations may have been restricted to suitably located upland areas.

In explaining the occurrence of Ice Ages through geological time, we must not only consider the role of continental drift and polar wandering, but also the associated endogenetic process of mountain-building. It has been seriously suggested that the building of the Tertiary mountain chains (North and South American cordillera, Himalayas and Alps, for example) fundamentally affected the pattern of global atmospheric circulation and led to upland glacierization. From this situation, positive feedback operated until glacierization spread to the high latitude lowlands and resulted in the growth of the Pleistocene ice sheets (Flint, 1957). Other related theories of Ice Age initiations appear in the extensive reviews by Brooks (1949), Charlesworth (1975), Embleton and King (1975), and Flint (1971). There is no space to review these theories here, except to reiterate that even on the extended scale of geological time essentially 'normal' mechanisms must have been responsible for the development of continental ice masses of whatever age. The multitudinous theories of Milankovich (1969), Ewing and Donn (1959), Fairbridge (1970), Simpson (1957) and Hughes (1975) are by no means incompatible; each Ice Age must have been initiated by a combination of circumstances involving complex galactic variables, interrelations between land configuration and altitude, ocean characteristics, polar position, and features of oceanic circulation (Steiner and Grillmair, 1973). The principle of uniformitarianism may apply in the youthful science of palaeoglaciology as in the other earth sciences.

Towards a model of world glaciation

From what has been said above, it now seems that glacier fluctuations in the long term are the result of factors similar to those which control short term ice margin oscillations. Essentially, we can see ice masses of continental proportions responding (albeit rather slowly) to changes in energy input. The only difference between an ice sheet and a corrie glacier is one of scale; the response time of the corrie glacier is so short that it can only realistically be interpreted in terms of minor climatic oscillations, whereas an ice sheet survives for so long that it can be seen to respond to endogenetic pulses of energy input (continental drift, mountain-building) as well as exogenetic ones (Menard, 1971; Damon, 1971). A major pulse of terrestrial energy, such as that experienced in Cenozoic time in the Alps and Himalayas, may result in such rapid uplift that fluvial processes cannot compensate. The upland mass may be lifted above the glacierization limit; glaciation will be initiated, and glacial and periglacial denudation will assist in the process of upland destruction. This is another way of looking at the influence of altitude and 'relief energy' as discussed in chapter 5; here we have to consider the effects of altitudinal change through time.

It is also possible to generalize concerning the horizontal dimension in the expansion and contraction of Ice Age glaciers. In Figure 7.9 an attempt is made to portray in simple terms how a continental ice margin fluctuates in space between glacial and interglacial stages, or between Ice Ages and 'Warm Ages'. It demonstrates simply how the ice margin retreats polewards at times when global climate is least suited to an extensive ice cover, and advances through a succession of latitudinal belts in response to stages of global cooling. Although this model is of course dangerously simplified, it is intended to demonstrate that the concepts

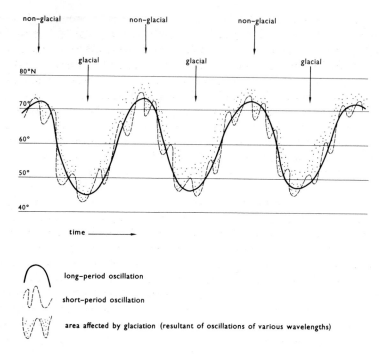

Figure 7.9
A model of continental ice margin fluctuations in the high and middle latitudes over time. This could be applicable at a variety of different time scales and for glacial oscillations measured on many different wavelengths. Note that the non-glacial intervals become less and less marked as they are traced into higher latitudes.

of glacial and non-glacial stages can have no universal meaning; if we simply consider glacial stages we can see from the diagram that in middle latitudes they may be so brief as to make little impact upon the landscape, while near the pole some areas may never have escaped from the grip of ice since the onset of Cenozoic cooling. The field evidence strongly supports the idea of latitudinal and altitudinal shifting (with compression of climatic belts and increasingly steep temperature and pressure gradients particularly adjacent to large ice sheets) during times of world glaciation. While the ice sheets move into the middle latitudes, ice cap glaciation replaces isolated upland valley glaciation in other mid latitude zones; regional glaciation appears in low latitude high altitude areas; and local glaciation becomes important in some low latitude areas with less elevated uplands. The full chain of linkages becomes enormous when one also considers the creation of permafrost, changes of oceanic circulation, and the adaptations required of animal and plant communities in the sea and on land (Flint, 1971).

    Among the points made on foregoing pages have been those concerned with the time-transgressive nature of glacierization, and with the normal climatological interactions responsible for large scale ice accumulation and degradation. Hence it may seem foolish to attempt to create a model of ice sheet growth and decay. Flint (1971) has attempted such models for the growth and decay of the Laurentide and Scandinavian ice sheets, postulating an original building of ice in the uplands and then a gradual coalescence of piedmont glaciers to form the ice sheets proper. Recent evidence points to a similar course of events in the development of the Antarctic ice sheet, with the initiation of glacierization in the Transantarctic mountains followed by the growth of first the east Antarctic ice sheet and then the west Antarctic ice sheet (Drewry, 1975). There is an alternative suggestion for North America, and Ives et al. (1975), Andrews and Barnett (1972) and Andrews and Mahaffy (1975), have suggested a much more complex pattern of Wisconsin Laurentide ice buildup and decay; they have argued that the ice sheet may have grown by the gradual broadening and expansion of small plateau glaciers and upland snowfields over a wide area. All these hypotheses stand as models, and models exist to be replaced when they have outlived their usefulness.

The following generalizations may apply in many cases:

(a) *Ice sheet growth.* Ice sheet growth must be the result of excessive precipitation, and the failure of the glacier system to achieve adequate throughput. Therefore the glacier must grow and increase its volume. It can only expand its storage capacity effectively by building up to a new equilibrium profile and expanding laterally. As yet there is still some uncertainty about whether there is a predictable sequence of growth. In upland areas the sequence may be as follows – cirque glaciers – névé domes – ice caps – ice sheet. In high latitude lowlands or plateau areas the sequence may be – snowpatches – small ice domes – coalescent domes – ice caps – ice sheet. In reality local conditions, and climatic oscillations during ice buildup, will have caused the sequence of development to be much more complex than this. However, the role of positive feedback probably becomes more and more important as the ice mass grows in size.

(b) *Maximum stage.* It is doubtful if ice sheets ever achieve a perfect equilibrium state, although probably they are stable and exert a certain influence over their own destiny once they are able to acquire their own climate. They are stable given a certain range of climatic oscillations; but there is no doubt that if climatic warming crosses a certain threshold (does every ice sheet have its own threshold?) then an ice sheet can become extremely unstable. The catastrophic disintegration of the Laurentide ice sheet illustrates this admirably. Glaciological instability leading to ice sheet surges is still a matter for debate (Hughes, 1972, 1975; Mercer, 1973).

(c) *The wastage process.* Once an ice sheet has achieved a state where it can dispose of all the precipitation it receives, it becomes vulnerable to either climatic warming or a reduction of precipitation supply. If both these events occur simultaneously, rapid wastage may follow. It is possible that an ice sheet may build up until the centre is too high and too remote from the coast to receive adequate precipitation. Once this occurs, ice sheet diminution becomes likely. The accumulation pattern for the present Antarctic ice sheet illustrates this idea perfectly, and there is debate about whether it is ready for collapse. Also, great significance must be attached to the fact that the Pleistocene Laurentide and Scandinavian ice sheets repeatedly expanded to almost identical southern limits and then collapsed. This must support the idea that each ice sheet has a certain maximum size beyond which it cannot expand without endangering its own existence. Once commenced, probably the process of disintegration has the characteristics of a positive feedback loop. If the equilibrium line rises above the ice sheet summit, wastage must become catastrophic, with the ice sheet downwasting and with little or no forward movement of the ice. If some supply is maintained, the ice sheet may waste by a gradual and regular withdrawal of its edge. The 'back-wasting versus down-wasting' debate still continues.

These generalizations raise as many questions as they answer, but gradually the answers to many of the imponderables of ice sheet behaviour are being clarified by field studies and computer models of the behaviour of the Weichselian Laurentide ice sheet (Andrews, 1973; Barry, 1973; Williams *et al.*, 1974), and we must hope that similar detailed work can be undertaken in the areas occupied by the other major Pleistocene ice sheets.

Further reading

ANDREWS, J. T. 1973: The Wisconsin Laurentide ice sheet: dispersal centres, problems of rates of retreat, and climatic implications. *Arctic and Alpine Res.* **5** (3), 185–99.

ANDREWS, J. T. and MAHAFFY, M. A. W. 1976: Growth rate of the Laurentide Ice Sheet and sea level lowering (with emphasis on the 115,000 B.P. sea level low). *Quat. Res.* **6,** 167–83.

CLIMAP Project Members, 1976: The surface of the Ice-Age Earth. *Science,* **191** (4232), 1131–7.

DENTON, G. H. and HUGHES, T. J. 1981: The last great ice sheets. John Wiley, New York (484 pp).

DREWRY, D. J. 1975: Initiation and growth of the east Antarctic ice sheet. *J. geol. Soc. Lond.* **131,** 255–73.

GRIBBIN, J. 1978: *The Climatic Threat.* Fontana/Collins, 206 pp.

HAYS, J. D., IMBRIE, J. and SHACKLETON, N. J. 1976: Variations in the Earth's orbit: pacemaker of the Ice Ages. *Science* **194** (4270), 1121–32.

HUGHES, T. 1980: Genes and glacial history: a letter to the editor. *Boreas,* **9,** 149.

JOHNSEN, S. J., DANSGAARD, W., CLAUSEN, H. B. and LANGWAY, C. C. 1972: Oxygen isotope profiles through the Antarctic and Greenland ice sheets. *Nature, Lond.* **235,** 429–34.

JOHNSON, R. G. and MCCLURE, B. T. 1976: A model for Northern Hemisphere continental ice sheet variation. *Quat. Res.* **6,** 325–53.

LORIUS, C., MERLIVAT, L., JOUZEL, J. and POURCHET, M. 1979: A 30,000-year isotope climatic record from Antarctic ice. *Nature,* Lond. **280,** 644–8.

LUNDQVIST, J. 1974: Outlines of the Weichsel Glacial in Sweden. *Geol. Fören. Förh. Stockh.* **96,** 327–39.

SPARKS, B. W. and WEST, R. G. 1972: *The Ice Age in Britain.* Methuen, London (302 pp).

TUREKIAN, K. K. (ed) 1971: *Late Cenozoic Glacial Ages.* Yale Univ. Press (606 pp).

WRIGHT, A. E. and MOSELEY, F. (eds) 1975: *Ice Ages: ancient and modern.* Seel House Press, Liverpool (320 pp).

Part III
Glacial erosion and its effects

8 The processes of glacial erosion

'Its stupendous unwieldly mass is dragged over the rocky surface, it first denudes it of every blade of grass and every fragment of soil, and then proceeds to wear down the solid granite, or slate, or limestone, and to leave most undeniable proofs of its action upon these rocks'. Since J. D. Forbes described the effects of a glacier in these words in 1843, there has been widespread acceptance of the reality of glacial erosion. Although admittedly there are known to be situations where a blade of grass might well be preserved beneath a glacier, most often a land surface is modified when traversed by a glacier. Indeed a variety of measurements point to the conclusion that glacial erosion is several times more potent than fluvial erosion (Embleton and King, 1975). However, an understanding of how this erosion takes place remains elusive. It is the purpose of this chapter to explore the fundamental processes involved in glacial erosion. Firstly, basic processes are discussed. Then the chapter goes on to consider variations in the intensity of glacial erosion, from place to place within single glaciers, between different glaciers and finally over time.

The basic processes

Traditionally it has been found helpful to recognize the two fundamental processes of glacial erosion as abrasion and plucking (quarrying). This subdivision was based partly on the interpretation of the smoothed and rough faces of features such as roches moutonnées (Figure 8.1), and partly on the interpretation of the two main components of glacially eroded debris –

Figure 8.1
The smoothed and plucked face of a small roche moutonnée on Storskär, Rödlöga Skärgård, Stockholm archipelago. Note the slices which have been sheared from the lee face.

rock flour and rock fragments. With improving understanding, it is perhaps helpful to sub-divide these basic processes further. In the first part of the chapter the basic processes are discussed under the following headings – abrasion, fracture of fresh rock, joint exploitation, meltwater erosion, and debris entrainment.

Abrasion

Generally speaking, abrasion describes the process whereby bedrock is scored by debris carried in the basal layers of a glacier. Evidence of such a process comes from several sources. Striations are etched into many rock surfaces affected by glaciers and reflect grinding by

Figure 8.2
The striated surface of a boulder caused by grinding at the base of the Ordovician ice sheet, Sahara Desert.
(Photograph by Anders Rapp.)

sand and rock particles moving under considerable ice pressures (Figure 8.2). Such surfaces have obvious analogies to surfaces attacked by artificial abrasives, as examination of the sand-papered surface of a piece of wood under a magnifying glass will show. The intricacy of striation patterns and their implications are discussed superbly by Chamberlain (1888). Another line of evidence concerns rock flour, a ubiquitous product of glaciers which con-tributes the characteristic blue-green colour to glacial meltwater streams. Consisting of fresh mineral fragments, generally smaller than $100\,\mu$ (Vivian, 1970), rock flour is difficult to explain except as a product of grinding. In addition the process of abrasion has been verified in the field; Veyret (1971) noted that during a recent readvance of the Glacier des Bossons, striations 0·4 m long, 10–20 mm across and 35 mm deep were abraded by boulders within a space of four weeks. Boulton and Vivian (1973) record the abrasion on blocks bolted to the rock underneath part of Breiðamerkurjökull in Iceland. After 9·5 m of basal ice studded with rocks had passed across the test blocks, both were striated and the marble block had been lowered by an average of 3 mm and the basalt by 1 mm.

Understranding of abrasion beneath glaciers has come from observations associated with subglacial hydro-electric tunnels (Galibert, 1962; Vivian, 1970), from theoretical considerations (Röthlisberger, 1968) and from laboratory work (Lister *et al.*, 1968; Hope *et al.*, 1972). From this one can construct a table of some important variables which are likely to affect the efficacy of the process (Table 8.1). (The role of bed shape is considered in the next chapter.)

Table 8.1   Tentative list of variables affecting the process of glacial abrasion.

Fundamental requirements for abrasion	(1) Basal debris (2) Sliding of basal ice (3) Transport of debris down towards bedrock
Other factors affecting rate and type of abrasion	(4) Ice thickness (5) Basal water pressure (6) Relative hardness of rock particles and bedrock (7) Particle characteristics (8) Efficient removal of rock flour

1   The presence and concentration of debris in the basal ice is an important requirement. Clean ice will not abrade solid rock (Röthlisberger, 1968), while high particle content will increase the efficacy of the abrasive. Presumably this is true so long as the debris content does not increase the viscosity of the basal layers sufficiently to prevent effective forward flow. However, in this context Peterson (1970) suggested that a frozen sheet of till being dragged along beneath cold-based ice is a most effective abrading agent. At the other extreme Gjessing (1965) drew attention to the possible abrading potential of a saturated till mass beneath warm-based ice.

2   The sliding velocity of the basal ice is important in that it affects the amount of abrasive glacier sole that can pass a given area of bedrock per unit time. Suppose there are two glaciers with equally dirty basal layers, one moving at 50 m per year and the other at 25 m per year. Clearly a bedrock surface under the first is passed by twice as much abrasive as the second.

3   Abrasion by rock particles is apparently ineffective unless particles are continually moving towards the rock bed of the glacier, and constantly renewing the abrasive. For example Lister *et al.* (1968) showed how sand particles in an experimental block of ice soon became polished when moved over sandstone and became increasingly ineffective abrading agents. Röthlisberger (1968) has examined the means by which particles are constantly renewed at the glacier sole. He concluded that melting of ice at the bottom of a warm-based glacier was the most effective way in which fragments could move down towards the bedrock surface (Figure 8.3a, overleaf). Calculations for a typical Alpine glacier showed downward velocities of 15 to 110 mm per year over wide areas but with locally high velocities of *c.* 2 m per year near meltwater streams. In addition, thinning of the basal ice layer (due to lateral deformation as the basal ice diverges round an obstacle)

(a)

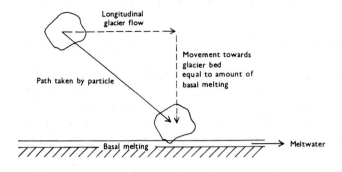

(b)

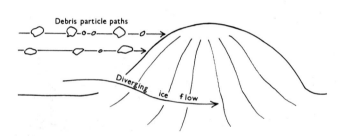

Figure 8.3
The renewal of abrasive material at the glacier base caused **(a)** by basal melting and **(b)** by the divergence of basal ice around an obstacle which brings particles into contact with the upstream side of the obstacle. Debris will also move round the obstacle.

was argued to have the effect of bringing stones down into contact with the top of the obstacle (Figure 8.3b). It is important to highlight the implications of these conclusions. The renewal of the abrasive material by basal melting can only occur beneath warm-based ice. The effect of ice deformation allows abrasion on the top of bumps and would presumably affect both warm-based and cold-based glaciers.

4   Ice thickness is important because of its effect on the vertical pressure on any particle beneath the glacier (Galibert, 1962). Assuming that there is no layer of water buoying up the glacier, the effective pressure exerted will increase with ice thickness. The effect on the amount of abrasion is shown diagrammatically in Figure 8.4. An increase in ice thickness increases the amount of abrasion until there comes a point, stressed by Boulton (1975), when the friction between the particle and the bed is sufficiently great to retard the movement of the particle, thus reducing the total amount of abrasion. A further increase in ice thickness accentuates the effect until the particle lodges, thereby crossing the threshold into chapter 11.

5   The presence of a layer of water at the base of a glacier can have important repercussions on the rate of abrasion because high water pressures have the effect of buoying up the glacier. This reduces the effective pressure applied by the ice on the bed and thus the

amount of abrasion achieved by a particle. On the other hand, by reducing the friction between the glacier base and the bed, sliding velocities tend to be locally increased (Boulton, 1972a).

6   The relative difference in hardness of the particles in the glacier and the bedrock affects the character of the abrasion. If the particles and bedrock are of equal hardness, one can assume equal amounts are abraded from each (Röthlisberger, 1968). If one is softer than the other then it will abrade faster. This is confirmed by Boulton and Vivian's experiment (1973), when marble was abraded three times as fast as basalt beneath the same basal ice. The most favourable conditions for rapid abrasion of the glacier bed are where the glacier sole leaves a hard rock armed with hard rock particles and passes on to a soft rock. The reverse situation would produce little abrasion of the glacier bed, though the soft rock particles in the glacier may be worn down.

7   Particle characteristics can have a marked effect upon the abrasion process. Other things being equal, one would expect an angular particle to scour more pronounced striations

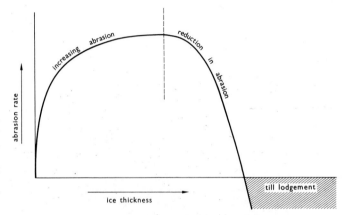

Figure 8.4
Schematic diagram to show how increasing friction associated with increasing ice thickness enhances abrasion until a critical ice thickness is reached. Then lodgement of particles begins. *(Boulton, 1974.)* The effect of basal water pressure is ignored.

than a rounded or flat particle. On the other hand, a flat particle may trap smaller particles beneath it and apply high pressures on them. Bigger blocks will exert more pressure on the bedrock than smaller blocks, and they will be capable of more efficient abrasion. This is because in the hydrostatic medium of ice, blocks exert a downward pressure proportional to their weight.

8   A final variable influencing the efficacy of abrasion is the removal of the fine debris produced at the ice/rock interface. Given a continued supply of tools from the basal ice it seems potentially more effective to have the fine debris removed from the zone of abrasion. In this context, the role of meltwater is important. Vivian (1970) recorded the presence of a film of meltwater at the ice/rock interface beneath the Glacier d'Argentière; the water film seems sufficiently thick to evacuate rock flour less than 0·2 mm in diameter into the main meltwater conduits under the glacier (Figure 8.5, overleaf). As a result of the removal of the fine material, grains coarser than 0·25 mm often support the ice and actively abrade, presumably until they too are sufficiently reduced to be washed away.

Much of this discussion of the variables affecting abrasion serves to emphasize the importance of melting and the presence of water at the base of the ice. Not only are both necessary

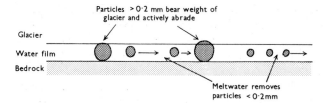

Particles > 0·2 mm bear weight of
glacier and actively abrade

Glacier

Water film

Bedrock

Meltwater removes
particles < 0·2mm

Figure 8.5
The removal of particles less than
0·2 mm in diameter in the layer of
meltwater at the ice/rock interface.

before effective abrasion can occur, but basal water has an important role in influencing
the velocity of basal slip. For these reasons one can state that, given the presence of basal
debris, conditions most favourable for abrasion occur beneath warm-based glaciers.

Fracture of fresh rock
The term is used here to describe the process by which ice or debris in the ice exerts sufficient
force on the bedrock to cause fracture. There are several lines of evidence pointing to such
a process. Firstly, there is the presence in glacier moraines of boulders frequently showing
freshly sheared faces. Secondly, the existence of friction cracks or chattermarks, which form
zones of crescentic grooves or cracks on some bedrock protuberances (Figure 8.6), is thought
to reflect fracture by the force of the overriding ice and the contained rock debris (Embleton
and King, 1975). Thirdly, Trainer (1973) analysed the jointing in bedrock in several parts
of the United States and recorded the occurrence of joints aligned at certain angles to former
ice movement. The consistent relationships were thought to demonstrate weakening by ice;
sometimes fractures followed pre-existing lines of weakness while sometimes they were new.

Figure 8.6
Crescentic fractures on quartzite, Simrislund, Sweden. *(Photograph by Sven Stridsberg.)*

Glen and Lewis (1961) have made a study of the forces likely to be available beneath a glacier. Since the yield stress in shear for ice is approximately 1–2 bars, there is an upper limit to the shear stress a glacier can apply on its bed. In a straight tug of war between ice and rock the forces involved are quite inadequate to shear fresh rock. However, if there are boulders in the basal layers of the ice then the situation is quite different. McCall (1960) calculated the downstream force associated with a 1 m cube of granite as approaching 50,000 kg and considered that with cautious assumptions this would be sufficient to shear a projection of granite with an initial surface of contact with the parent rock of 16 cm². The forces do not increase appreciably with increasing depth because of the inherent weakness of glacier ice. Even more significant, and in contrast to rivers, the force does not increase materially with increasing ice velocity. This is because the force on a single boulder increases only as the $\frac{1}{4}$ power of velocity (Glen and Lewis, 1961). However, the bigger the boulders in the ice the greater the potential force available. These calculations assume regular deformation of ice at the base of a glacier. However, in view of the evidence of a jerky stick-slip motion, it now seems likely that much higher forces are available, at least momentarily. It is likely that the slip phases accomplish the bulk of geomorphological work. In this context, it is interesting to recall that the spacing of friction cracks and chattermarks has long been recognized as suggestive of an irregular or vibratory force such as stick-slip.

Boulton (1974) has suggested that differences in pressure between the upstream side of a protuberance where ice is pressing hard on the rock and the downstream side where the pressure is less, may be sufficient to cause rock fracture. Such a process operates best where the ice only just retains contact with the lee side of the bump. For this reason the process is favoured by high velocities or by the presence of an irregular bed. Also small obstacles are clearly more susceptible to fracture than large ones.

Joint exploitation

The previous section was concerned with the fracture of fresh unjointed rock. Where there are lines of weakness, then clearly the available forces may remove much larger blocks of rock. Thus, a crucial part of glacial erosion involves any process which loosens or weakens blocks of rock beneath a glacier. Broadly speaking there are two groups of theories. First are those that resort to weakening of bedrock before it is covered by glaciers. Second are those processes of weakening thought to operate beneath a glacier.

The importance of the weakening of bedrock prior to glaciation has been stressed above all in the continental European literature. It has been suggested that periglacial conditions during the onset of a glacial phase can result in the loosening of rock in front of a glacier by frost shattering (Tricart and Cailleux, 1962). Another view, stressed above all by Bakker (1965), Feininger (1971), and Thomas (1974), points to the fundamental importance of the deep weathering of bedrock before a glaciation. In places, susceptible rocks may effectively be weakened to depths of over 100 m. In view of this, it is probably instructive to bear in mind Linton's reconstruction of deeply rotted rock profiles in Britain, resulting from chemical weathering (Linton, 1955) (Figure 8.7, overleaf). It is possible that such conditions prevailed in most of the humid areas of the world before the onset of glaciation.

A third mechanism of rock weakening before glaciation is associated with the idea that unloading of the land surface due to erosion can lead to dilatation joints in the surface layers of the bedrock. In this process, rocks crack parallel to a face of unloading (Ollier, 1969). Such a process has been long known to quarrymen when a rock face or quarry bottom bursts

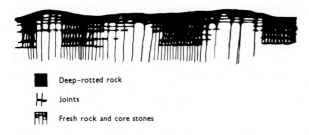

■ Deep-rotted rock

⊢ Joints

⫟ Fresh rock and core stones

Figure 8.7
Schematic representation of deeply rotted rock profiles which were probably widespread before glaciation. Chemical weathering has exploited joints. *(Linton, Geogr. J. 1955.)*

and is physically dislodged from the parent rock. It is on record that such bursts have thrown machines weighing one ton off railway tracks (Bain, 1931). Apparently such bursts are most frequent in the hardest and most compact rocks. The expansion of rocks when the constraining pressures are reduced is illustrated by the way rocks close in on drills. Galibert (1962) has investigated the process in association with hydro-electric tunnels in French mountains where it is found to depths of 200 m, and is convinced of its fundamental importance to glacial erosion. Birot (1968b) mentions that, due mainly to such dilatation jointing, French dam engineers find many rocks in nature to be 50–100 times as weak as indicated in the laboratory. In France much of this jointing is thought to lie parallel to pre-glacial valley slopes, suggesting it is pre-glacial in age (Galibert, 1962). Similar conclusions have been reached in other glaciated landscapes where terms such as pseudo-bedding and topographic sheeting have been used for similar features (Jahns, 1943; Waters, 1954; Chapman and Rioux, 1958; Sugden, 1968).

If block weakening could occur only before glaciation of an area, it would be difficult to explain the depths of troughs up to 3,000 m deep which occur, for example, in the east Greenland fjord region. It is for this reason that particular interest is focused on any process of block weakening which might occur beneath glaciers. The periglacial and deep rotting

Figure 8.8
Sheeting developed parallel to a glacial facet near Søndre Strømfjord airfield, west Greenland. The sheeting cuts across strongly foliated gneisses.

theories mentioned are unlikely to play much part except in areas of marginal glaciation or of moderate erosion. However, there are reasons to believe that dilatation jointing can operate beneath active glaciers. Several workers have noted that sheeting or jointing can frequently be observed parallel to glacial facets, implying some relationship. Battey (1960) and Lewis (1954) described examples in Norway, and Linton (1963) described examples from the Cuillins in Scotland. In the fjords of Greenland, for example Søndre Strømfjord, there are spectacular examples of sheeting parallel to trough walls, regardless of strong structural trends in the rock (Figure 8.8).

The problem is whether dilatation occurs beneath a glacier or whether it develops after a glacier withdraws because of the removal of the overlying weight of the glacier. Although

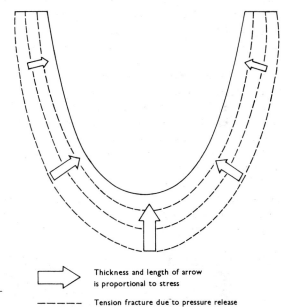

Figure 8.9
Tension cracks caused by dilatation beneath a glacier.

⇨ Thickness and length of arrow is proportional to stress

– – – – – Tension fracture due to pressure release

the latter will almost inevitably have had an effect, there are grounds for thinking that dilatation jointing can occur beneath ice. Suppose that a glacier begins to erode a depression into a pre-glacial surface. At first the increased weight of the ice on the rock will counteract any dilatation effect, and erosion will need to occur by other means. Eventually, however, the bedrock may be deepened to such an extent that the pressures exerted at the base of the ice are reduced in comparison with the original pre-glacial pressures exerted when the depression was occupied by rock. This is because ice is lighter than rock with a density approximately one third that of rock. With further deepening, dilatation will occur, probably increasing progressively as the deepening increases. Since rocks are subjected to hydrostatic pressure at even quite modest depths, the pressures will act perpendicular to the glacial facet and thus favour sheeting parallel to the facet (Figure 8.9).

The stresses can be calculated for an example which serves to give an idea of their order of magnitude. In that part of Søndre Strømfjord near the Sukkertoppen ice cap in west Greenland some 1,000 m of deepening by ice has occurred. After extrapolating an ice surface profile and allowing a generous ice thickness of *c.* 1,500 m above the trough bottom when it had been excavated, this represents an effective unloading at the bottom of the trough equivalent to a thickness of 500 m of rock. This is equivalent to a pressure of about 136 bars. The potential importance of this sort of figure can be appreciated when one compares Bain's observation that rock bursting in Vermont marble quarries occurs when only 6–7 m of rock is removed (Bain, 1931). Yet dilatation does not necessarily explain the occurrence of joints perpendicular to the ice/rock interface, which are necessary to produce blocks (Trainer, 1973). These would only occur by dilatation on an irregular surface where there is room for lateral expansion.

Another long favoured mechanism for weakening rocks beneath a glacier is freeze–thaw. Water beneath the glacier penetrates the rock joints, before freezing to widen and thus loosen the joints. On small glaciers, a process of meltwater flowing down the rock walls to freeze beneath the glacier has been advanced as a possible theory (Lewis, 1940). To be effective this will require basal ice below the pressure melting point and a supply of meltwater. The theory and observation of regelation in the lee of rock protuberances offers a mechanism of shattering (Figure 2.13). Since the observations of Carol (1947), this latter mechanism has seemed an attractive explanation of the block loosening in the lee of roches moutonnées. In this case, basal ice would need to be just at or near the pressure melting point.

Observers of conditions beneath Alpine glaciers have recently tended to deny the possibility of freeze–thaw. According to Vivian (1970), the temperature of the air in cavities beneath the ice is usually positive by several tenths of a degree while rock temperatures are $1\cdot5^{\circ}$C or above. Obviously this is not conducive to freezing. Further, to illustrate the impossibility of freeze–thaw, it is suggested that temperatures of -20°C are necessary before frost can shatter fresh blocks of granite (Corbel, 1968; Vivian, 1970). Whereas from the evidence presented the process is unlikely to be important under certain Alpine glaciers, it would be premature to dismiss it as a possibility beneath other glaciers. If there are joints in the rock, then minor fluctuations around the freezing point may be sufficient for freezing water to exploit the weakness and loosen blocks.

Meltwater erosion

Meltwater action is discussed fully in chapter 15. At this stage it is only necessary to mention that it plays an important part in eroding bedrock beneath warm-based glaciers. This is over and above the effect of evacuating the products of abrasion. Meltwater erosion takes place both in subglacial channels and also in intervening areas where there is a film of water at the ice/rock interface. The greater the quantity of subglacial meltwater and the greater its extent, the greater will be its contribution to erosion beneath a glacier.

Debris entrainment

Effective erosion is unlikely to occur unless the debris produced at the ice/rock interface is evacuated. The main variables affecting the efficacy of the process are the mechanisms by which ice entrains debris, and assuming this takes place, the velocity of the glacier and the thickness of the basal debris layer.

For material to be incorporated within a glacier it is necessary for ice to exert a tractive force on a particle. In many situations where an isolated block rests on the glacier bed, ice may simply deform round the block and surround it. The greater the surface of contact between ice and rock, the greater will be the tractive force. Eventually, if this force is sufficient to overcome any frictional resistance offered by the glacier bed, the particle will move. Ice thickness will influence the efficacy of the process in that depths greater than c. 20 m are required to cause effective ice deformation round the block, whereas under great thicknesses of ice the tractive force may never overcome resistance with the bed. Particle shape and size will also influence the process. Probably the process is responsible for the entrainment of large erratics such as that illustrated in Figure 8.10.

In certain situations enormous erratics of the order of hundreds or thousands of metres across may be entrained by glaciers. Price (1973) records that the largest known erratic in Germany has dimensions 4 km by 2 km by 120 m. Boulton (1972a) suggested that such erratics may be moved by cold-based glaciers if the 0°C isotherm is relatively close to the glacier bed; if the friction at the frozen ice/rock interface exceeds that on a lower unfrozen plane of weakness in the rock, then the whole block may be dragged along.

Figure 8.10
A large erratic in Kjove Land, east Greenland.

Another means of entrainment which is only effective for small particles is regelation. Debris has frequently been observed in regelation layers (Boulton, 1970b; Kamb and LaChapelle, 1964; Gow, 1970) and is thought to be frozen onto the glacier bottom in the lee of prominences (Figure 8.11, overleaf). In a warm-based glacier the regelation layers which build up in the lee of a bump tend to be destroyed by pressure melting associated with the next bump of comparable size (Figure 8.11a). For this reason the zone of debris-bearing regelation ice cannot grow in thickness and debris is transported in a relatively thin layer less than 0·5 m in thickness. Within this layer the debris commonly comprises 5 per cent of total volume, but figures as high as 55 per cent are recorded (Boulton, 1975). Although the layer is thin, it tends to move relatively rapidly because of the high sliding velocities associated with warm-based ice. Far greater thicknesses of layers of regelation ice and associated debris can be frozen on to a glacier which is cold-based near its margin and warm-based in the interior (chapter 2). The meltwater and debris from the interior may freeze on to the glacier

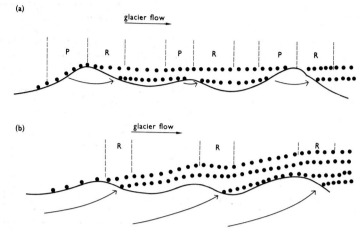

Figure 8.11
Debris incorporated into regelation ice (R) in the lee of rock bumps. **(a)** In a warm-based glacier, the layers tend to be destroyed by pressure melting (P) associated with other comparable bumps and remains thin. **(b)** Beneath a cold-based glacier, the freezing-on of successive regelation layers can build up to considerable thicknesses. *(Boulton, Inst. Br. Geogr. Sp. Pub. 1972a.)*

base layer by layer to form considerable thickness (Figure 8.11b). Such a process has been inferred by Boulton (1970a) beneath Makarovbreen in Svalbard, where successive freezing of debris by regelation has caused the sequence of rock types in layers in the snout to reflect the sequence of rocks crossed by the glacier. Regelation requires sensitive thermal conditions which allow a change from basal melting to freezing.

Assuming conditions are such that material can be incorporated into the base of a glacier, then clearly any upward movement of the material or ice will increase the thickness of the basal layers and the glacier's total transporting power. Röthlisberger (1968) pointed out that differential movement within the lower layers of the ice causes contained fragments to rotate and this causes them to rise from the bed. But most important of all is upward ice flow, associated in particular with zones of compressive flow (Nye, 1952b). Whether this upward movement takes place by thrusting or creep, it is potentially a powerful way of increasing the thickness of the load at the base of a glacier. Evidence accumulating from the study of sedimentary deposits suggests that thrusting can increase the thickness of the layer of debris-bearing ice at the bottom of a glacier in a spectacular manner, at least under certain conditions (Moran, 1971). The topic is examined in more detail in chapter 11.

Variations in the intensity of erosion within and between glaciers

Having discussed the basic processes involved in glacial erosion, it is clear that the factors affecting the efficacy of erosion *in toto* vary in a highly complex manner. Figure 8.12 attempts to portray the combination of variables that influence the rate of erosion at any one point. Whereas the shape of the bedrock channel of a glacier is left over for discussion in the next chapter, it is perhaps helpful to summarize the probable role of the main glacier and bedrock variables and the way their effects vary both within and between glaciers.

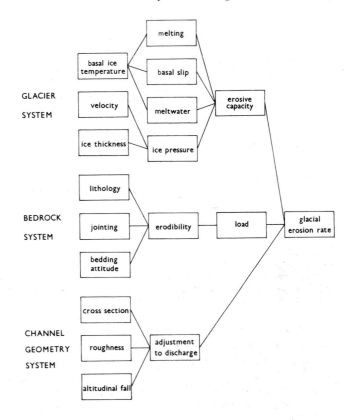

Figure 8.12
The main variables influencing the
rate of glacial erosion.

Glacier variables
Thermal regime   Table 8.2 attempts to summarize the influence of basal ice temperature
regimes upon each of the individual processes discussed. A few of the processes, such as dilatation, jointing, fracture of fresh rock, and abrasion caused by ice divergence round an obstacle,
can occur beneath cold- and warm-based ice. Processes such as abrasion, meltwater transport

Table 8.2   The role of basal temperature regime in influencing the
efficacy of various processes of erosion.

Process	Basal ice temperature regime		
	warm	fluctuating	cold
Abrasion (due to basal melting)	✓	✓	—
Abrasion (obstacle related)	✓	✓	✓
Fracture of fresh rock	✓	✓	✓
Joint exploitation – freeze-thaw	✓	✓	—
Joint exploitation – dilatation	✓	✓	✓
Debris entrainment – ice pressure	✓	✓	✓
Debris entrainment – regelation	✓	✓	—
Meltwater – erosion	✓	✓	—
Meltwater – evacuation of debris	✓	—	—

and regelation are most effective beneath warm-based ice, while freeze-thaw is probably most important when temperature conditions at the bed fluctuate around the pressure melting point. Boulton (1972a) has produced a useful schematic summary of these effects by distinguishing zones of differing basal regime (Figure 8.13). He distinguished a Zone A where there is net melting and a Zone B where there is a balance between melting and freezing. Here abrasive and meltwater processes dominate. In Zone C where water freezes on to the glacier sole abrasion is still important, but freeze-thaw processes become possible and meltwater activity is reduced. In Zone D the glacier sole is frozen to the bed and as a result abrasion is negligible and meltwater activity impossible.

The processes listed as operating under cold-based ice in Table 8.2 would seem to contradict the widely held view that erosion does not occur when ice is cold-based. For this

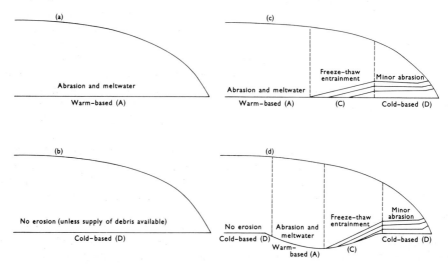

Fig. 8.13
Zones of contrasting basal ice temperatures and associated erosional processes. Capital letters refer to the labelling used by Boulton (1972a).

reason it is worth giving a fuller explanation. As discussed in chapter 2, at temperatures below the pressure melting point the adhesion at the ice/rock interface is greater than at other levels in the ice. Thus deformation in a cold-based glacier takes place in a zone immediately above the bed rather than between ice and rock. At first sight it is reasonable to expect no erosion under these conditions. However, several forms of erosion seem possible, especially if there are boulders in the basal ice which can penetrate through the stationary layer of ice at the ice-rock interface. Such boulders might perhaps be derived from a cliff overhanging a valley or cirque glacier. Alternatively there may be loose blocks on the glacier bed sufficiently prominent to thrust through to the zone of effectively moving ice just above the bed, or they may be inherited from ice passing from a warm to cold-based regime. Under such circumstances, rock fracture and abrasion may take place when the basal ice thins to bypass an obstacle. This may help to explain the striated boulders found beneath the cold-based Meserve glacier (Holdsworth and Bull, 1970) and in the moraines of cold-based glaciers

in the Transantarctic mountains (Mercer, 1971). If such processes can be effective in deepening a valley initially, then it is possible to envisage dilatation also taking place as an additional
process. The possibility of effective and widespread erosion beneath cold-based ice would
seem to depend largely on the presence of debris in the basal ice. If it is present then various
processes can begin to operate; if not, then it is difficult to envisage any effective erosion.
For this reason many cold glaciers seem to be devoid of any debris.

In view of the importance of zones of contrasting basal ice temperatures on erosion, it
is useful to discuss any regular spatial variation in their occurrence. Boulton (1972a) has
suggested several characteristic associations (Figure 8.13). As discussed in chapter 2, it is
reasonable to expect many glaciers in temperate or maritime environments to be warm-
based, and abrasion and meltwater activity to be relatively important (Figure 8.13a).
This might help to explain the great emphasis placed on these processes by workers in the
French Alps. Glaciers in cold continental areas are unlikely to have much effect unless they
are thick. At Camp Century where the ice is thin, basal ice temperatures are below
freezing and, as apparently confirmed by Goldthwait (1960), erosion is unlikely (Figure
8.13b). On the other hand, the deeper Byrd core demonstrated basal melting and the debris
in the bottom layers of regelation ice suggested erosion beneath that part of the Antarctic
ice sheet. On glaciers which are warm-based beneath their interiors and cold-based near their
peripheries, there may be an inner zone of abrasion and an outer zone where meltwater
and debris flowing from the centre freezes onto the cold ice of the periphery. Beyond this
may be a zone of minor erosion (Figure 8.13c). Topography may influence the distribution
of zones in any particular glacier. In otherwise cold-based ice caps there may be local patches
of warm-based ice overlying bedrock depressions where abrasion and meltwater erosion can
proceed (Figure 8.13d). Alternatively a thick warm-based ice sheet in a cold environment
may have local patches of cold-based ice over bedrock highs where ice thicknesses are
reduced.

It is important to note that the pattern of zones in a glacier can affect the type of erosion
in any one zone. For example, ice passing from a warm-based regime to a cold-based regime
may be armed with basal debris and achieve considerable abrasion in the succeeding cold-
based zone.

Basal velocity   Basal ice velocity is a critical variable affecting erosion, partly because of its
effect on the amount of debris evacuated or dragged across the glacier bed, and partly because
of rock fracture associated with differential pressures as the glacier flows across a rough bed.
Boulton (1975) discussed how the velocities of the basal ice and particles within it may be
calculated. In particular, he emphasized the importance of high basal water pressure in
reducing friction between a glacier and its bed. In places the effect is to double the sliding
velocity (Vivian, 1970). Localities favourable for such high water pressures are bedrock hollows in impermeable rock and the deeper central part of any glacier channel. An increase
in sliding velocity in both locations has many suggestive implications.

Ice thickness   Ice thickness is a crucial glacial variable because of its influence on the friction
between basal debris and the bed. So long as a particle is not materially retarded by friction,
the thicker the ice the greater the potential for abrasion and rock fracture. This helps to
explain the apparent efficacy of abrasion in rock basins, where the ice is thicker than average

(Galibert, 1962), and suggests another means of perpetuating topographic hollows beneath glaciers.

Bedrock variables
Bedrock can influence the efficacy of erosion in several ways through its influence on:

a the type and quantity of basal debris,
b the susceptibility of the glacier bed to erosion,
c the roughness of the bed and
d the permeability of the glacier bed.

There is no attempt to pursue this topic in any depth in this chapter. But as an example of the complexity of the possible interrelationships, Table 8.3 attempts to list those bedrock characteristics most favourable for various individual processes of erosion. The most effective tools for erosion would appear to be large resistant blocks, such as are commonly derived from deep erosion of crystalline rocks. Erodibility of the bedrock base is influenced by such characteristics as rock hardness, jointing, bedding and permeability.

Table 8.3   Bedrock conditions most favourable to the effective operation of the various processes of erosion.

Abrasion	Large blocks in basal ice
	Tools harder than bedrock
Rock fracture	Large blocks in basal ice
Joint exploitation	Hard, jointed rock
	Suitable bedding attitude
Entrainment – ice pressure	large fragments
Entrainment – regelation	small fragments
Meltwater erosion	impermeable bedrock
	soluble bedrock
Meltwater evacuation	impermeable bedrock

Variations in the intensity of erosion over time

In general there has been little attempt to examine the possible variations in erosive activity related to the time or stage of a glaciation. European workers have frequently stressed the importance of bedrock or regolith 'preparation' before a glaciation, implying that most glacial erosion takes place during an advance. The readily loosened material is immediately available for transport, and can also be used as an abrasive. Such a view is typified by Lliboutry (1964) and Tricart and Cailleux (1962). Warnke (1970) has extended and developed these ideas in relation to the Antarctic ice sheet where rates of erosion are thought to be reflected in the type and rate of sedimentation in the surrounding oceans. In the case of the Antarctic, Warnke suggested that most erosion took place at an early stage and has subsequently tailed off (Figure 8.14). He discussed a combination of three reasons: in part it may reflect an early period of maximum erosion which phased out when easily eroded preglacially weathered material was no longer available, in part an achievement of some sort of equilibrium after a certain amount of erosion, and in part long term changes in the boundary conditions at the ice/rock interface.

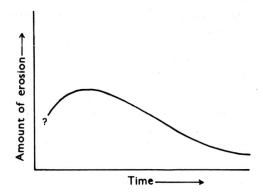

Figure 8.14
Schematic representation of the
erosive capacity of glaciers over
time. *(Warnke, Am. J. Sci. 1970.)*

Whereas the first two possibilities are highly attractive, the third is a little more problematical for there are several directions in which changes may occur. Robin (1972) has pointed out that the buildup of an ice sheet in polar latitudes might be accompanied by a change from cold- to warm-based conditions as ice thickness increases. This might lead to a peak of erosion at some time after the initiation of glaciation. If large surges occur as envisaged by Hughes (1975) then the ice may return to cold-based conditions after the thinning associated with a surge. However, in the case of Antarctica another possibility exists. As Mercer (1968b) and Drewry (1972) pointed out, the earliest Antarctic glaciers in the Transantarctic mountains were probably warm-based. They may have remained so until global refrigeration caused a change to cold-based ice. Whatever the possible reasons and however uncertain the time scales, Figure 8.14 poses many questions. These are immediately highlighted if one applies the graph to present glaciers or indeed to the Pleistocene ice sheets of temperate latitudes.

Further reading

BOULTON, G. S. 1972: The role of thermal régime in glacial sedimentation. In Price, R. J. and Sugden, D. E. (eds) Polar geomorphology, *Inst. Brit. Geogr. Spec. Pub.* **4,** 1–19.

BOULTON, G. S. 1974: Processes and patterns of glacial erosion. In Coates, D. E. (ed) *Glacial geology*, State Univ. of N.Y. Binghampton, 41–87.

CHAMBERLIN. T. C. 1885: The rock scorings of the great ice invasions. *US geol. Surv. Seventh Annual Rept,* 147–248.

HALLETT, B. 1976a: Deposits formed by subglacial precipitation of $CaCO_3$. *Bull. Geol. Soc. Am.* **87,** 1003–1015.

RÖTHLISBERGER, H. 1968: Erosive processes which are likely to accentuate or reduce the bottom relief of valley glaciers. *Int. Ass. scient. Hydrol.* **79,** 87–97.

VIVIAN, R. 1975: *Les glaciers des Alpes Occidentales*. Allier, Grenoble (513 pp).

There are many excellent articles on the processes of glacial erosion in *Symposium on glacier beds: the ice-rock interface. Ottawa, 15–19 August, 1978. J. Glaciol.* **23** (89) (445 pp), 1979. Also in *Proceedings of the symposium on processes of glacial erosion and sedimentation, Geilo, Norway, 25–30 August, 1980 Annals of Glaciology,* **2** (192 pp), 1981.

9 Landforms of glacial erosion

Landforms of glacial erosion can be viewed as modifications carried out to a glacier bed in such a way that the hydraulic geometry of the bed becomes more efficient for the evacuation of glacier ice (Evans, 1969). Such a framework is helpful in that it focuses attention on the links between process and form. It encourages the recognition of equilibrium forms as well as directing attention to the main variables affecting process/form relationships. In the field of glacial erosion the main variables are the basal ice conditions beneath the glacier, the character and shape of the underlying bedrock surface and the time available for modification by glaciers (Table 9.1). The study of glacial erosion will have achieved much when

Table 9.1 Some variables affecting the character of glacially eroded landforms

Glacial	morphology of ice mass
	ice thickness
	direction of ice flow
	velocity of basal ice
	temperature regime of basal ice
	character and amount of basal debris
Bedrock	structure
	spacing and character of joints
	lithology
	degree of pre-glacial weathering
Topography	relative relief
	relief shape
	alignment in relation to ice flow
	altitude
Time	length of glaciation
	change of any of the above over time

it can express the importance and role of these variables in quantitative terms. At present, however, in spite of an extensive literature, surprisingly little is known. An understanding of the processes is only now becoming based on a sound theoretical footing backed by field observations. It is only too common to find that forms are described in a generalized imprecise way, while interpretation has frequently stressed only certain of the variables involved. For example, von Engeln (1937) claimed that the 'grossness' of glacial erosion completely dominated structural variation, while Zumberge (1955) drew attention to structurally controlled glacial relief developed under two radically different directions of ice flow. Davis (1900) stressed the evolution of glacial landforms through time while other workers, especially in the Tatra mountains of Poland, have stressed the control of pre-glacial relief on glacial landforms (Klimaszewski, 1964).

Table 9.2.   Classification of features of glacial erosion

PROCESS	RELIEF TYPE	RELIEF SHAPE	Micro m⁻² (1 cm)	m⁻¹ (10 cm)	m⁰ (1 m)	m¹ (10 m)	m² (100 m)	m³ (1 km)	m⁴ (10 km)	m⁵ (100 km)	m⁶ (1,000 km)	Macro m⁷ (10,000 km)
	Eminence	Streamlined				←— Whaleback —→		Streamlined ←— spur —→				
							←— Rock drumlin —→					
Areal ice flow		Part-streamlined		←— Roche moutonnée —→			←— Flyggberg —→			Landscape of Areal Scouring		
	Depression	Streamlined	←— Striae —→			←— Groove —→						
		Part-streamlined		←— P-form —→								
Linear flow in rock channel	Depression	Streamlined				←— Rock basin —→			←— Trough —→	↑	Landscape of Ice Sheet Linear Erosion	
							←— Alpine trough —→					
Interaction of glacial and periglacial	Depression						←— Cirque —→				Valley glacier landscape	
	Eminence					Residual summit or ←— horn —→					Nunatak landscape	

$$SCALE$$

Given this background it would be helpful if this chapter could present an accurate description of landforms of glacial erosion and a summary of the role of the different variables which are highlighted in Table 9.1. However, this is not possible at the moment, and so the chapter stresses the huge gaps in understanding. The writers agree with Haynes (1968a) and Andrews (1972a) that one of the most serious gaps seems to be the lack of accurate observations on form. Without this, theoreticians are deprived of one source of highly significant information with which to constrain their theories.

A classification of features of glacial erosion is given in Table 9.2. The subdivision on the vertical axis reflects the contrast between forms fashioned by fundamentally different processes. These include *areal flow* where the ice flows across the landscape in a relatively unconfined way, *linear flow* where a major stream of ice flows in a rock channel, and an interaction of *glacial and periglacial* activity. Where necessary, further subdivision is made on the basis of whether a form is an eminence or a depression and on its degree of streamlining. Identification of any landform is largely a function of its size and thus the horizontal axis is a logarithmic scale of dimensions. Such an arrangement allows an appreciation of both the range of landforms involved and also, by scrutinizing the gaps, of the landforms which ice does not create. There are many problems in plotting a particular landform on such a diagram. Not only are many terms used in different ways by different authors, but also considerable subjectivity is required to subdivide what is often a continuum of forms. However, whatever the imperfections, the classification affords a basis for further discussion. In the pages that follow, each major subdivision is treated separately. For each, the forms and relationships are described and then the links between form and process are discussed. The chapter concentrates on medium scale landforms. Smaller scale features such as striae and crescentic gouges have already been mentioned in chapter 8, while whole landscapes are discussed in chapter 10.

Forms created by areal ice flow

Eminences

At a scale of tens to a few hundreds of metres, a family of streamlined *whaleback* forms has been recognized (Flint, 1971). These do not seem to have been measured morphometrically and are thus difficult to describe in anything other than tentative terms. The forms are well displayed in coastal areas of Sweden. Here individual hillocks tend to have smoothed rock surfaces on all sides, and are moulded into a shape which is longer than it is wide (Figure 9.1). In spite of the rounding, smoothed slopes may attain angles as high as 40°. Heights range from less than one metre to tens of metres. Jointing seems important in determining the boundaries of individual hillocks. Although any correlation remains to be demonstrated, the hillocks seem to occur most often on crystalline rocks like granite and gneiss.

At a scale varying from tens of metres to a few kilometres, there exists a family of streamlined hillocks and hills, for which Linton (1963) has suggested the use of the term *rock drumlin*. Again it is difficult to generalize about their form. In several situations, rock drumlins occur as peripheral members of fields of drumlins, most of which are built of drift (Charlesworth, 1957). In such cases the rock drumlins are similar in shape to the classic drift drumlins with a blunted upstream side and a tapering downstream side. Length to breadth ratios commonly vary between 2:1 and 4:1, while heights vary from 5–50 m. In other situations the rock drumlins may comprise hard rocks in areas of softer rocks; examples are the quartzite rock drumlins surrounded by schist near the Sound of Jura in Scotland (Linton, 1963). Elsewhere they coincide with areas of ice convergence; examples several kilometres long near the Bay of Húnaflói in northern Iceland are mentioned by Linton (1963). In still other situations, upstanding hills in areas of generally limited glacial erosion may be streamlined; examples include isolated mountains rising clearly above the plateaux of central Sweden (Rudberg, 1954).

Probably the largest streamlined landforms are tapered spurs and interfluves. The view that such large features may be due to ice erosion has been somewhat unfashionable in recent years. However Linton (1962) has described how, in central Scotland, 10 km long spurs of sandstone, flagstone and shale lie in the lee of more resistant mountain or hill masses, such as the Monteith hills. Streamlining is thought to have been accompanied by the erosion of at least 100–120 m of the less resistant rocks in places. On interfluves in the Finger Lakes

Figure 9.1
Whaleback forms, Rödlöga Skärgård, Stockholm archipelago.

region of New York State, Clayton (1965) describes moulding of a similar scale and type. In both cases the smoothness of slopes over several kilometres is thought difficult to explain in terms of normal fluvial erosion, whereas the alignment in relation to known ice movements supports the idea of ice as the chief moulding agency.

Roches moutonnées comprise a family of partly streamlined forms generally regarded as the hallmark of glacial erosion. Notwithstanding some uncertainties of usage in the literature (Flint, 1971), the term roche moutonnée seems well established and is understood to describe an asymmetric hillock or hill with one side moulded and the other side steepened and often craggy. A wide range of feature sizes is encompassed by the one term. At the smallest scale of 1 m or less it is common to find convex ice smoothed surfaces sharply truncated by individual joint planes. At a scale of 10–20 m roches moutonnées are very similar to whaleback forms except with one face truncated along series of joint planes (Figure 8.1). Larger roches moutonnées measure several hundreds of metres across and their dimensions are closely

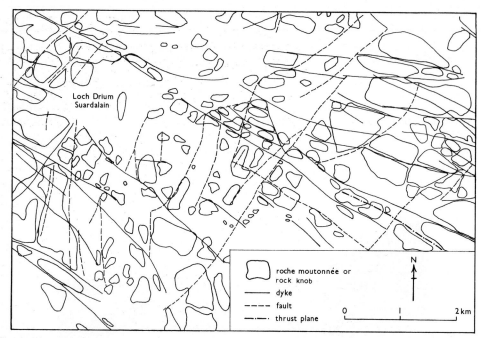

Figure 9.2
The outlines of large roches moutonnées and their relationship to structure on Lewisian gneiss near Lochinver, northwest Scotland. *(By permission of K. G. Pike.)*

related to the distribution of faults and dykes (Figure 9.2). On their backs are smaller roches moutonnées measuring tens of metres across. Most descriptions of roches moutonnées in the literature suggest that the features are best developed on well jointed granites and other crystalline rocks.

The asymmetry of larger hills measuring a few hundred metres to several kilometres across has also been related to ice action. Perhaps the best known examples were carefully docu-

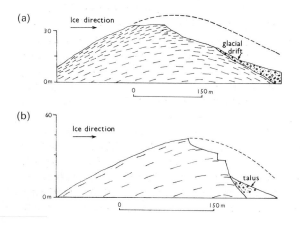

Figure 9.3
The relationship of roche moutonnée hills to preglacial sheeting in granite. **(a)** Fletcher Hill, New England. *(Jahns, J. Geol. 1943.)* **(b)** A spur north of Loch Avon in the Cairngorm mountains, Scotland. Dashed lines are a diagrammatic representation of the sheeting and the dotted line is the reconstructed preglacial form.

Figure 9.4
Flyggberg or steep sided hills in
Västerbotten, Sweden. Tjälberget
and Bergvattenberget near
Svanabyn (*above*) and (*below*)
the Fredrika area. (*Rudberg, Geographica 1954.*)

mented by Jahns (1943) in New England. Here the tops of granite hills 300–1,300 m in
length and 30–50 m high have been steepened on their downstream sides to form craggy
bluffs (Figure 9.3). By examining the pre-glacial sheeting structures in the granite, Jahns
deduced that up to 33 m of granite had been removed from the lee side of the hills by ice
action, compared to 3–4 m from the smoothed upstream sides. Rudberg (1954) described
how isolated mountains 1–3 km across and 100–350 m high, which rise above the plateaux
of central Västerbotten in Sweden, have steepened facets. He gave them the local Swedish
name 'flyggberg' (flygg=steep rock face). The facets lie either parallel to or on the lee side
in relation to ice movement (Figure 9.4). None face into the ice. As in New England, these
are thought to be essentially pre-glacial hills steepened on one side by ice.

Depressions
Grooves comprise a family of medium scale streamlined depressions. In the Mackenzie valley,
Canada, individual grooves may be 12 km long, 100 m wide and 30 m deep (Smith, 1948).
Usually, however, they are smaller by an order of magnitude (Figure 7.7). In situations,
grooves are straight and excavated parallel to the direction of ice movement, often indiscri-
minately across the underlying bedrock structure. In other situations, especially at smaller

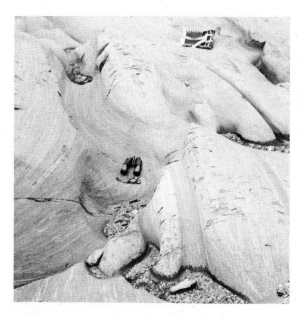

Figure 9.5
Glacial grooves on a rock surface
in Patagonia. *(Photograph by J.
H. Mercer.)*

scales, grooves may be slightly sinuous with soft flowing outlines (Figure 9.5). Striae often run parallel to the sides. Sometimes the lip of a groove may be overhanging.[1]

Rock basins measuring from a few metres to many hundred kilometres in size owe at least part of their excavation to the action of ice. Not only do such depressions and the surrounding slopes bear extensive evidence of ice abrasion, but the overdeepening which they represent is difficult to explain by other agencies of erosion.

One useful aid to the morphometry of such depressions is the shape of lakes occupying the whole or part of the depressions (Figure 9.6). The overwhelming impression is that the depressions within any one lowland area tend to be sub-parallel and elongated. The elongation is even more marked than it first appears, since a line of small lakes may be picking out only the deepest parts of what is a more or less continuous depression extending for many tens of kilometres across the land surface. The depth of the depressions seems to be moderate in relation to their length. For example, Zumberge (1952) noted that many lakes 1·5–13 km long, which are cut into slates in the Rove area of Minnesota, have depths of about 30 m. The slopes surrounding the depressions are often highly irregular, a feature brought out clearly by the often tortuous course of individual lake shorelines.

There are clear relationships between the location of the depressions and lithological and structural factors. Linton (1963) and Nougier (1972) noted that depressions are aligned along

[1]Some of the latter features have been classified as *p-forms* (plastically sculptured forms) (Dahl, 1965). Considerable controversy surrounds the origin of p-forms, some researchers attributing them to the action of meltwater (Dahl, 1965), some to the action of saturated till (Gjessing, 1965) and some to normal processes of abrasion (Boulton, 1974). In such cases of uncertainty it frequently turns out that one classification category incorporates features of different origins. Believing this to have happened in the case of p-forms, streamlined linear features are regarded as primarily due to abrasion and are considered here. Other p-forms are considered in chapter 15.

faults, joints and dykes (Figure 9.6a), while in Sweden, Rudberg (1973) found that fracture frequencies beneath the depressions are higher than beneath rock on either side. In Minnesota, Zumberge (1952) noted that the lakes occupy depressions which follow banding in gabbro while on granite they follow joints. In areas of interbedded slate and diabase sills they lie on the former (Figure 9.6b). In part of the eastern Canadian shield, Brochu (1954) recorded how lakes in New Quebec are preferentially located at the junction of different rock types, either adjacent and parallel to or astride the junction. In addition, Brochu noted that different rock types tend to support lake hollows to varying extents. Thus 60 per cent of the surface underlain by schists is covered by lakes while only 1 per cent of the gabbro

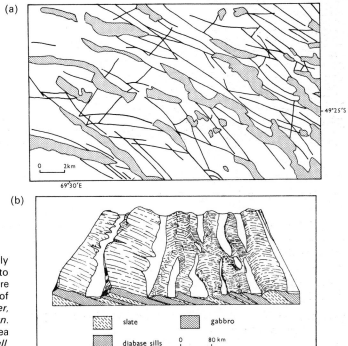

Figure 9.6
The relationship of glacially eroded lake depressions to structure. (a) Lakes and fracture pattern in the central plateau of the Kerguelen Islands. *(Nougier, Rev. Géogr. phys. Géol. dyn. 1972.)* (b) Lakes in the Rove area of Minnesota. *(Zumberge, Bull. geol. Surv. Minn. 1952.)*

is similarly affected. The largest lakes are in the least resistant rocks. Indeed, Brochu was able to rank the rock types in order of their suitability for lake formation as follows: dolomite-→schist→andesite, rhyolite and basalt→granite→gabbro.

Although lithology and structure often have a dominant influence on the depressions, it is important to recognize the role of other factors. There is frequently a relationship between the direction of ice flow and the alignment of the main depressions, as for example Virkkala (1952) has recorded in parts of Finland. In such cases it is assumed that the ice has preferentially exploited joints subparallel to its direction of flow. Also the role of pre-glacial relief may be important, especially in influencing the alignment of the larger depressions (Zumberge, 1955; Rudberg, 1954; 1973).

A special category of rock depression occurring at a continental or subcontinental scale may be due to glacial erosion exploiting structural features at this scale. Davis (1920) and

Tricart and Cailleux (1962) have drawn attention to the depressions eroded at the junction of crystalline rocks and younger sediments. In Canada, Great Bear lake and Great Slave lake are two of many lakes lying at the margin of the Canadian shield forming what White (1972) has termed an arc of exhumation. The Gulf of Finland is a comparable European example. Similar depressions have been observed on continental shelves in glacierized areas (H. Holtedahl, 1958; O. Holtedahl, 1970).

Relationship to processes of areal ice erosion
In spite of the imperfections in understanding even the most elementary forms created by areal flow of glacier ice, it is clear that there are three basic features to be explained: streamlining of eminences, lee side roughening, and the creation of depressions.

(a)

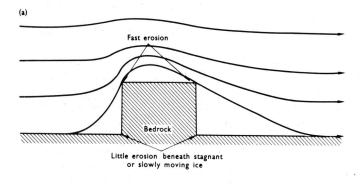

(b)

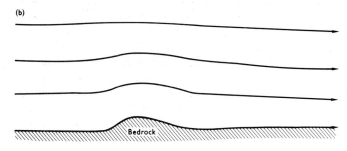

Figure 9.7
Two dimensional representation of the process of streamlining. **(a)** Streamlines develop over a bump and lead to a contrast in velocity between the top of the bump and the adjacent hollows. **(b)** Abrasion wears down the bump until there is little differential movement and an equlibrium form results.

Since *streamlined forms* are smooth and since striae may exist on well preserved examples, the basic mechanism involved in their formation is likely to be abrasion, aided to some extent by rock fracture of larger blocks. As seen in the previous chapter, such a mechanism is most effective where the basal ice is at the pressure melting point and there is a constantly replenished supply of rock tools. The actual transformation of a presumed irregular form into a streamlined form can be envisaged in terms of streamlines which are the result of enhanced basal creep. Figure 9.7 shows an irregular rock surface with ice streamlines reconstructed. Presuming the ice is equally armed with tools, the surfaces exposed to greater ice movement will be subjected to a greater amount of abrasion and thus will wear down faster than surrounding surfaces subjected to slower ice movement. The equilibrium shape will thus be a smooth surface parallel to the streamlines.

Another factor which may reinforce the process of streamlining is the pressure differential over the obstacle. The forward movement of the ice will locally enhance pressures at irregularities exposed to moving ice and thus increase the rate of abrasion at these points.

The actual shape of the streamlined form will reflect many variables. Other things being equal, blunter forms are likely to be favoured by strong variations in bedrock resistance, frequent and large irregularities in the pre-glacial topography, or a weakly moving ice mass. Elongate streamlined forms are likely to be favoured by minor variations in bedrock resistance, few or small pre-glacial irregularities or rapidly moving ice. Streamlining will also be affected by the length of time available. An initially irregular form will be roughly streamlined quite quickly. Thereafter, in the longer term, the streamlined form is likely to become more elongate but at a progressively slower rate of change. This is because the more streamlined the obstruction becomes the smaller the differential between ice flow over the obstacle and that over the surrounding ground.

The form of the roughened lee side faces of *roches moutonnées* suggests that the basic mechanism is block removal. The upstream side, however, is often streamlined with evidence of abrasion. The problem resolves itself to one of explaining why under certain circumstances the lee side of a streamlined feature is subjected to block loosening and removal rather than to abrasion. As discussed in chapter 8 the two most favoured mechanisms of block loosening are freeze-thaw and rock fracture. Both mechanisms operate best when ice pressures are high on the upstream side of the bedrock obstacle and when the ice either rests lightly on the lee side or separates from the rock to form a cavity. Once loosened, blocks may be removed by the flow of ice round them. If there is a cavity in the lee of the roche moutonnée then the blocks may fall to the floor of the cavity before being picked up. In this context it is important to note that cavities may fluctuate in size depending on seasonal and other variations in sliding velocity. At times of low sliding velocity ice may encroach and obliterate cavities altogether, thus allowing any loose blocks to be evacuated. Such a process has been observed beneath the Glacier d'Argentière (Boulton, 1974).

Since block loosening is favoured by low ice pressures on the lee side of a bump, it is likely that there is a threshold condition distinguishing roches moutonnées from wholly streamlined forms. If the ice is able to mould itself round the hill and exert sufficient pressure on the lee side to cause abrasion then streamlined forms occur; if not, then processes of block loosening cause asymmetry. Conditions favouring the development of low lee side pressures are thin or fast moving ice and prominent obstacles, whereas streamlined forms are favoured by thick or slow moving ice and low or small obstacles.

It is difficult to assess the role of the different variables affecting the shape of roches moutonnées largely because it is not certain whether they are equilibrium forms. Although a long held view stresses that they are the characteristic product of glacial erosion (von Engeln, 1937; Lewis, 1947), another view is that they represent irregularities in the pre-glacial relief which are being eroded away by glaciers, and which will eventually disappear (Davis, 1909). It may be that both views are correct but applicable at different scales. At scales of tens of metres or less the very abundance of roches moutonnées suggests that they are typical products of glacial erosion and it is likely that they relate to the interplay of glacial processes and certain lithological conditions, provided apparently by jointed crystalline rocks. At scales of hundreds of metres and above the situation may be different and it is notable that the form of the pre-glacial relief is frequently mentioned as a major reason for the initial existence of an eminence. This is clearly the case in the hill and mountain examples cited above but is also a widely held view for the fields of roches moutonnées occurring so widely in Scandinavia, North America and Greenland (Bird, 1967; Rudberg, 1973; Sugden, 1974).

So far the idea of the process operating in the lee of an obstacle has been viewed in profile. Rudberg (1954; 1973) has drawn attention to its probable role in plan, with plucking taking place on the lee side of spurs thrusting laterally out into the ice stream. This lee side effect in the horizontal plane is thought to produce truncated spurs and also the steepened facet of a 'flyggberg' which is parallel to the direction of ice flow. In such situations it would clearly be wrong to assume that the cliffed facets have formed in the lee of the hill as a whole; indeed, this would suggest an ice direction at right angles to the one responsible.

Currently there are two schools of thought as to which processes are responsible for *depressions*. One attributes their excavation to plucking, and the depressions are thought to be underlain by closely spaced joints while eminences are underlain by widely spaced joints (Flint, 1971). The assumption that more quarrying has taken place in areas underlain by closely jointed bedrock is not unreasonable bearing in mind the way elongated depressions are obviously related to joints, faults and fracture lines. However, a problem concerns the process of block loosening, for the ice over depressions would be thicker and conditions less suitable for freeze-thaw or rock fracture mechanisms.

The other view considers that abrasion is the most effective process in creating rock basins because of the greater thickness of ice and the presence of high water pressures favouring higher than average rates of basal sliding. If abrasion is important, it is interesting to consider why basins in general seem more irregular and less streamlined than their convex counterparts. One reason may be that the basin as a whole offers less resistance to the passage of ice than an upstanding hill. Not only are basins less exposed to the passage of ice but processes such as divergence and pressure melting, which are important in bringing rock tools into contact with the rock, will be less effective. One might expect rock basins to be asymmetric to a certain extent, for abrasion is likely to be most severe on the downstream side of the basin since this is where the ice is forced to ascend.

On a smaller scale, Boulton (1974) has drawn attention to the ability of abrasion to create grooves such as that depicted in Figure 9.5. Assuming that the ice is carrying only a thin layer of debris and that it diverges round an obstacle due to enhanced basal creep, then the abrasive tools will be concentrated on either side of the obstacle. This process will perpetuate linear depressions on a relatively small scale so long as most debris bypasses the obstacle.

Troughs: forms created by glaciers flowing in channels

Forms

Troughs are perhaps the most spectacular and well known landforms associated with glaciers. Their morphology and indeed their existence reflects a situation where all or a major part of the glacier discharge was in a rock channel. This includes not only mountain valley glaciers but also ice streams within ice sheets and ice caps.

At the outset, it is helpful to recognize three main types of trough. *Alpine troughs* are those cut by valley glaciers and the major characteristic is that their areas of ice accumulation are overlooked by higher ground (Linton, 1963). A second type of trough, closed at one end by a trough head, has been termed *Icelandic* by Linton. The trough has been cut by ice spilling over the trough head and is associated with erosion beneath ice caps or ice sheets. Such troughs are common not only around such small ice caps as Jostedalsbreen in Norway

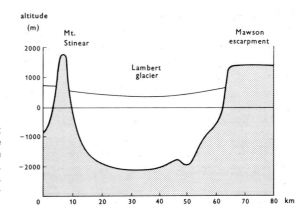

Figure 9.8
Cross-section of the largest glacial trough in the world. The Lambert glacier trough is 50 km wide and about 3,400 m deep. *(Morgan and Budd, J. Glaciol. 1975 by permission of the International Glaciological Society.)*

and local ice caps in Iceland, but magnificent features on a continental scale are associated with outlet glaciers in the Antarctic (Figures 9.8 and 3.8). A third type of trough is open at both ends and forms a through valley. Frequently such troughs breach watersheds and there are many magnificent examples in Scotland, the Scandinavian mountains and in the fjord region of east Greenland. Again, most appear to have been eroded beneath ice sheets. Troughs of different types have many similar characteristics and thus it is most convenient to discuss different aspects of trough morphology in turn.

The use of the term U-shape to describe trough cross-sections is rather optimistic with regard to the steepness of the sides (Figure 9.9). Not only has Rudberg (1973) noted that asymmetric troughs are the most common form in parts of central Sweden, but Svensson

Figure 9.9
Part of Geiranger fjord, Norway, showing steep trough walls. The sinuous shape of the trough and the benches representing a former high level river valley are also clearly seen. *(Photograph by Normanns.)*

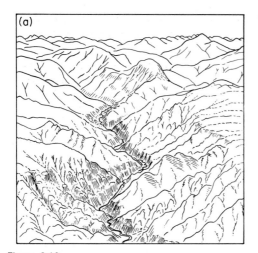

Figure 9.10
A 'before and after' reconstruction of Yosemite valley. Apart from widening the preglacial valley, the ice has deepened it by *c*. 500 m at the upper end and *c*. 200 m at the lower end. *(Matthes, US geol. Surv. Prof. Pap. 1930.)*

(1959) and Graf (1970) have found that even in well developed symmetrical troughs a parabola approximates to the shape more closely. They suppose that the cross-section of the trough takes the form of the curve $y=ax^b$. Each half of the trough is treated separately with y=vertical distance, x=horizontal distance from the origin in mid valley, a=the coefficient and b=the exponent. Methods of curve-fitting have been described fully by Doornkamp and King (1971). 'Good' troughs are true parabolas with exponents of about 2. Since a parabola is an endless curve additional measures are needed to define the cross-section. Graf used

Figure 9.11
A trough head sculptured beneath a former Scottish ice sheet, Loch Avon, Cairngorm mountains, Scotland.

a simple form ratio (Fr) in which the $Fr = D/W_1$, where D=depth, and W_1=valley top width. In the Beartooth Mountains of Montana and Wyoming, form ratios vary between 0·242 and 0·445 (Graf, 1970).

A term often used to describe long profiles of glacial troughs is that they are 'over-deepened'. This of course is a relative term comparing their profiles with those of river valleys. Equally appropriately, most river valleys are 'underdeepened' when seen from a glacial view-point. Nevertheless, it is clear that many glacial troughs have a 'down-at-heel' profile with a steep gradient near or at their heads and a gentler slope, sometimes a reverse slope, towards their mouths (Linton, 1963). This feature is perhaps best illustrated by an early and careful

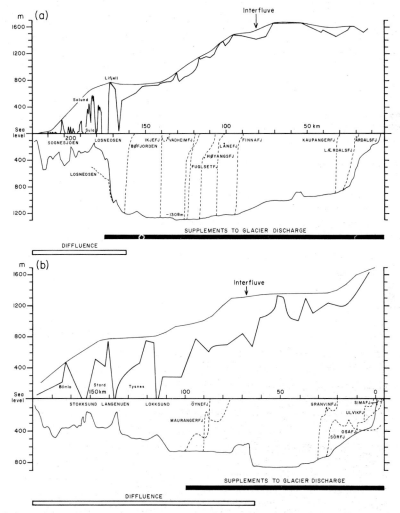

Figure 9.12
Longitudinal profiles of Sognefjord **(a)** and Hardangerfjord **(b)** showing the 'down-at-heel' long profile. Rock basins are associated with the confluence of tributary glaciers and rock bars with diffluence. *(Holtedahl, Geogr. Annlr 1967.)*

comparison of the Yosemite trough with the reconstructed pre-glacial river canyon profile (Figure 9.10). The same is true of *Icelandic* troughs. Here the overdeepening takes place dramatically at the trough head and contrasts with flat or basin forms immediately down-stream (Figure 9.11). At a far larger scale typical long profiles are demonstrated by two Norwegian fjord troughs (Figure 9.12). The long profiles of open troughs are different in that the floors are highest at some mid point and decline in altitude from this point in both directions.

Irregularities in the form of rock bars (riegels) and rock basins, or steps, are superimposed on these major long profile characteristics. Sometimes it has been suggested that the irregu-larities are the hallmark of glacial erosion (King, 1970). However, viewed at the scale of the trough as a whole, it is probably fairer to regard them as relatively minor irregularities superimposed on the general characteristics described above. These irregularities are well shown in fjord profiles (Figure 9.12) while King (1959) described some land-bound examples in Austerdalen, Norway.

An individual trough is often described as 'straightened'. Again this is in comparison to river valleys. The contrast between the two is effectively illustrated by the before and after bird's eye view of Yosemite canyon (Figure 9.10). However, it would be misleading to suggest that all troughs are straight. It is possible to highlight many examples of sinuous troughs, for example Geiranger fjord in Norway (Figure 9.9), and Nordvestfjord in east Greenland (Figure 5.8).

Relationship to linear ice flow

There are indications that glacial troughs are equilibrium forms related to the amount of ice discharged. This has long been suspected in the case of valley glacier troughs and indeed Penck (1905) termed the relationship as the law of adjusted cross-sections. He pointed out, for example, that although hanging valley floors were obviously discordant, the surfaces of the glaciers which cut them were not. The discordance of the valley floors was simply a reflection of the contrast in ice discharged in each trough. It seems that the size of icecap troughs is also related to ice discharge. Haynes (1972) found that there was a close correlation of trough cross-sectional area with drainage area and ice activity around the Sukkertoppen ice cap in west Greenland. The conclusion that discharge and trough size are mutually adjusted is an important step from which to examine the processes involved.

The cross-section of a glacial trough is analogous to a river channel, though of vastly greater proportions. The need for a larger channel is simply a function of the slower flow of ice. It is reasonable to assume that the most efficient shape for ice evacuation is a semicircle since this affords minimum frictional resistance in relation to a given volume of ice. Any irregularity in the cross-section will tend to be obliterated either by lee side plucking (Rud-berg, 1973) or by abrasion as discussed in relation to streamlined forms, assuming of course that all variables are equal. However, in the real world all variables are not equal and they all influence the actual cross-profile.

A parabolic rather than semicircular profile is probably favoured by ice flow, at least when the glacier is warm-based. As demonstrated in Figure 3.10b, basal ice velocity is highest in the middle of the valley where the ice is thickest. This favours both abrasion and the ability to transport eroded debris, and will tend to deepen the middle of the valley more rapidly than the sides. Another comparison of interest is with sewers. Apparently these are often made parabolic in cross-profile because this is one of the most efficient forms for discharging variable quantities of sewage. In other words the shape is relatively efficient for low and high discharges. By extrapolation one wonders whether a parabolic shape is not also the most natural trough shape for glaciers whose thickness and discharge has fluctuated widely over time. If one considers a situation where a trough is periodically full of ice but only partly filled at intervening times, then the trough bottom is subjected to erosion beneath glacier ice for longer than parts nearer the sides. In the longer term this would have the effect of creating a parabolic profile.

The actual shape of the parabola is influenced by several variables. With regard to the activity of the glaciers concerned in the cutting of a trough, Graf (1970) noticed a tendency for greater activity to be correlated with deeper troughs. This agrees with the conclusions of Rudberg (1954) who noticed that in an area of relatively minor modification by ice sheet erosion in Västerbotten, troughs have been widened rather than deepened. Clearly this relationship needs further testing, but, assuming that it can be established, it remains to ask why it should take place. If there is a tendency for a glacier to erode more in the middle of its bed than at the sides then perhaps it is reasonable to argue that the greater the total amount of erosion by the glacier the greater the differential will be. However, the type of ice stream involved may also have an effect on the cross-profile. It is not known whether troughs formed beneath ice sheets tend towards the same shape as troughs cut by valley glaciers under subaerial conditions. After all, in the case of the latter there is likely to be more debris available for erosion at the sides of the glacier.

Another variable influencing cross-section shape is rock type. It has frequently been observed that troughs are deep and narrow where the bedrock is resistant, and wider and shallower where the bedrock is less resistant (Matthes, 1930; King, 1959). Such a relationship is also hinted at by the commonly expressed view that trough forms are best developed in areas of hard rock (Pippan, 1965; Veyret, 1955). The reasons for this relationship are not wholly clear. Perhaps it simply reflects the greater ability of strong rocks to support high and steep trough walls. Thus one would expect steeper trough sides on granite than on shales simply because the shales cannot withstand the high stresses associated with steep slopes.

Pre-existing topography is likely to affect cross-sections in several ways. It is common to find that troughs, especially fjords, are narrow and deep where they cross mountains or plateaux but wider and shallower when flowing through less constricted topography. This is well illustrated by a series of cross-profiles of Inugsuin fjord on the east coast of Baffin Island (Figure 9.13, overleaf). If the height of the trough walls above sea level is also taken into consideration then the contrast is all the more striking. Another relationship with pre-existing topography is illustrated by Rudberg's suggestion (1973) that the asymmetric cross-profile common in much of central Sweden is the result of the removal of a pre-existing fluvial valley spur from one or other side of the trough. Finally one may note that Lewis (1947) and Veyret (1955) believed that V-shaped cross-sections might be associated with steep gradients.

The 'overdeepening' of a glacial trough compared to a river valley is explained by the relative position of the equilibrium line along a trough which usually determines the location of the zone of greatest ice discharge. Here ice velocity or thickness will tend to be greater than elsewhere, thus favouring higher than normal rates of erosion. On land based glaciers this zone of maximum erosion will be at some distance up glacier from the snout and cause relative overdeepening. Where the equilibrium line is at or near sea-level, then one would expect a progressive increase in the amount of deepening towards the coast.

The trough head which marks the commencement of overdeepening in the long profiles of ice sheet troughs is difficult to explain. Its very occurrence suggests the existence of a threshold where for some reason the ice ceases to flow as a sheet and suddenly flows as an ice stream in a rock channel. Perhaps the threshold is related to basal ice conditions which determine whether or not the ice slips over its bed. If for some reason the basal ice reaches the pressure melting point at some point in a glacier long profile, then this might allow the excavation of a trough to begin. Once initiated, the trough would tend to concentrate increasing amounts of ice into it and thus enhance its permanence. Probably the trough heads retreat up the trough (Souchez, 1967; 1966). If this view of the origin of trough heads is accepted, then the initial threshold might be related to many factors, for example, to a pre-existing valley step or convergence of ice, causing a critical ice thickness to be reached, or to lithological variations. The relative importance of these can only be worked out as a result of field

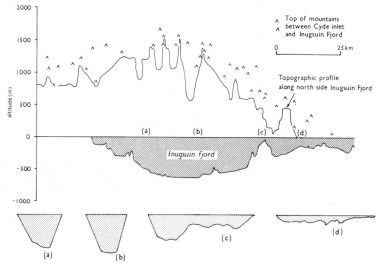

Figure 9.13
Diagram to show how the depth of Inugsuin fjord in Baffin Island is deeper and narrower in the vicinity of high mountains but becomes wider and shallower towards the coast. (**a, b, c, d** refer to position of cross-sections.) *(Løken and Hodgson, Can. J. Earth Sci. 1971.)*

testing. At the moment understanding is limited particularly through a lack of information about trough heads, namely their frequency of occurrence, distribution, amplitude and other characteristics.

The basin and riegel or stepped irregularities in troughs can be explained in terms of what is known about glacier flow along a glacier. There are two aspects to the problem: (a) how does ice erode existing irregularities? and (b) how are such irregularities initiated? The mechanisms of erosion which perpetuate hollows have already been discussed in chapter 8. In confined channels there may be an upper limit to the concavity of the basin depending on the attitude of curved slip lines under compressive or extending flow (Nye and Martin, 1968). The attitude and curvature of the latter reflect the large scale geometry of glacier flow, which is itself largely influenced by the size and surface slope of the glacier in relation to the underlying channel. Figure 9.14 shows how concavities tend to be eroded to conform with the slip lines. Nye and Martin also showed that on convex stretches where extending flow applies, there is no upper limit to the radius of curvature. This implies that rock basins will tend to curve more gently than intervening bars or steps, a feature which seems to occur in practice.

The initiation of irregularities in a trough is another problem and depends on many factors. Variations in rock type or jointing have been shown to be related to basin and bars in some situations (Matthes, 1930; King, 1959). The changing volume of a glacier caused by con-fluence or diffluence of ice may also have dramatic effects (Veyret, 1955). Holtedahl (1967) noted how the deeper reaches of Hardanger fjord occur where the trough narrows. In other situations the irregularities may be inherited from the pre-existing form of the relief (Rud-berg, 1954). Similar views have been applied to steps in the Tatra mountains (Klimaszewski, 1964) and the Alps (Bakker, 1965).

The creation and overall morphology of a glacial trough depends on a complex interrela-tionship of topography, ice and bedrock conditions. Since there is no reason to expect glaciers

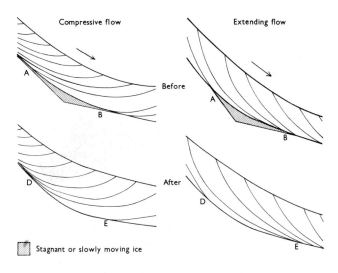

Figure 9.14
Diagram to show how concavities in a glacier bed are eroded so as to conform to slip lines. Actively eroding ice associated with slip lines bypasses stagnant or slowly moving ice with little erosive capacity in the concavity AB. Then, erosion associated with active ice proceeds until the whole concavity is lowered to conform to a slip line DE. *(Nye and Martin, Int. Ass. scient. Hydrol. 1968.)*

to meander, it is generally assumed that a sinuous trough represents the exploitation of a pre-existing river valley by a glacier. This is most obvious in the case of alpine troughs where modifications arise from the widening of the valleys to accommodate the glaciers (Figure 9.10). In ice sheet situations the relationship of troughs to pre-existing valleys is less straight-forward. The existence of a former river valley has often been inferred from old traces on the flanks of a trough or from the sinuous course of the trough itself (Figure 9.9) (Gjessing, 1966; Holtedahl, 1967; Funder, 1972). Troughs of this type seem to occur most frequently when they lie approximately parallel to the direction of ice movement. Initiation of a trough under such circumstances is favoured because basal ice over such a site is likely to be the first to rise to the pressure melting point, permitting basal slipping. This is partly because of the greater ice thickness over the valley (Haefeli, 1968) and partly because of the frictional heat released by the flow of ice down the valley if it is favourably orientated. Also, the presence of ice moving longitudinally down the valley is capable of effectively evacuating any debris eroded from the base. This view implies that in certain circumstances valleys unsuitably orientated for ice evacuation are unlikely to be exploited. This has been confirmed by field studies for example in Scotland (Linton, 1963; Sugden, 1968) and in North America (Clayton, 1965). The often dramatic contrast between unmodified fluvial valleys and glacial troughs in these areas suggests that a sharp threshold is involved.

There are several situations when troughs have been argued to be essentially new features. This seems true of most open troughs and especially those which breach imposing pre-existing watersheds. Often 'new' and open troughs can be related to distinctive lines of structural weakness (Bretz, 1935; Funder, 1972). Perhaps the process of erosion involves two stages. First areal ice erosion picks out and selectively erodes a linear depression. The more successful the deepening the greater the chance of channelling flow in the depression and creating an ice stream, which in turn creates a trough. Assuming the process is roughly as outlined above, it is not difficult to explain the existence of new troughs cut in the direction of ice flow. Thus in northwest Scotland where a succession of troughs breach the main watershed they can be attributed to the overall flow of ice across the watershed (Dury, 1953). However, it is more difficult to understand the alignment of apparently new troughs transverse to the overall direction of ice flow, especially when they are many tens of kilometres in length as in east Greenland. In such a situation it is tempting to suggest that the basal ice in the depressions was diverted either way down the troughs by major massifs. The long profiles of the east Greenland troughs which are higher in the centre than at either end (Funder, 1972) would seem to lend some support to this hypothesis.

Forms created by the interaction of glacial and periglacial activity

Whereas the landforms discussed in the two previous sections are primarily due to the action of ice alone, there are a number of distinctive landforms loosely described as 'glacial' whose morphology reflects a combination of glacial and periglacial activity.

Glacial cirques

Few landforms have caught the imagination of geomorphologists more than the *glacial cirque* (corrie) (Figure 9.15). In spite of this its form is far from clear. For example, a study of over 100 cirques by Haynes (1968b) in Scotland showed that the two 'classical features mentioned in many texts – a rock basin and an L-shaped break between headwall and basin – are uncommon and indeed occurred at opposite ends of the spectrum of shapes. The problem of definition was well illustrated at a meeting on cirque morphometry by a sub-group of the British Geomorphological Research Group, where it took a whole day to evolve a definition even reasonably acceptable to the participants (Evans and Cox, 1974). It was agreed that a cirque can be defined as 'a hollow, open downstream but bounded upstream by the crest of a steep slope (headwall), which is arcuate in plan around a more gently sloping floor. It is "glacial" if the floor has been affected by glacial erosion while part of the headwall has developed subaerially, and a drainage divide was located sufficiently close to the top of the headwall for little or none of the ice that fashioned the cirque to have flowed in from outside.' Normally the headwall is cliffed.

There have been several attempts to measure cirque shape. In the Cairngorm mountains, Scotland, where cirques are cut into a plateau of uniform granite, semicircles closely fit the cliff tops surrounding the basins and thus apparently agree with long held views concerning the simple arcuate shape of the ideal cirque (Sugden, 1969). On the other hand, cirques in steep mountain relief are often more angular in outline. For example, Galibert (1965)

Figure 9.15
A cirque close to sea level near Aðalvik, northwest Iceland. *(Photograph by permission of H. R. Barðarson.)*

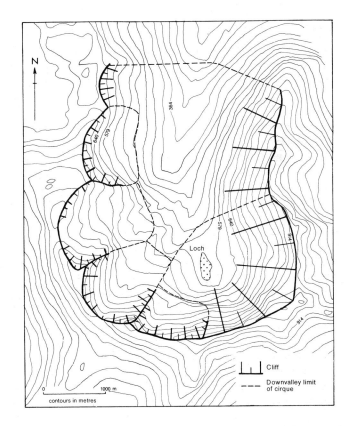

Figure 9.16
The morphology of cirque-in-cirque forms, Coire Uaine, Kintail, Scotland. *(By permission of J. Gordon.)*

noted the frequent occurrence of rectangular and triangular shapes in the Alps. Other complications are introduced by cirque-in-cirque forms which bite into the overall curve of the backwall (Derbyshire, 1968; Gordon, 1975) (Figure 9.16). The long profiles of cirques have been best described by logarithmic curves of the type $y=(1-x)e$ ' which combine a curved arcuate section at the base of a backwall with a basin and steeper headwall (Haynes, 1968b). By introducing the constant k so that $y=k(1-x)e$ ' the curves allow for varying concavity. Where $k=2$, the basin is deep and the headwall is steep; where $k=\frac{1}{2}$ the basin is poorly developed (Figure 9.17).

The absolute dimensions of cirques are not well known. Whereas there is similarity between the widths of cirques in the Cairngorms and part of Baffin Island ranging from c. 300–1300 m (Andrews and Dugdale, 1971), widths in the Alps and Antarctica are often considerably larger. Lengths of corries vary from somewhat less than 200 m to several kilometres. Again, Antarctic examples seem longer than average. Depths vary from a few tens of metres to 1,300 m in the Fauteuil des Allemands in the Alps (Galibert, 1965) and an estimated 500–2,700 m in parts of the Antarctic.

The morphology of the rock surface with a cirque is generally accepted as reflecting a

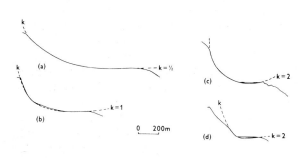

Figure 9.17
Fitting k-curves to cirque long profiles in Scotland. 81 per cent of the long profiles sampled resembled k-curves closely. Some, as in **(d)**, did not. **(a)** Calbach, Cranstackie-Beinn Spionnaidh, Sutherland. **(b)** Duail, Foinne-bheinn, Sutherland. (c) na Poite, Applecross. **(d)** an Lochan Uaine, Ben Macdhui, Cairngorm mountains. *(Haynes, Geogr. Annlr 1968b.)*

contrast between glacial erosion of the floor and base of the headwall and subaerial frost action in the upper headwall. Haynes (1968b) noted that in Scotland the slope of the ice moulded part of the headwall varies between 19° and 36°. Upper backwall angles may vary from c. 30° to over 70° depending on the rock type involved.

The equilibrium cirque shape is far from certain. A common view is that once a semicircular armchair hollow is formed it will enlarge itself in plan without change of form (Linton, 1963). The recognition of a characteristic form in a wide variety of environments and rock types would seem to be sufficient confirmation of such a view. Whereas there is little evidence that cirques progress from simple to complex forms, there are suspicions that the regular semicircular shape is an oversimplification (Gordon, 1975). It is interesting to record that, after observing the characteristics of cirques cut into volcanic cones in the Antarctic, Andrews and LeMasurier (1973) considered that given sufficient time cirques will tend to lengthen rather than widen their basins and thus become more elongate. However, this situation might be more accurately envisaged as a cirque of constant width extending a valley by backwearing.

The behaviour of a cirque glacier confined to the cirque basin has been admirably demonstrated by the detailed work of Lewis and associates on Veslskautbreen in Norway (Lewis, 1960). Here with an equilibrium line approximately 3/5 the way between the snout and upper limit the glacier can be envisaged as a foreshortened valley glacier. Similarly the cirque long profile can be seen as a foreshortened trough with maximum erosion occurring in the vicinity of the equilibrium line. In addition, a rotational component of flow is introduced by the normal downward velocities above the equilibrium line and upward velocities below. Since the net balance gradient of cirque glaciers is high and since the glacier is short, then the vertical components become sufficiently dominant for the whole glacier to resemble a rigid rotating body. This again favours the excavation of a rock basin. In situations where the cirque has been occupied by the head of a valley glacier and the equilibrium line is some way downglacier, the cirque can be viewed as a rock basin forming in sympathy to slip lines (Nye and Martin, 1968). The relatively large size of a cirque in relation to the glacier responsible for its excavation may be attributed to the tendency for cirque glaciers to be dependent on drifted snow and thus to be more active than adjacent larger glaciers.

The main process occurring on the backwall of a cirque is widely assumed to be frost-shattering, and certainly in temperate and sub-polar latitudes there is ample evidence of frequent freeze–thaw cycles (Tricart, 1970) and of rock falls on cliffs (Rapp, 1960). Apparently freeze–thaw may also be significant in areas like central Antarctica where air temperatures remain well below zero for all the year. Voronov (1968) has noticed that in the Conrad mountains of Queen Maud Land, sunny weather causes snow to melt at air temperatures of $-17°C$ and that meltwater fills cracks in the bedrock before refreezing under cloudy conditions. Other examples have been mentioned by Andrews and LeMasurier (1973). In some mountainous areas the headwall is shaped by thin glaciers which are responsible for steep abraded slopes common in couloirs (Figure 4.16). Such slopes may attain angles of 45–50° in the Alps, 70–80° in the Himalayas but only 40–45° in less humid environments like east Greenland (Galibert, 1965).

It is difficult to assess the role of different variables in cirque morphology, because of uncertainty about the duration of cirque glaciations. The probable major contrast in the length of glaciation between the Antarctic and temperate latitudes may do much to explain the apparent large size of Antarctic cirques in comparison to most others. At a smaller scale in the Cairngorm mountains, one of us was led to the conclusion that size differences between cirques could best be explained in terms of different ages and thus differing durations of ice occupancy (Sugden, 1969).

With regard to ice characteristics, a cirque glacier which lies wholly in a cirque is more likely to erode a rock basin than one which feeds and merges with a valley glacier (Penck, 1905; Evans, 1969). This reflects the position of the equilibrium line in relation to the cirque, and means that cirques close to the glaciation level will tend to form deeper basins than those at higher altitudes. Also ice activity is likely to be associated with cirque volume, with the largest cirques in areas of greater snow input, although this has not yet been demonstrated convincingly. Rock variables are apparently important at different levels. Rock structure can influence the overall shape of cirques (Galibert, 1965) as well as their detailed form (Haynes, 1968b). Pre-existing relief is important above all in determining the characteristics of the site in which a cirque develops. Altitude of the mountain, the presence of a suitably orientated hollow for snow collection, and the suitability of the mountain shape for snow drifting are all variables mentioned in many regional studies.

Horns and arêtes
Mountains that are surrounded by glaciers tend to have a characteristic form and display

steep straight slopes. Where isolated, such mountains may form upstanding horns with three or four distinct faces. Sometimes they may be linked by ridges or arêtes with similar steep straight slopes (Figure 9.18). Such features occur both in mountain chains like the Alps and in massifs or nunataks which stand above ice sheet surfaces.

Steep rectilinear slopes are fully consistent with a situation where debris removed from the mountain slopes is prevented from accumulating at the foot of the slope by the constant flow of ice (Bakker and Le Heux, 1952). The angle of the slopes is probably related to several factors. The initial steepening is related to back- and downwearing by glaciers, whereas the

Figure 9.18
Arêtes in basalt rocks in east Greenland. *(Reproduced with the permission (A421/74) of the Geodetic Institute, Denmark.)*

angle will reflect a balance between the processes operating on the rock face and the strength of the rock mass. Other variables being equal, the relative positions of cirque glaciers which are attacking the horn will influence the number of faces. In ice sheet and ice cap situations, the configuration will likewise reflect the flow of the surrounding glaciers. There is a tendency for the horn or nunatak to be elongated in the direction of ice flow.

Linton (1963) suggested that horns are extremely stable equilibrium forms. If, as he argued, dilatation of the rock parallel to each face is an important mechanism of glacial erosion, then the process will become decreasingly effective as the horn gets smaller, for rock expansion can then take place evenly in all directions rather than loosening sheets of rock parallel to one face. In other words the smaller the horn, the more resistant to erosion it becomes. Arêtes may be relatively stable forms also. Not only will the importance of rock dilatation decrease as the arête narrows, but the activity of a cirque glacier will diminish as an arête is lowered. This is because the snow-fence role of the arête in trapping drifted snow will decrease. Under such conditions the headwall attack by the cirque glacier may progressively decrease. This view may help to explain the observation of White (1970) that the very frequency of occurrence of arêtes suggests that cirques cannot easily remove them.

Further reading

ANDREWS, J. T. 1972: Glacier power, mass balances, velocities and erosion potential. *Z. Geomorph.* **13,** 1–17.

GORDON, J. 1981: Ice-scoured topography and its relationship to bedrock structure and ice movement in parts of northern Scotland and west Greenland. *Geogr. Annlr. Stockh.,* **63A** (1–2), 55–65.

JAHNS, R. H. 1943: Sheet structure in granite: its origin and use as a measure of glacial erosion in New England. *J. Geol.* **51** (2), 71–98.

LINTON, D. L. 1963: The forms of glacial erosion. *Trans. Inst. Brit. Geogr.* **33,** 1–28.

MATTHES, F. E. 1930: Geologic history of the Yosemite valley. *US geol. Surv. Prof. Pap.* **160** (137 pp).

NYE, J. F. and MARTIN, P. C. S. 1968: Glacial erosion. *Int. Ass. scient. Hydrol.* **79,** 78–86.

RASTAS, J. and SEPPÄLÄ, M. 1981: Rock jointing and abrasion forms on roches moutonnées, S.W. Finland. *Annals of Glaciology* **2,** 159–163.

Several interesting papers on cirques and troughs are contained in *Geografiska Annaler Stockholm Series* A, **59** (3–4), 1977.

10 Landscapes of glacial erosion

The aim of this chapter is to describe and explain the types of landscape resulting from glacial erosion. The emphasis is on the search for order in the regional associations of landforms rather than on individual landforms themselves. As can be seen from Table 9.2, in effect the chapter is dealing with the results of glacial erosion at a larger scale than in the previous chapter, with different landscape categories representing the effects of different types of glacier flow. At the outset it must be admitted that the chapter can barely attempt the beginnings of an explanation. As is so often the case in geomorphology, large scale features of the landscape represent the results of processes operating over long time periods. This means that any discussion of these processes must involve a good deal of speculation.

Classification

Before any analysis of landscapes of glacial erosion can proceed, it is necessary to derive some form of classification. In the past classifications have tended to take several forms. Some writers have suggested that there is an orderly evolution of landscape through time (Davis, 1900; Linton, 1963); it follows from this that landscapes may be classified in terms of the stage of evolution they have reached. Others have used classifications based on various broad topographic characteristics, and landscapes of glacial erosion have been discussed under such headings as *areas of low relief* (e.g. Price, 1973; Embleton and King, 1975), *glaciated plains* (Davies, 1969) and *coastal environments* (Embleton and King, 1975). Still others have distinguished zones on the basis of the degree of glacial modification to pre-existing land surfaces (Clayton and Linton, 1964; Clayton, 1974). All such classifications emphasize the role of one variable affecting the process of glacial erosion, usually topography or length of glaciation, and are thus difficult to relate to the actual processes of glacial erosion involved. Table 10.1 is an attempt to classify landscapes in terms of their morphology and their genesis. It is a development of Table 9.2 in that the categories are related to three main types of process – areal ice flow, linear ice flow, or a combination of glacial and periglacial processes. The first two categories relate to the type of flow within ice sheets or ice caps while the third

Table 10.1 A classification of landscapes of glacial erosion

Glacier system	Glacier type	Process	Landscape type
Ice sheets and ice caps (unconstrained by topography)	Ice domes	Sheet flow	Landscapes of little or no erosion
			Landscapes of areal scouring
	Outlet glaciers	Stream flow	Landscapes of selective linear erosion
Glaciers constrained by topography	Valley glaciers	Stream flow and periglacial processes	Alpine landscapes
			Cirque landscapes

category refers to the effects of glaciers which are constrained by topography. It must be emphasized that the classification is intended to be no more than a structure around which to hang a discussion of possible links between process and form.

Ice sheet and ice cap landscapes
Landscapes with little or no sign of erosion include areas known to have been covered by ice sheets or ice caps but which bear no obvious sign of the ordeal. Jameson Land in east Greenland is typical of the type (Figure 10.1). The landscape is essentially fluvial in character

Figure 10.1
Landscape with little or no sign of glacial erosion, Jameson Land, east Greenland. *(Reproduced with the permission (A421/74) of the Geodetic Institute, Copenhagen.)*

with smooth slopes and a considerable thickness of regolith. The area is dissected by an intricate network of stream valleys. The main problem in recognizing this landscape type is to demonstrate that the area was covered by ice. Although there are still doubts in certain localities, in others there is good evidence of a former ice cover. For example, in Jameson Land there have been discoveries of erratics derived from interior Greenland which show that the whole area was covered by the main Greenland ice sheet at some stage (Funder, 1972). In the Queen Elizabeth Islands in northern Canada, where there is similar relief, evidence of post-glacial uplift seems to imply that the islands were covered by ice during the last glaciation (Blake, 1970).

Landscapes of areal scouring everywhere bear signs of glacial erosion (Figure 10.2). The main characteristics were clearly described by Linton (1963), who used the term 'knock and lochan' topography. Joints, faults and dykes are the master features and are scoured to form irregular depressions often with lakes occupying the deeper basins. The bosses in between the depressions are scraped by ice. In places the lee sides of the knobs have been roughened

Figure 10.2
Landscape of areal scouring just north of Søndre Strømfjord airfield, west Greenland. *(Reproduced with the permission (A 421/74) of the Geodetic Institute, Copenhagen.)*

into the form of roches moutonnées. In west Greenland where this type of landscape is well displayed, there is often a dramatic contrast in the scenery depending on whether you look eastwards or westwards. In one direction you see a rugged blocky view of the lee sides of the knobs and in the other a smooth, vegetated stoss side view. In other places, all sides of the knobs may be smoothed, a feature of some significance in the islands of the Baltic archipelagos where sunny sunbathing slopes are often on the lee side! Most descriptions of this topography stress the relatively limited relief amplitude which is generally less than 100 m. In some situations and especially on soft rock outcrops, areal scouring may be represented by large scale smoothing, such as occurs in parts of central Scotland (Linton, 1962).

Landscapes of selective linear erosion describe situations where ice erosion has been concentrated in a trough or series of troughs and has left the intervening slopes or plateaux unmodified (Figure 10.3). Such selectivity of glacial erosion has been described in a number of regional studies in North America and Europe (Bretz, 1935; Linton and Moisley, 1960; Clayton, 1965; Sugden, 1968; 1974). There seems general agreement that such scenery develops beneath an ice sheet. Often the trough sides may be characterized by striations or grooves right up to the cliff top, showing that they were once completely filled with ice.

Figure 10.3
Landscape of selective linear erosion, Nordvestfjord, east Greenland. *(Reproduced with the permission (A421/74) of the Geodetic Institute, Copenhagen.)*

The intervening plateau areas may be regolith covered and devoid of glacial erosional forms. Fragile rock remnants like tors of possible pre-glacial age may survive on such interfluves (Figure 10.4, overleaf). One of the most striking features of such relief is the abruptness of the junction between the troughs and the unmodified plateau surface. In east Greenland one can often stand on the edge of an essentially unmodified pre-glacial surface and peer into a trough over 1500 m deep. Trough patterns may be complex. Figure 10.5 (overleaf) shows a dendritic network of troughs and fjords in Iceland. Similar networks occur between Ellesmere and Axel Heiberg islands (Figure 10.6a, p. 197), and in the Aligegheny plateau of New York State (Clayton, 1965). In other situations, as for example in east Greenland (Figures 10.6b and 5.8) there is, in addition to a dendritic tendency, a series of straight transverse troughs.

Figure 10.4
This tor in the western Cairngorm mountains, Scotland, has survived inundation by at least one ice sheet.
Troughs were created in the immediate vicinity. Erratics occur around the tor.

Figure 10.5
A well developed dendritic trough pattern in the Akureyri area of northern Iceland. *(Reproduced by permission of NASA; ERTS no. 8139212185.)*

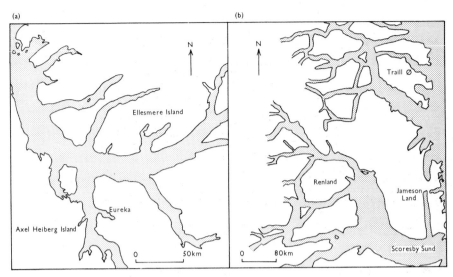

Figure 10.6
(a) A dendritic trough pattern forming the strait between Ellesmere Island and Axel Heiberg Island, Arctic Canada, and **(b)** a dendritic pattern with the additional complexity of straight transverse troughs in east Greenland.

Landscapes with both areal scouring and linear erosion can be recognized in many areas and serve to demonstrate that any classification is an attempt to order what is in effect a continuum of forms. In these cases there are distinctive troughs but the intervening interfluves are affected by typical areal scouring. The cliff top of the trough is often rounded. Examples of such landscapes are discussed by Dahl (1963) and John (1972a).

Landscapes of glaciers constrained by topography
Alpine landscapes have been widely described and recognized in the classic literature. The essential feature is a dendritic network of troughs with steep, often precipitous slopes (Figure 4.13). Such networks are reminiscent of stream networks and can be ordered in the same way as streams. Generally speaking it is rare to find a trunk trough above third or fourth order (Evans, 1969). Where the relief type is well developed, the upper slopes may form arêtes and horns. Frequently the relief is asymmetric, with the steepest slopes associated with the most active glacierization which in turn is influenced by aspect (Derbyshire and Evans, 1975). In north temperate latitudes, the steepest slopes tend to face north and northeast, while in south temperate latitudes the preferred orientation is south and southeast (Evans, 1972). In many Antarctic massifs the steepest slopes face towards northern sectors. In equatorial latitudes asymmetry is weak.

Cirque landscapes include cases where essentially discrete cirques are set into a hill massif (Figure 9.15) The overall landscape appearance depends on the density of cirques, the shape of the hill mass and the size, shape and orientation of individual basins. As with steep faces in alpine scenery, there are similar preferred aspects for cirques.

Composite erosional landscapes

In many parts of the world, landscapes contain elements related to ice sheets, ice caps and valley glaciers in various combinations. Two important landscape types are involved: (a) those landscapes which are or have been moulded by ice sheets, ice caps and valley glaciers contemporaneously; and (b) those landscapes subjected to successive phases of valley glacier and ice sheet or ice cap erosion.

Figure 10.7
Nunatak landscape on the central ice cap of Ellesmere Island. *(Original photograph supplied by the National Air Photo Library, Canadian Department of Energy, Mines and Resources.)*

The first category includes areas subjected to alpine valley glaciation, but which are also traversed by ice sheet forms, normally troughs cut by outlet glaciers. Such a situation is exceedingly common around the margins of the present ice sheets of Greenland and Antarctica. Perhaps the best contemporary example is the Transantarctic mountain range where peaks are sculptured into alpine forms and yet the whole range is traversed by a succession of outlet glaciers draining the east Antarctic ice sheet (Figure 3.8). There is a progression from vast ranges such as this to a landscape of individual nunatak peaks (Figure 10.7).

Figure 10.8
The Loch Torridon area, western Scotland, showing cirques which have subsequently been subject to areal scouring beneath an ice sheet which submerged the whole landscape. *(Copyright Aerofilms Ltd.)*

The second category includes landscapes with signs of both valley glacier forms and ice sheet forms which cannot have evolved contemporaneously. Such landscapes abound on massifs which have been submerged beneath mid latitude ice sheets (Figure 10.8). Frequently, cirques occur on massifs which are known from other evidence to have been submerged beneath thick ice sheets. Since there is no reason to suppose that cirques form beneath ice sheets, such situations have long been accepted as representing different phases of mountain and ice sheet glaciation.

Relationship of landscape to process

Ice sheet and ice cap landscapes
From what has been written in earlier chapters it is reasonable to suppose that effective erosion beneath an ice sheet can only take place when the basal ice is at the pressure melting point. Under such conditions, assuming a topographic alignment suitable for ice flow, one can expect basal sliding to occur. Following arguments given in more detail elsewhere (Sugden, 1974), it seems reasonable to regard the main landscape types as part of a continuum of forms related to the areal extent of basal ice at the pressure melting point. Thus areal scouring with evidence of sliding on all rock surfaces is likely to relate to areas where all the basal ice is at the pressure melting point. In contrast, areas with no sign of erosion are probably related to areas where the basal ice is below the pressure melting point and where there is no movement between ice and bedrock. Landscapes of selective linear erosion form an intermediate category where basal ice is at the pressure melting point only in troughs; presumably the basal ice remains below the pressure melting point over the sites of intervening interfluves.

Assuming that the extent or presence of basal ice at the pressure melting point is the critical factor in determining the type of landscape eroded by an ice sheet, it is possible to survey the main variables involved. Of the many ice characteristics which affect basal ice temperatures the three most important are ice thickness, ice surface temperature and ice velocity (chapter 2). Since ice temperatures rise with depth, conditions are more favourable for basal melting where ice is thicker and surface temperature higher. Velocity is important because of the heat which is released by internal deformation in the basal layers. These relationships mean that, other variables being equal, basal melting is likely beneath ice sheet centres where the ice is thickest. Also basal melting is important where an ice sheet is nourished under a maritime regime for this favours high surface temperatures and high ice velocities. The inverse argument is that ice sheets are likely to be frozen to their beds when the ice is thin, especially where it is nourished in a cold continental climatic regime.

Topography is an important variable which directly affects ice thickness and thereby basal ice temperatures. Beneath an ice sheet, basal melting is more likely over topographic lows in the bedrock than over adjacent highs. Topography also influences basal temperatures through its influence on the horizontal flow of an ice mass. A topographic situation which favours convergence of ice streams is likely to increase ice velocities and thus the amount of heat released by internal deformation. Topography favouring divergence is likely to have the opposite effect. Thus other variables being equal, situations of convergence will tend to raise basal ice temperatures more than situations of divergence. The role of bedrock variables is not clear, but following Weertman (1966) and Boulton (1972) it is likely that water will be more abundant at the ice/rock interface when the rock is impermeable. Since the amount of water is likely to affect the amount of basal slip, and thus erosion, one can postulate that, other variables being equal, erosion will be more effective on impermeable rocks and less effective on permeable rocks.

The spatial distribution of landscapes cut by ice sheets lends much support to these views. It is notable that areal scouring is characteristic of areas once submerged thickly by ice of the central zones of the Scandinavian and Laurentide ice sheets. It is also common along the maritime edges of past and present ice sheets. This is true of much of southwestern Greenland (Figure 10.9), eastern Canada, western Scandinavia, and the exposed peripheries of east Antarctica (Figure 10.10, p. 202). A relationship between areal scouring and impermeable rocks may be indicated in Arctic Canada where Bird (1967) notes that the relief type is best developed on crystalline shield rocks, although many other factors may also be involved in this relationship. On a smaller scale, landscapes of areal scouring are associated with areas of ice convergence or ice streaming, for example in Scotland (Sissons, 1967) and in north and east Greenland (Sugden, 1974).

As noted earlier, selective linear erosion is associated with uplands where the ice cover would have been thin, for example in east and north Greenland, the Allegheny plateau area of North America, Scandinavia and some Arctic islands such as eastern Svalbard. Though rarer in the Antarctic, similar landscapes can be recognized in the Prince Charles mountains near the Lambert glacier and in the higher parts of the southern Antarctic Peninsula (Figure 10.10). Another association, recognizable in Greenland (Figure 10.9), Scotland and Scandinavia, is that selective linear erosion is more common in areas removed from the maritime western edge of an ice sheet. In a maritime climatic environment, the interfluves may be subject to areal scouring, as in western Norway, western Scotland and parts of western Greenland.

Landscapes with little or no sign of erosion occur in cold dry continental environments.

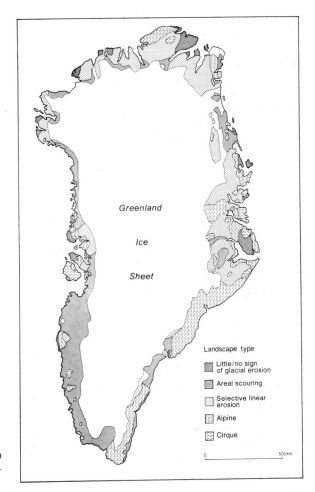

Greenland

Ice

Sheet

Landscape type

Little/no sign
of glacial erosion

Areal scouring

Selective linear
erosion

Alpine

Cirque

0   500km

Figure 10.9
Landscapes of glacial erosion in
Greenland. *(Sugden, Inst. Br.
Geogr. Spec. Pub. 1974.)*

Furthermore, in north and east Greenland they occur on peninsulas where ice may be
expected to have diverged and also where the underlying bedrock is permeable (Figure 10.9).
   The role of the variables mentioned above changes according to the size of area being
considered. Variables related to the nature of the ice sheet reflect the size of the ice sheet.
Thus in Antarctica, ice conditions vary on a continental scale and in Greenland on a subcontinental scale. However, at the other extreme, in Scotland variations related to the ice sheet
change on a regional scale and thus the contrast between areal scouring in the west and
selective or no erosion in the east takes place within a distance of 80–150 km. Topographic
variables can operate on a subcontinental scale where major contrasts in bedrock altitude
occur between mountain ranges and adjacent lowlands, as in Greenland and Antarctica.
But topography also plays a role at much smaller scales where one hill-mass or one depression
a few tens of kilometres across can influence the resulting landscape type. Lithological variations are most important at smaller scales though a large scale effect may be present between
shield and non-shield rocks.

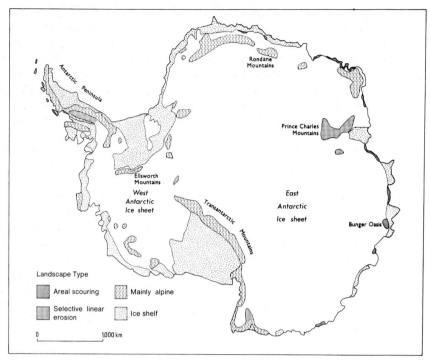

Figure 10.10
Landscapes of glacial erosion in Antarctica. A preliminary map compiled from various sources.

Landscapes of glaciers constrained by topography

The existence of landscapes created by valley glaciers reflects two main situations: (1) topographic conditions where regional relief is too steep to accommodate the equilibrium profile of an ice cap or ice sheet, as for example occurs in narrow mountain chains or uplands close to the edge of an ice sheet and (2) climatic conditions which are unable to provide sufficient snow for the buildup of an ice cap or ice sheet. In both cases the glaciers occupy hollows in the pre-existing relief, and in essence the contrasts in relief type reflect the varying levels of success of the glaciers in transforming the original relief. Several variables seem relevant.

The glacial climate is important in determining how much of the pre-existing relief is modified by glacial erosion. In a marginal glacial climate, only a few hollows favourably sited for snow accumulation will be eroded by glaciers. In a more favourable glacial climate a greater number and variety of sites will be eroded. The controls vary from place to place. In mid latitude areas a strong constraint is imposed by insolation and wind-drifting, as is illustrated by the tendency for cirques to be confined to the shady, leeward side of mountains. In polar latitudes other constraints may apply. On Disko Island, Greenland, in addition to a northeast orientation of cirques there is a secondary peak reflecting a large number facing north-northwest. This may reflect the importance of katabatic winds in blowing snow off the Greenland ice sheet at times when the ice sheet was more extensive. This has had the effect of broadening the range of suitable sites for cirque erosion. Again, Markov *et al.*

(1970) suggested that the restriction of cirques to north facing slopes in Antarctica is related to the fact that it is only on sunny slopes that backwall sapping take place. However, other factors involved in the northward aspect may be that mountain flanks are higher on the downstream north facing side of any massif rising above the Antarctic ice sheet, that they face moisture bearing winds, or that the aspect is suitable for the accumulation of snow drifted by katabatic winds. Climate also has an important influence on the altitudinal distribution of cirque landscapes. Not only is there the widely accepted decline in cirque altitude from the equator towards the poles, but cirque altitudes are lower near continental coasts than they are in continental interiors (Figure 5.6). This gradient is best known in mid latitudes where cirque floor altitudes rise dramatically from a maritime west coast towards the east (Ljunger, 1948; Linton 1959; Sissons, 1967).

Pre-existing relief plays an obvious role in determining the frequency of suitable sites for glacier accumulation and erosion. One would suppose that valley frequency, orientation and arrangement would be reflected in the glacial landscape but this has yet to be examined in depth. Rudberg (1954) noted how the lack of cirques in part of the central Scandinavian peninsula can be attributed to the occurrence of plateaux which favoured the build-up of ice caps rather than valley glaciers. It is to be expected that the role of pre-glacial relief will be more obvious and important in areas of marginal glaciation. For this reason, it is perhaps no surprise to find that the school of thought which emphasizes the dominant role played by pre-glacial relief in determining the type of glacier and thereby the resulting landforms, prevails in the marginally glaciated Tatra mountains (Klimaszewski, 1964).

Landscape models

It is helpful to follow the discussion of the last few pages by attempting to construct models of typical landscape associations and evolution. These are intended as no more than broad generalizations which invite testing and modification.

Models of process/form relationships
Figure 10.11 (overleaf) is an attempt to relate landscape type to different glacial processes. The models are confined to ice sheet situations, and landscape associations are discussed in terms of ice characteristics and topography. Figure 10.11a is an upland continent covered by an ice sheet, Figure 10.11b a lowlying continent covered by an ice sheet and Figure 10.11c an undulating continent covered by an ice sheet. The left side of the diagram represents a warm maritime environment while the right side represents a cold dry continental environment. Bedrock variables are assumed to be constant in all cases.

In Figure 10.11a alpine scenery occurs at the peripheries of the ice sheet where uplands protrude above the ice sheet surface. The mountains are more jagged in the maritime environment than in the continental environment. Selective linear erosion occurs in a zone inland from the mountains where ice cover is thin. In the maritime area ice activity may be sufficient for areal scouring to occur on the interfluves. On the continental side there is no areal scouring on the interfluves. Beneath the ice sheet centre the ice is sufficently thick for areal scouring to occur. However, sliding velocities in the centre may be low and prevent effective erosion.

In Figure 10.11b which is a lowlying continent, there is no alpine scenery. Instead there is areal scouring in the maritime and central areas. Probably there is a gradation from rock

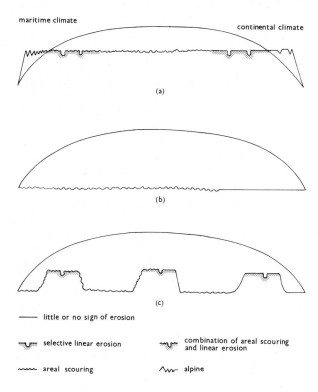

maritime climate

continental climate

(a)

(b)

(c)

——— little or no sign of erosion

⌣⌣ selective linear erosion

combination of areal scouring
and linear erosion

⌢⌢⌢ areal scouring

⋀⋁⋀ alpine

Figure 10.11
Models of landscape associations
beneath idealized ice sheets. See
text for explanation.

drumlin forms in the centre to roches moutonnées near the maritime periphery, since reduced ice thickness and increased velocity towards the periphery favour lee side block removal. On the continental side of the ice sheet, areal scouring gives way to areas where there is no sign of erosion.

In Figure 10.11c the undulating relief causes a corresponding variation in landscape types. Areal scouring occurs in lowlying areas in the maritime and central portions but is replaced by areas of no sign of erosion towards the continental side. Selective linear erosion occurs on all topographic highs. In the maritime areas it is combined with areal scouring while on the continental side it is not.

To distinguish three models is not to suggest that there are three types of ice sheet. The models are highly idealized and clearly a portion of any model can be combined with portions of any of the others. Also it is important to remember that they ignore lithological variations and any changes in topography or ice conditions through time. Finally, they are speculative and have not been tested.

Models of landscape evolution
One major problem which impedes understanding of landscapes of glacial erosion concerns uncertainty over their evolution through time. The questions asked about any particular landscape will vary depending on whether it is regarded as being in equilibrium or merely a transient stage on the way to some other ultimate equilibrium form. Before considering various models of evolutionary development, it is important to note that calculations of

amounts of glacial erosion point to the clear ability of glaciers to change a landscape profoundly. This is true of deductions of the amount of overdeepening from evidence of trough deepening, hanging valleys and other forms, as for example mentioned by Davis (1909), Matthes (1930), Bretz (1935), Pippan (1965), Sissons (1967) and others, and also of extrapolation of estimates of current and past rates of erosion. For example, Markov *et al.* (1970) note that Evteev's (1960) modest estimate of the rate of lowering beneath the east Antarctic ice sheet is sufficient to have removed a layer of rock 1 km thick from the whole area. Other more rapid estimates elsewhere in the world (Embleton and King, 1975) carry even more dramatic implications. It seems reasonable to suppose that at least some landscapes have had time to achieve some form of equilibrium.

Valley glacier landscapes have long been discussed in terms of evolution. In 1926 Hobbs, although using different terms, envisaged an evolution from isolated cirque basins in a massif to an alpine landscape including arêtes and horns as individual cirque basins expanded over time. Davis (1900) believed in a similar evolution and equated the isolated cirque stage with youth, and the horn and arête landscape with maturity. He postulated a post-mature stage of more subdued relief created when the horns and arêtes were removed and covered by ice. Such a sequence raises several problems. Whereas it is probable that individual cirques expand with time and in some situations lead to sharpened mountain relief, it would be false to argue that all sharpened landscapes have evolved this way. The shape of the interfluve between cirques varies with many other factors, such as slope steepness, cirque density, rock type, etc. For example, the contrast between the sharpened Cuillins and the subdued Cairngorm mountains of Scotland can be attributed to variations in cirque density since both sets of cirques are similar in size.

Davis's mature landscape may be an equilibrium form. The landscape type is displayed in a wide variety of environments which, in view of past climatic and ice sheet fluctuations, presumably incorporates landscapes of widely different ages. Theoretically there seems no reason why the landscape type should not be an equilibrium form. As argued in the previous chapter, horns and arêtes are self preserving in the sense that their removal causes the adjacent glaciers to lose the benefit of a snow-fence effect and thus they become deprived of snow. Also a pyramidal or ridge form is conducive to long term stability. The overall altitude of any mountain chain will be maintained in the long term by isostatic uplift compensating for the amount of rock removed. However this argument does not exclude the possibility that on time scales of over 30 million years or so there may be another longer term equilibrium landscape type.

Long term changes in landscapes created by ice sheets are difficult to discuss, mainly because of lack of information. However, a number of disparate pieces of evidence point tentatively to the conclusion that the landscapes described do represent equilibrium forms. In the first place, the correlation between trough size and ice discharge noted in the previous chapter carries the implication that troughs cease to grow in size once they have attained a size suitable for ice discharge. In the second place, there are observations that in mid latitude environments the last ice sheet carried out little erosion in comparison to earlier ice sheets (Ahlmann, 1919; Rudberg, 1954). One possible explanation is that forms suitable for ice evacuation were already present for the last ice sheet. However, there are alternative ways of interpreting the observations and thus it is necessary to discuss the various situations on a more theoretical level.

Superficially, landscapes of areal scouring seem to represent a balance between areal ice erosion and the characteristics of the underlying bedrock. Structural weaknesses are picked out and determine the relief and the dimensions of the intervening bumps. Hence there may be no reason to expect any particular change with time, other than overall lowering. Such an argument would agree with the view of White (1972) that the areal scouring of Canada and Scandinavia represents glacial erosion on a massive scale. On the other hand some writers have argued that the landscape type represents at least in part the exhumed weathering front of a landscape subjected to deep rotting under pre-glacial conditions (Feininger, 1971). If this is true then continued glacial erosion might change the landscape either way and

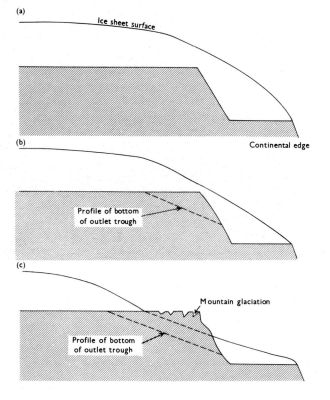

Figure 10.12
Evolution of glacial landscapes in mountains near the peripheries of ice sheets.
(a) The ice sheet covers unmodi-fied plateau and scarp near the ice periphery. The ice surface is steepest where the ice is thin.
(b) Selective erosion cuts troughs through the plateau edge where the ice is thin and lowers the ice sheet surface.
(c) Further deepening of the out-let trough lowers the adjacent ice sheet surface until the pla-teau edge is exposed. Moun-tain glaciation dissects the outer side where (1) the mountains are longest exposed and (2) the relative relief is greatest.

produce a smoother or rougher surface. In view of the arguments that landscapes of areal scouring in many areas represent relatively limited modification of pre-glacial fluvial land-scapes, the question would seem to be wide open. Such views have been put forward for Scandinavia (Rudberg, 1970), Scotland (Godard, 1965), Arctic Canada (Flint, 1971; Bird, 1967) and Greenland (Sugden, 1974).

Landscapes of selective linear erosion pose many interesting problems. In situations well beneath an ice sheet it is possible to envisage an equilibrium situation whereby little erosion takes place once the troughs can cope with the discharge of ice. In uplands near ice sheet margins, on the other hand, a very different situation may occur (Figure 10.12). As a trough develops by backward extension or by deepening, it becomes an outlet glacier for the main ice sheet. This in turn will lower the overall ice sheet surface in the vicinity of the outlet

Figure 10.13
A progression from clear arête and horn forms in the foreground to a landscape of selective linear erosion nearer the ice sheet, Blosseville Kyst, Greenland. *(Reproduced with the permission (A421/74) of the Geodetic Institute, Copenhagen.)*

and may expose plateau remnants on either side. These in turn develop into areas of cirque and alpine glaciation. One can go further and say that alpine glacier erosion will lead to local isostatic recovery of the upland, which in turn will cause the outlet to incise further, exposing more of the mountains above the ice sheet surface.

Speculative as this view is, it receives support from some suggestive evidence. Nunataks, consisting of flat plateaux separated by deep glacier-filled troughs, are common round the upper flanks of outlet glaciers, for example, the Prince Charles mountains near the Lambert glacier (Figure 10.10) and near the outlets draining into Scoresby Sund in east Greenland. Secondly, there are many instances of peripheral mountains dissected by outlet glaciers whose forms show a progression from clear horns and arêtes at the ice sheet periphery to unmodified plateau remnants nearer the ice sheet centre. This pattern is particularly clear in many peripheral areas of Greenland (Figure 10.13) and is also apparent in the Vestfirðir area of northwest Iceland and the Transantarctic mountains. Thirdly, there is evidence that peripheral mountain ranges like the Transantarctic mountains have undergone considerable uplift since they were inundated and traversed by outlet glaciers of the Antarctic ice sheet

(Grindley, 1967; Bull and Webb, 1973). Some of this may prove to be isostatic recovery responding to the erosion of the mountains by alpine glaciation.

Any discussion of landscape evolution by glacier action must involve consideration of Linton's model of divide elimination developed in a number of papers (1963a; 1964) and applied to valley glacier and outlet glacier situations. Linton argued that the widening of glacial troughs would progressively truncate and shorten pre-existing divides (Figure 10.14). The troughs widen from their source areas as far as the equilibrium line in response to the increase of ice discharge. In areas like west Spitsbergen and parts of the Antarctic Peninsula, divide elimination was argued to have occurred on a scale of tens of kilometres and to have literally removed mountains. This model is not in conflict with any of the views discussed above if it is assumed that the process of trough widening will cease or slow down once the glacier trough is big enough for ice discharge and some type of equilibrium obtains. In such a case Linton's model can be seen as one applicable to the change from a fluvial to glacial equilibrium. The intensity of the change will depend on the original fluvial characteristics as well as the ice conditions.

(a)   (b)

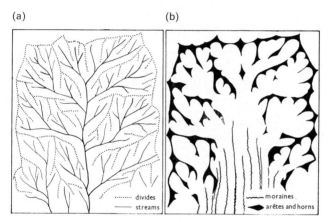

Figure 10.14
Linton's model of divide elimination. **(a)** Idealized pattern of streams and divides in an area some 50 km across. **(b)** The same area after and during glacierization with the equilibrium line near sea-level. *(Linton, Trans Inst. Br. Geogr. 1963.)*

One major problem concerning the evolution of any landscape of glacial erosion concerns the often dramatic changes in ice cover and conditions over time. Two particular sets of problems are posed. In Antarctica and Greenland there has been at the very least a change from the glaciation of certain mountain chains to an ice sheet cover which has on several occasions fluctuated in thickness and extent. The effects of such variations on the overall landscape differ from place to place and are discussed in many regional examples, for example in the Transantarctic mountains by Drewry (1975), Mercer (1968b), Calkin and Nichols (1972), and in the Ellsworth mountains by Rutford (1972).

In mid latitude areas the fluctuations between full ice sheet cover, no glaciers, and mountain glaciers have been more complete and interpretation is far from easy. The presence of ice sheet forms and valley glacier forms (particularly cirques) for a long time was interpreted in terms of the hemicycle: cirque glaciation – ice sheet glaciation – cirque glaciation. Examples are particularly common in the early work of the Officers of the Geological Survey of Scotland and the idea has persisted until recently. However, as argued in chapter 7, it is unlikely that important cirque activity accompanies the decay of an ice sheet. Probably as suggested by several regional studies the cirques represent a succession of phases of mountain glaciation occurring in between or at the commencement of ice sheet or ice cap glaciations (Rudberg, 1954; Miller, 1961; Sugden, 1969).

A conclusion reached in many regional studies is that forms of cirque and alpine glaciation may survive intact beneath ice sheets. The evidence in support of the selectivity of ice sheet erosion goes some way to explain this phenomenon. Cirques are likely to occur on mountain massifs which, by virtue of their bulk and altitude would probably be covered by thin and diverging ice when submerged beneath an ice sheet. Such conditions are unlikely to lead to much modification. On the other hand, cirques in maritime environments tend to be lower and may be submerged by thick and active ice sheet ice. In such situations considerable modification of the cirques may take place.

The probable disappearance of mid latitude ice sheets between glacials also raises the issue of the contribution of subaerial processes to 'glacial landforms'. The role of subaerial slope processes in modifying and widening troughs, for example, has not been examined in any depth in the literature.

Further reading

BRETZ, J. H. 1935: Physiographic studies in east Greenland. In Boyd, L. A. (ed) The fiord region of east Greenland, *Am. Geogr. Soc. Spec. Pub.* **18,** 161–266.

DERBYSHIRE, E. and EVANS, I. S. 1976: The climatic factor in cirque variation. In Derbyshire, E. (ed) *Geomorphology and Climate*, Wiley, London.

GORDON, J. E. 1979: Reconstructed Pleistocene ice-sheet temperatures and glacial erosion in northern Scotland. *J. Glaciol.* **22** (87), 331–44.

HAYNES, V. M. 1977: The modification of valley patterns by ice-sheet activity. *Geogr. Annlr. Stockh.* **59A** (3-4), 195–207.

LINTON, D. L. 1964: Landscape evolution. In Priestley, R., Adie, R. J. and Robin, G. de Q. (eds) *Antarctic research*. Butterworths, London, 85–99.

RUDBERG, S. 1973: Glacial erosion forms of medium size – a discussion based on four Swedish case studies. *Z. Geomorph.* **17,** 33–48.

SUGDEN, D. E. 1974: Landscapes of glacial erosion in Greenland and their relationship to ice, topographic and bedrock conditions. In Brown, E. H. and Waters, R. S. (eds) Progress in geomorphology, *Inst. Brit. Geogr. Spec. Pub.* **7,** 177–95.

SUGDEN, D. E. 1978: Glacial erosion by the Laurentide ice sheet. *Journal of Glaciol.* **20** (83), 367–91.

THORÉN, R. 1969: *Picture atlas of the Arctic.* Elsevier, Amsterdam (449 pp).

Part IV
Glacial deposition and its effects

11 The processes of glacial deposition

This chapter is closely related to chapter 8, which deals with the basic principles of erosion by ice, and looks at the variations in erosional intensity which occur in both the space and time dimensions for glacier systems. Whereas chapter 8 is largely concerned with the manner in which bedrock is broken down and entrained in the glacier, this chapter follows the history of detrital particles via glacial transport to glacial sedimentation as till (Figure 11.1, overleaf). Hence its main theme is the analysis of the processes involved in the transport and deposition of glacial rock detritus.

This chapter concentrates mainly upon the lower parts of the glacier system. There are various types of ice marginal environments in which glacial deposition can occur. These are as follows:

a   The 'ice-bed margin', on the floor and flanks of a trough in the case of a valley glacier, and on the undulating rock and sediment bed in the case of an ice cap or ice sheet.
b   The 'ice-atmosphere margin', or glacier surface.
c   The 'ice-bed-atmosphere margin', around the whole glacier periphery. In the case of a simple ice cap the whole margin will have 'snout characteristics', but in a confined valley glacier situation there will be a much greater range of ice edge conditions between the upper and lower limits of the glacier.
d   The 'ice-water margin', where the glacier snout or ice sheet edge is grounded in relatively shallow water.
e   The 'ice-water margin', where a glacier snout or ice shelf is afloat or partly afloat in lake or sea water. The critical margin here is the glacier base, from which material is dropped or released into suspension as a consequence of bottom melting or ice breakup.

Each of these types of margin can be referred to as a threshold, but there are distinct differences between them. Whereas (c), (d) and (e) are real *output situations* in which deposits may be given permanent lithological characteristics and where lasting landforms may be created, (a) and (b) are *throughput situations*. Here, with the exception of lodgement till, most deposits are of a temporary nature, being made of material which is still in transit. Final deposition cannot be said to have occurred until there is no ice (active or stagnant) in contact with the depositional site, and even then the action of periglacial or fluvial processes may lead to the quite rapid modification of till characteristics and morainic landforms (Figure 11.1). In the cases represented by (d) and (e), processes associated with slumping and sliding, wave action, tidal scour or turbidity currents may modify the original characteristics of the sediments (Kuenen, 1951; Lundqvist, 1965).

When studying the remainder of this chapter it will be as well to bear in mind overall glacier system relationships, and some of the above principles of debris release. After a consideration of the various processes of till genesis on land and in water, the chapter goes on to refer briefly to rates of sedimentation, and ends with an examination of some till characteristics. Fluvioglacial sedimentation is excluded here, but is treated in detail in chapter 16.

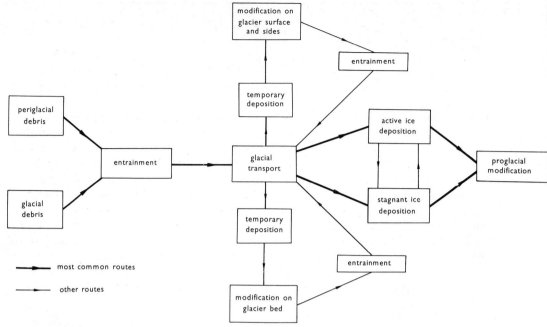

Figure 11.1
Flow diagram showing the input, throughput and output relations of detrital particles in the glacier system. Most particles follow a simple route: they are entrained, transported, and then deposited from either active or stagnant ice. Some particles follow more complicated routes: they are deposited temporarily on the glacier bed, surface or sides, then entrained again, then deposited finally.

Processes of glacial deposition on land

The product of terrestrial glacial sedimentation is generally referred to as till, and this is the term used everywhere in this book in preference to the older terms 'drift' and 'boulder-clay' (Charlesworth, 1957) and the new and potentially useful term 'glacial diamictite', suggested by Flint (1971). Here it is accepted that *till* is a deposit which can be analysed by sedimentological techniques, whereas *moraine* is an accumulation of glacial deposits having an independent surface expression. Whereas till is more the concern of the glacial geologist, moraine is the particular concern of the glacial geomorphologist. In the only English language volume concerned specifically with till, Goldthwait (1971) refers to its diagnostic features as follows:

1   it is poorly sorted, and often has clasts of many sizes (including boulders) in a variable finer matrix;
2   it tends to be massive in structure, without smooth lamination or graded bedding;
3   it is composed of mixed minerals and rock types, some of which are far-travelled;
4   it generally has a proportion of striated stones and micro-striated grains;
5   it may have a common orientation of elongated particles;
6   it may be more compact than neighbouring sediments, due to the great pressures exerted during deposition;

7   it may rest upon a striated rock or sediment basement;
8   its component clasts are predominantly sub-angular, due to frequent breakage during transport and partial smoothing by abrasion.

Flint (1971) refers to other characteristics of till also, but it will become clear from the following text that the range of till types is enormous, and many tills encountered in the field will fail to satisfy one or more of these idealized indicators of glacial genesis (Francis, 1975).

In recent years there have been considerable advances in the understanding of the processes involved in till formation. Sedimentological, stratigraphic and structural studies of ancient tills have been made over the last century or more, and have now reached a sophisticated level in many countries (Dreimanis, 1971); but often these studies have been inadequately supported by detailed knowledge of current ice marginal environments or by theoretical work on the forces involved at the various types of ice margin threshold. Although there is a long history of studies of till genesis from contemporary glaciers (Garwood and Gregory, 1898; Todtmann, 1932), the last decade has seen a great increase in detailed studies of the processes involved (Boulton, 1968; 1970a; Mickelson, 1971). Theoretical studies of glacial sedimentation have also begun to appear in the literature (Nobles and Weertman, 1971; Boulton, 1972a; 1975). From this promising position the following paragraphs discuss the main processes of till deposition on land as far as they are known at present.

Subglacial lodgement
This is the process by which lodgement till or 'comminution till' (Elson, 1961) is deposited from the base of the glacier. At its simplest, the process involves pressure melting beneath active moving ice, allowing small detrital particles to be freed and lodged or plastered onto the glacier bed. Alternatively, basal drag may be invoked as causing the gradual reduction of basal ice velocity leading to shearing over successive layers of ice-debris mix at the glacier sole (Figure 11.2, overleaf). The till surface is gradually built up by a process of accretion, and indeed the term 'accretion till' has been frequently used in the literature. Subglacial lodgement can occur anywhere beneath flowing ice, but it may be most common beneath a warm-based glacier where the ice is gradually losing its erosive capacity and releasing its load of detritus. Lodgement is not necessarily associated with glacier retreat, and indeed Goldthwait (1958) has shown that much Late Wisconsin lodgement till in Ohio is associated with the expansion phase of the last Laurentide ice sheet.

Recent work makes it possible to elaborate somewhat on this simple hypothesis of the lodgement process. Boulton (1970b; 1971) has observed the lodgement process on a small scale beneath the Svalbard glaciers of Nordenskioldbreen and Aavatsmarkbreen. In one case (Figure 11.3, p. 217), lodgement till was seen accumulating on the upglacier flank of a roche moutonnée. Hollows and even minor irregularities in the bedrock surface were filled in with till, providing a more or less smooth subglacial profile in contact with the glacier sole. Where the debris-rich sole was moving in contact with bedrock or previously deposited till, pressure melting led to the gradual release of detrital particles. This is entirely in agreement with glaciological theory (chapter 2). Some of the material, in a thin layer less than 0·5 cm thick, seemed to be moved along by shearing beneath the glacier sole. The rate of movement was about 7 mm per day or more than 200 mm per month. Since fresh till was also observed accumulating above a striated bedrock/till surface, it was concluded that pressure melting is not the only lodgement process, but that direct plastering on the underside of the shearing plane may be locally very important. This plastering occurs where the frictional drag on detrital particles being moved over the bed equals the tractional force exerted on it by the glacier ice (Boulton, 1972a). Particularly in subglacial zones where there is net basal melting which is gradually lowering the glacier down onto its bed, frequent contacts between particles and the bed will cause high rates of lodgement (see Fig. 8.3a). If the bed is rough or if the overburden pressure increases (due to greater ice thickness) the frictional drag will increase and so will the rate of till lodgement. If, however, the rate of bottom melting is high,

Figure 11.2
Debris-rich basal ice exposed on the edge of the Greenland ice sheet near Søndre Strømfjord. There is one major shear-plane in the centre of the photograph.

meltwater may occur at the interface between the glacier sole and its bed and reduce effective normal pressure and frictional drag. Hence it would also reduce the rate of till lodgement, unless the bottom material (either till or bedrock) were permeable enough to evacuate excess meltwater.

The creation of lodgement till may be greatly affected by thrusting and shearing within the glacier on its sole and in till which has already been deposited on the bed (Figure 11.2). Boulton (1970b) has described the overriding of debris-rich basal ice at Nordenskioldbreen by relatively more plastic debris-free ice.

As mentioned in chapter 2, shearing creates heat, so that the glacier sole may be melted from above and from below, by friction and geothermal heat respectively. Till is gradu-

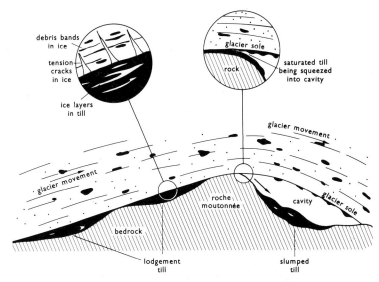

Figure 11.3
Types of subglacial till accumulating at the base of Svalbard glaciers (modified after Boulton). Note that the creation of a lee side subglacial cavity plays an important part. This is a composite schematic diagram.

ally released in this way onto the glacier bed and meltwater is expelled by pressure, making the till more and more compact until it is affected by brittle fracture in the form of low angle shear planes and high angle transverse joints (Figure 11.4). These and other structures in lodgement till are discussed later in this chapter, but here it is worth noting that shearing of sediments in some form (as a result of direct glacier pressure) may be considered an essential lodgement process (Moran, 1971). In addition to the shearing which might orientate particles at the till-ice interface and just below it, shearing stresses may affect particles several metres

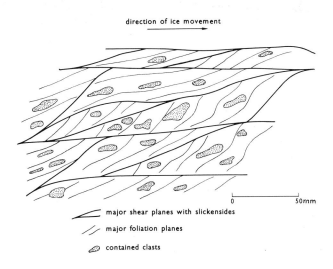

Figure 11.4
Schematic diagram showing the shear structures in lodgement till which may result from stress beneath moving ice. *(Boulton, J. Glaciol. 1970b by permission of the International Glaciological Society).*

below the surface of an accumulating till sheet, giving rise to both foliation planes and fine foliation.

Conditions suitable for the operation of lodgement processes exist beneath glaciers where basal melting takes place. Debris-rich ice may be let down onto the bed at a rate of up to 20 mm per year even where the ice is stagnant, as a result of the geothermal heat flux (Robin, 1955). Where the ice is moving, the frictional heat may be sufficient to let down a further 20–40 mm per year. Thus even beneath the warm-based parts of the Greenland and Antarctic ice sheets of today, debris-charged ice may be let down onto the bed at the rate of up to 6 m per century, allowing the possibility of considerable accretion of a lodgement till layer. In other environmental situations, lodgement will be favoured in zones of net basal melting or approximate net balance between melting and freezing. These zones will be variously located in different glacier systems, and in any case lodgement processes may be highly local-ized if bed conditions favour erosion rather than deposition. Here the scale factor comes strongly into play, and local variables may determine the precise location of areas of lodge-ment. Nobles and Weertman (1971) argue from a consideration of bedrock-ice temperature gradients for active glaciers that rates of deposition (by lodgement) must be greater in topo-graphic hollows than on hills and ridges. This is of course amply supported by field observa-tions of the manner in which till sheets tend to mask pre-existing topographic irregularities (Norris and White, 1961). To generalize, we can say that while subglacial erosion tends to exaggerate pre-existing relief, subglacial deposition tends to mask earlier irregularities and create a more even glacier bed.

Subglacial flowage
The subject of subglacial till flowage has caused some debate in the literature. Gripp (1929) suggested that some englacial moraine bands in Spitsbergen glaciers were formed from water soaked till being squeezed upwards into basal crevasses by the pressure of overlying ice, and Hoppe (1952; 1957) used the same hypothesis in his explanation of certain hummocky morainic landscapes. Boulton (1970a) questioned this idea, but later (1971; 1972) was more inclined to accept it as a possible process where considerable thicknesses of unfrozen sub-glacial till exist, as beneath warm ice.

Even in subglacial lodgement tills, localized stress conditions can cause till fabrics to reflect local bedrock configuration as well as the broad direction of ice movement. If the till is wet, it may experience fabric reorientation to depths of 10 m or more (MacClintock and Drei-manis, 1964). Where cavities exist either on the lee side or upglacier side of bedrock protu-berances or large boulders on the glacier bed, unfrozen lodgement till may move under dif-ferential overburden pressures into the spaces, providing of course that it is soft enough to be mobile. Since subglacial till surfaces are naturally rough they are often characterized by high stress contrasts over small areas. Under these circumstances, there may be a great deal of secondary deformation along subglacial pressure gradients which may radically change till structure and fabric (Gravenor and Meneley, 1958). The till may even qualify for Elson's (1961) term of 'deformation till' if deformation structures are common enough. Sometimes fluted lodgement till surfaces may result from the secondary movement of par-ticles, and this is a topic discussed in more detail in the next chapter.

Subglacial till flowage under pressure is inadequately understood. The process sometimes follows the original deposition of particles by lodgement, even if only by a short while. The widespread flowage of till and other sediments under grounded active ice has been discussed by Hoppe in a number of papers since 1948, and it appears to have been common below the highest shoreline around the Gulf of Bothnia. The 'kalix till' of Sweden, consisting predominantly of stratified sands and silts with inclusions of coarse material, has closer affinities with fluvioglacial and glacio-marine materials than with true till, and it is thought to have been deposited by slow moving sheets of basal meltwater close to an ice margin. Water depths at the margin may have been 200 m or more. If there is no evidence for the subglacial penetration of lake or sea water, then a closed system environment in which subglacial meltwater is temporarily trapped may be suitable for till squeezing and flowage. Under conditions of advanced ice wastage, the till may be squeezed into basal crevasses, and this process has been much discussed by such authors as Andrews and Smithson (1966), Hoppe (1952), Elson (1969) and Price (1970). Mickelson (1971) has shown the process in operation on Plateau glacier, Alaska. Here, and indeed on many glaciers elsewhere, there is 'free flowage' of till into 'closed system' cavities beneath the glacier sole, for example on the lee side of roches moutonnées. Boulton (1971) has described how wet, thin till oozed out like toothpaste from the ice-rock contact illustrated in Figure 11.3 to slump and slide down on to the surface of a detrital accumulation below. Hillefors (1974) had assumed that some 'lee side moraines' in western Sweden formed in a similar fashion.

Bulldozing and recycling

Under certain conditions where ice is highly active at the snout of a glacier, perhaps as a result of the arrival of a kinematic wave or a surge, ablation may not be able to compensate and the ice edge will advance. Characteristically the ice margin will be steep and highly crevassed (Kalin, 1971; Garwood and Gregory, 1898). While some glacier advances have virtually no effect upon the ground over which they pass, from the point of view of deposition the most important effects of an advancing snout are often as follows:

a   the bulldozing of previously pro-glacial material, including frozen blocks of silt and sandur sands and gravels, peat and slabs of bedrock;
b   the recycling of material (mostly till) being delivered to the snout englacially and supraglacially by the normal mechanisms of transport.

Spectacular push-moraines may be created, and these are discussed in the next chapter. Concerning the till produced during the advance, two main groups may be recognized. Both of these groups can display widespread contortions, and would be classified as 'deformation till' by Elson (1961).

a   In the case of a polar glacier frozen to its bed and moving across permafrost, the till will retain most of the sedimentological characteristics of the overridden material, be it fluvioglacial outwash material, peat, lacustrine silts and clays, or an earlier till. However, the material will be distorted by shearing, faulting and folding as slabs of frozen detritus are bulldozed forward by the glacier (Gripp, 1929). The degree of distortion will depend upon the compactness of the material and the extent to which it is bound by its permafrost cement (Moran, 1971). Frozen sands and gravels, for example, may have a high density of shearing and faulting structures compared with a compact till. As in the spectacular Thompson glacier push-moraine on Axel Heiberg Island (Kalin, 1971), surface material may be redistributed by melting, sliding and flowing during the summer, and perhaps also by frost-heaving and frost-cracking during the winter. The end product – the 'push till' – will display a somewhat chaotic structure (Banham, 1975).

b   Different kinds of structures will result if the bulldozed material is unfrozen and is being moved by a glacier which advances by sliding over its own bed. Detritus may be faulted and thrusted, as in frozen sediments, but folding and other violent distortions will be more common, and because the environment is a 'warm' one there will be wholesale redistribution of material by flowage, sliding, secondary sedimentation in temporary lakes, and by fluvioglacial processes (Bayrock, 1967).

Melt-out processes

These are the processes by which englacial detritus is gradually added to a till layer, either subglacially or supraglacially, by the melting of ice. The fabric characteristics of the resultant

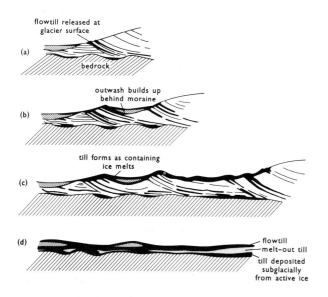

Figure 11.5
Diagram showing how different types of till may be interstratified after ice wastage, producing a bipartite, tripartite or even multi-layered till sequence relating to *one* glacial phase. *(Boulton, Inst. Br. Geogr. Spec. Pub. 1972a.)*

till will resemble those of the parent ice fabric unless some reorientating process acts at the boundary of the ice. High concentrations of englacial detritus are particularly associated with those parts of glaciers where regelation layers have been added to the glacier bed or where thrusting has carried basal material up into the ice, as in Boulton's zones C and D (Figure 8.13). This material is released as melt-out till by ablation processes, which are more characteristic of peripheral ice cap zones than of the central zone of streamlining and lodgement. Nevertheless, because of the complex glacial dynamics even of wasting ice caps, melt-out till may in some situations be deposited above lodgement till and below flow till, replacing the common bipartite till sequence by a tripartite sequence (Figure 11.5). Regarding the precise mechanics of melt-out till deposition, these appear to be determined above all by the provision of heat at a controlled rate.

Surface melting Melting on an upper ice surface is most likely to be gradual and 'controlled' beneath a cap of ablation till where summer heat slowly penetrates the overburden and achieves thawing of the surface of debris-rich ice. In Iceland, southern Greenland, and Scandinavia, the summer thaw may be able to penetrate a till cap of several metres (Østrem, 1959), but in polar situations, where the depth of thaw may be less than a metre, melt-out tills accumulate much less thickly. However, in Svalbard and other polar areas, if part of the overburden is removed, thawing will be permitted to proceed at the surface of the underlying ice, and more melt-out till will be added to the base of the overburden (Figure 11.6). Thick sequences of melt-out tills may develop beneath supraglacial lakes, where heat for melting is transferred downwards at a controlled rate from the base of the water-body.

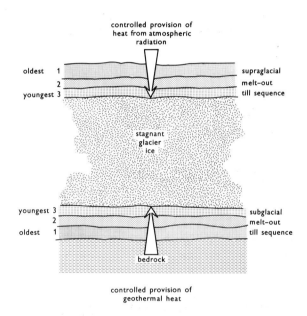

Figure 11.6
Melt-out till sequences in supraglacial and subglacial situations,
showing the age relationships of
the layers added.

Lacustrine or fluvioglacial materials may accumulate above the melt-out sequence, as demonstrated by Boulton (1970b). It is extremely rare for a supraglacial melt-out till to retain exactly the structural characteristics of its parent ice. If ice foliation is approximately horizontal, and if the melting surface is also horizontal, it may indeed be possible, but rare, for fresh melt-out till to be created with minimum disturbance of the fabric (Figure 11.7a, overleaf). But if the foliation planes in the ice are dipping (as they usually are), then some particle disturbance must take place during melting as a result of settling and contact between previously separated particles (Boulton, 1970b). The result may be that the melt-out till retains clear foliation but with a reduced dip (Figure 11.7b). On a gently sloping thawing surface, meltwater evacuation may be efficient and there may be minimum disturbance of the melt-out till fabric. But above a steeply dipping ice surface, meltwater lubrication at the ice-overburden interface may induce the whole mass to slide downslope.

Another factor which will produce instability in the accumulated cap of melt-out till is rapid melting at the thawing surface, either because of an exceptionally warm ablation season

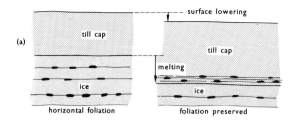

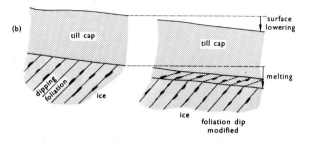

Figure 11.7
Diagram showing how foliation may be preserved or modified during the formation of melt-out tills. In **(a)** both the ice surface and foliation layers are horizontal. In **(b)** both are dipping, leading to much greater modification during melt-out.

or because part of the overburden has been removed by sliding, shearing or some other process, thus allowing deeper than usual penetration of thaw into the stagnant ice beneath. If the thawing surface is horizontal or only gently sloping the meltwater produced cannot be efficiently evacuated. The whole melt-out till cap may become saturated, and as high pore pressures build up at its base, instability, flowage or the creation of injection features may be the inevitable result of overloading (Boulton, 1971). Similar features will also occur in melt-out tills formed beneath standing water-bodies, for the sediments may be completely saturated or subjected to loading pressures from overlying sands, silts and clays as they accumulate.

Basal melting Turning to the creation of subglacial melt-out tills, a special circumstance is the controlled provision of geothermal heat (Mickelson, 1971). This heat is largely responsible for the basal melting process where there is little ice movement, and it may cause the very gradual sedimentation of 'ideal' melt-out tills (Figure 11.6). On the other hand a new set of special circumstances is encountered which might disturb the till structure. Where a glacier is stagnant, the geothermal heat flux can cause melting upwards through the basal ice layers. As long as the meltwater created can be expelled laterally, the annual layers of melt-out till, as they accumulate, will retain more or less the fabric characteristics of the dirty parent ice. However, this is a delicate equilibrium situation which is all too easily disturbed. If meltwater is not evacuated efficiently, the melt-out till can become saturated and lateral flowage will occur under the pressure of overlying ice. If lateral flowage is impeded, then injection features may well be created in the till mass (Hartshorn and Ashley, 1972). And if the water accumulates and cannot escape the ice surface will 'lift off' from a small part of the till surface. This will create a basal cavity, and further particles released from the ice will be dropped onto the till surface, perhaps with the creation of a new fabric (Marcusson, 1973).

From the above it may be appreciated that the creation, and especially the preservation, of melt-out tills requires a rather special combination of environmental conditions. Such tills may often be converted into flow tills (supraglacially or subglacially), and in some situations they are difficult to distinguish from lodgement tills or deformation tills. Hence there is a vast range of melt-out till types, some of which are virtually indistinguishable from tills formed in other ways. However, because melt-out tills are common in Pleistocene till sequences, the mechanisms involved in their formation require careful study.

Supraglacial flowage

Most of the ablation tills which accumulate on and near the snouts of glaciers are eventually affected by flowage, unless special circumstances allow melt-out tills to form and survive. It has long been known that a great deal of ablation till is modified by sliding and flowing on a downwasting ice surface (Lamplugh, 1911; Tarr, 1909). The present authors have noted, to their considerable discomfort, the instability of till masses on irregular downwasting ice surfaces in Greenland and Antarctica. Following on the work of Gripp (1929) and other pioneers, Hartshorn (1958) used the term 'flowtill' for such materials, which are highly variable in composition and mode of occurrence. Flowtills are described in the course of studies of ice-disintegration features or 'glacier karst' by Gravenor and Kupsch (1959), Clayton (1964), and many other authors. Perhaps the most influential recent studies of the processes involved in the present day emplacement of flow tills are those of Boulton (1968; 1971). From studies of Svalbard glaciers, Boulton has shown how a great deal of ablation till melted out at the surface of debris-rich ice is redistributed by various types of flowage. An example from Iceland is shown in Figure 11.8. The major factors influencing flow are the gradient of the ice surface beneath the till cap, its roughness, the amount of meltwater available for

Figure 11.8
A tongue of flowtill resting on the surface of Kaldalon glacier, northwest Iceland. The till is derived from an upstanding ridge of ice cored moraine off the picture to the right.

till lubrication, and the porosity of the till itself. This last factor is affected partly by the particle-size characteristics of the till.

Till which has just been released from the ice surface, and which is saturated by meltwater, may flow at rates of several metres per minute downslope on the flanks of wasting hummocks of ice. This type of flow is referred to as *mobile liquid flow* by Boulton (1971), and it is illustrated in his 1968 paper. It is generally restricted to the surface layers of waterlogged till, being up to 20 cm thick. Boulders and stones tend to settle through flows of this type, and there may be normal water sorting which leads to a simple stratification.

Flow which is slower, but still discernible with the naked eye, is called *semi-plastic flow* by Boulton. It occurs on gentler or more stable slopes where the till may be redistributed with ample, but not excessive, meltwater lubrication. The slip-plane may be the surface of contact between unfrozen and frozen till, or else between unfrozen till and ice. Boulton refers to this type of flow as occurring after slope failure along an arcuate slip-face, but this is a somewhat restrictive definition, for semi-plastic flow can occur in many different situations, and almost continuously during the ablation season, in many areas of downwasting ice. On the margin of the King George Island ice cap near Three Brothers Hill there is an extensive area of dead-ice topography in which semi-plastic till flowage is common. Drainage is chaotic, and much of the flow till collects in hollows as unstable, saturated masses. As these hollows fill up with liquid till, they are fed by slumping on the flanks of the hollow, and eventually an overflow is affected which evacuates at least some of the material to an adjacent lower hollow. Commonly, as described by Boulton (1968), ablation till is provided from the top of an arcuate scar; it slumps down over a wall of clean melting ice into the centre of a small amphitheatre, and then escapes downslope as a lobate flow. The material thus emplaced may show signs of stratification and the settling of the largest stones and boulders to the base of the flow. Washing by meltwater may also produce laminations, and these are often useful for the recognition of flow structures which can be diagnostic in ancient flow tills (Banham, 1966).

The slowest type of flow is referred to as *downslope creep*, where materials not actually exposed to the atmosphere move slowly downslope. Little meltwater lubrication is involved and rates of flow may be less than 50 mm per month. Nevertheless, the till fabric is subject to some adjustment. Although the altered till will probably have no fold or thrust structures, it may acquire a slight stratification or foliation and it will be distinguishable from the other types of flow till by its compact and massive appearance. Often this type of flow till will be simply a modified supraglacial melt-out till.

It should be recognized that this classification of flow tills and flowage processes is extremely arbitrary. There are no clear dividing lines between the sedimentary types mentioned, and anyone who has wandered through an area where ablation till is currently being modified by slumping and flowage will recognize what a hazardous task it is to attempt any classification of flow tills.

Processes of glacial deposition in water

These processes have been inadequately studied, and are therefore little understood at the present day. This is unfortunate, because sedimentation from grounded and floating ice is vastly important today on the peripheries of the Antarctic ice sheet and in northern hemisphere fjords occupied by glacier tongues. And such sedimentation was of major significance in late-glacial times during the wastage of the major continental ice sheets. Glacio-marine sediments cover much of the floors of the Arctic and Antarctic Oceans and the North Sea and Norwegian Sea (Goodell *et al.*, 1968; Olausson, 1972). Many of the uppermost tills of the Hudson Bay region date from the time of catastrophic ice sheet disintegration about 7,800 BP when the ice was lifted from its bed by impounded meltwater (Andrews, 1970), and much of the till cover of eastern Sweden and Finland was deposited from grounded ice which remained active even during its retreat. In many upland areas also, there are problems associated with the dumping of till into ice marginal or even subglacial water-bodies (Lundquist, 1967; Gjessing, 1960), and the real processes of sedimentation have never been adequately investigated from contemporary grounded and floating ice margins. This is understandable, for, as noted by Holdsworth (1973), ice fronts are none too safe when viewed by the intrepid fieldworker from either a supra-glacial or subglacial position.

Another problem is one of definition. Are glacio-marine and glacio-lacustrine sediments really *glacial* sediments? Many materials, such as the Swedish 'kalix till', varved silts and clays, and bedded deltaic sands and gravels, are transported and may be deposited by melt-water. Hence they have no claim to the term 'glacial' either from a genetic, morphological or sedimentological point of view. They may not look in the least like tills. But some materials dumped by ice into water-bodies look so much like land-deposited tills that it is virtually impossible to tell them apart (Flint, 1971); and it is these deposits which concern us here. As in the above section, a somewhat artificial distinction is made between tills, fluvioglacial, lacustrine, and marine deposits; often they occur in close association, and interstratification is common especially in environments of rapid ice wastage (Francis, 1975).

The 'starting process' for glacial deposition in water is, of course, melting or ablation. Except in some ice shelf situations, where material may actually be added to the bottom of the shelf by freezing (Swithinbank, 1970), ice in contact with water is melted at a controlled rate. This will lead to the release of englacial detritus, which must first be transported and then deposited. If the ice is grounded, much detritus may be transported and sorted by slow moving sheets of meltwater, and it may be deposited before it reaches the glacier snout (Figure 11.9a, overleaf). If the ice is afloat, the released detritus must settle through standing water to the lake or sea bed (Figure 11.9b). The heaviest particles (boulders and stones) will plunge directly to the bottom, but sands, silts and clays will settle more slowly from suspension, with settling rates determined by the temperature and salinity characteristics of the water and its turbulence. Some sorting may therefore occur, but if the input of debris into the water is more or less continuous the end product may be almost identical to a land based till. If turbidity currents operate on the bed, they may well disturb the bed deposits and create secondary structures such as shear planes, folds and involuted structures (Hartshorn and Ashley, 1972). These may be difficult to distinguish from the features which are created in terrestrial lodgement or melt-out tills as a result of shear stress or loading. At another extreme, turbidity and other powerful deep-water currents may be capable of creating substantial forms which are difficult to distinguish from moraines (Holtedahl and Sellevoll, 1972), especially if they are built of redistributed glacial materials. On the other hand, various clues to the real origin of glacio-marine tills may be found in occasional lenses of stratified drift, and in the occurrence of layers of biogenic matter and marine organisms in their growth positions. Also, freshly deposited glacio-marine sediments have larger void ratios and lesser bulk densities than terrestrial tills (Easterbrook, 1964).

Sediment characteristics will be strongly influenced by the nature of glacier calving. Since grounded and floating ice is often active, even when it is in an overall state of retreat, it may achieve the throughput of much englacial material right to the snout. If the glacier is really afloat it is here, at the calving snout, that by far the greatest amount of debris release will occur, with relatively little from the floating glacier base. However, considerable sedimentation may occur within and at the grounding line of an ice shelf or a glacier grounded in a calving bay, such as those envisaged for parts of northern Sweden (Hoppe, 1948). If the glacier snout is standing in very shallow water, most ice output will be achieved by the production of small brash ice fragments which may release the greater part of the englacial

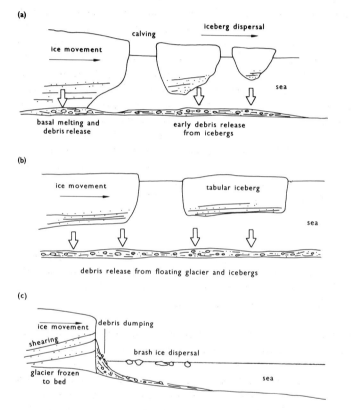

Figure 11.9
Three types of sedimentation from floating ice or ice fronts standing in water.
(a) A case where a warm-based glacier is standing in deep water in a calving bay.
(b) A case where a cold glacier or ice shelf is afloat in very deep water.
(c) A case where a cold-based glacier is standing in shallow water or on a beach.

detritus in the immediate vicinity of the snout (Figure 11.9c). On the other hand, a snout which is afloat in deep water will produce large numbers of bergs (tabular bergs in the case of an ice shelf) which may disperse sediment over a wide area (Conolly and Ewing, 1965; Holtedahl, 1959). In locally shallow areas which cause icebergs to ground and release their debris, there may be 'nests' or clusters of erratic boulders and patches of till which have been deposited without being greatly modified by settling through standing water (von Engeln, 1918).

The thermal characteristics of grounded and floating ice will also affect the nature of sedimentation. Warm-based glaciers whose snouts are in contact with standing water may well experience basal pressure melting conditions upglacier where they are in contact with bedrock. This means that the glacier base will be a surface of net ice loss rather than accretion, leaving englacial material for the most part near the glacier base where it is available for early release. Warm-based ice, which is subject to rapid surface ablation as well as loss by calving, will have meltwater streams which produce large quantities of fluvioglacial materials into the adjacent water-body, particularly during the summer months. This means that the nature of the sediment being let down into the sea bed from suspension will show a seasonal variation. In addition, the injection of the meltwater itself, perhaps in great quantities, will lead to variations of temperature, salinity and hence density, which will affect settling rates. Under such conditions the lithological characteristics of the accumulating bottom sediments will be highly variable; there may be complex interbedding of till and fluvioglacial material, and turbidity currents will frequently disrupt and rearrange the deposits (Price, 1973). In addition, if the snout is advancing, the bottom sediments will be unfrozen and easily distorted and piled up by subaqueous bulldozing. In contrast, cold-based glaciers may carry less englacial material. They release relatively little fluvioglacial material into the water-body, and there will be less seasonal variation in the pattern of sedimentation. Englacial till, located relatively high within the floating glacier as a result of past shearing, thrust-stacking and accretion of regelation layers on the sole, may be immune from immediate release when large icebergs are calved. Hence this debris is much more likely to be widely dispersed than that of calving warm glaciers. Carey and Ahmad (1961) considered some of the theoretical grounds for variations in sedimentation from floating glaciers with different basal ice regimes, and this discussion was continued by Andrews (1963). The influence of floating and calving ice upon the formation of moraines is described in the next chapter.

Rates of sedimentation

Rates of till sedimentation are highly variable, being controlled largely by the precise mode of ice wastage, the rate of wastage, and the detrital content of ice. It may well be that flowtills are deposited most rapidly, melt-out tills at variable but slower rates, and lodgement tills most slowly; but local conditions lead to very great variations in rates of deposition by all these processes.

Flowtills can accumulate at rates of several metres per year, although negative feedback often operates to ensure that high rates of flow are not maintained for more than a few years (Figure 11.10, overleaf). There are, so far as the authors are aware, no prolonged direct measurements of rates of flowtill accumulation, although this would make a useful and interesting study. Meltout tills at a glacier surface may form at rates of 200 mm per year or more, depending on the rate of heat receipt at the ice surface (Østrem, 1959; 1965). On the other hand basal melt-out tills accumulate much more slowly, and Mickelson (1971) has calculated rates of 5 and 28 mm per year for the Burroughs glacier, southeast Alaska. From the work of Boulton in Svalbard, it seems that similar rates of lodgement may occur under present day conditions, but Goldthwait has estimated that under Pleistocene ice sheets, rates as high as 200 mm per year may have occurred at the close of the last glaciation. Where a variety of depositional processes are involved, and where an active glacier is heavily charged with debris (as during the terminal phase of a surge), several metres of till may be created in a single year. Flint (1971) mentioned the example of the Sefström glacier in Spitsbergen, which deposited 30 m of till in the space of only ten years. Where the squeezing process is involved, localized sedimentation rates of several metres per year may be commonplace.

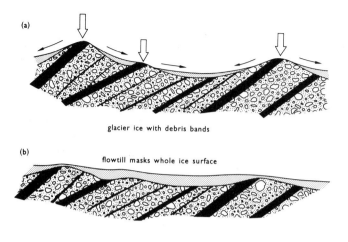

glacier ice with debris bands

(b)

flowtill masks whole ice surface

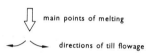

main points of melting

directions of till flowage

Figure 11.10
Diagram illustrating the negative
feedback which operates to slow
down rates of flowtill formation.

Till characteristics

Pleistocene tills have been studied more and more intensively during the present century, and there is now an enormous literature on till texture, lithology, structure, and fabric, as well as on various engineering properties such as degree of compaction and plasticity. A whole variety of criteria are employed in different areas to describe and correlate tills, including pebble fabric, particle size distribution, heavy mineral and carbonate concentrations, pebble lithologies, colour, and weathering characteristics (Goldthwait, 1971). Many of these characteristics have now been studied for tills adjacent to present day glaciers, and the new data are making it possible to identify, with much greater certainty than before, ancient tills which have accumulated by the lodgement, melt-out and flowage processes referred to above (Francis, 1975).

The two till characteristics which have perhaps been most intensively studied are macro-fabric and particle-size distributions, and the techniques employed in the field and laboratory have now become sophisticated if not standardized. Influential papers on till fabric are those of Richter (1936), Holmes (1941), Harrison (1957), Wright (1957), West and Donner (1956), and Young (1969). In addition to the papers referred to above, Boulton has also provided a great deal of valuable fabric information from Svalbard, and has made a major contribution by comparing the fabrics of englacial material and recently formed tills of various types. Particle-size distributions are analysed, for example, by Murray (1953), Kelly and Baker (1966), Järnefors (1952), Dreimanis and Vagners (1971), and Beaumont (1971). Both of these methods of till investigation, and a number of others, are extensively reviewed in the standard texts (e.g. Flint 1971; Embleton and King, 1975; Goldthwait, 1971; Price, 1973). There is therefore little to be gained by attempting to repeat the exercise here. However, a number of points can be added on the basis of recent work on the sedimentology, structures and stratigraphy of tills. These points contribute to the foregoing discussion on the mechanics of till deposition.

Sedimentology

While till is generally considered to be a non-sorted agglomeration of particles of all sizes, glacial processes do in fact set certain limits to the character of till as a sediment. In the last few years progress has been made towards the definition of the largest clast sizes which can be incorporated in till; and at the other end of the scale the definition of the smallest particles which may be typical of certain types of till.

The dimensions of boulders and erratics carried by ice are often spectacular (Figure 8.10). Flint refers to a number of vast erratics, one of which weighs 13,700 metric tons and is derived from a bedrock outcrop 7·2 km distant. On the Norfolk coast there are large erratics of chalk and masses of glacial drift thought to have been moved as vast frozen blocks (Banham and Ranson, 1968). Little is known about how vast erratics are transported and emplaced in till. Meneley (1964) has calculated that the physical properties of ice allow boulders up to

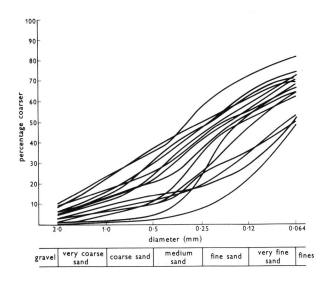

Figure 11.11
Typical particle-size curves for tills from an area of variable glacial deposition in north Pembrokeshire, Wales. Note the wide range of curves.

25 m in diameter to be stably supported and transported, whereas tabular slabs up to 8 m thick have a theoretical maximum length of 330 m if they are to be carried intact by a glacier. More work is needed on this topic, for all we know at present is the simple fact that breakage and crushing of clasts are essential processes which operate during glacial transport (Holmes, 1960; Drake, 1968).

However, it is somewhat unwise to generalize about the breakdown of till clasts without having any detailed information about where, when and how breakage occurs. It is known that ablation till has far more angular and broken clasts than basal till (Drake, 1971), and that abrasion and crushing do occur when till particles are in contact with one another or with bedrock (Boulton, 1975). But for much of the time that particles are entrained in the ice, they are effectively protected from modification unless they happen to be located on a thrust plane or other surface or zone of differential ice movement. Does a till acquire its lithological character during transport, or during deposition? How much destruction of till clasts occurs during deposition, particularly in the case of lodgement till? And how much

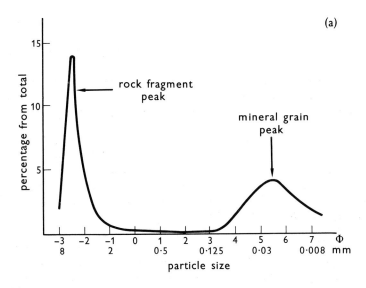

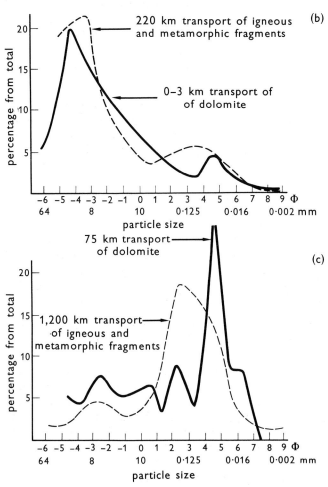

Figure 11.12
Bimodal distributions of particle sizes in sediments. *(Dreimanis and Vagners, Ohio State Univ. Press 1971.)*
(a) Experimentally crushed quartz. *(Gaudin, Trans Am. Inst. Min. and Metall. Engineers 1926.)*
(b) Till matrix composed mainly of rock fragments.
(c) Till matrix composed mainly of mineral fragments after prolonged glacier transport.

can a till lithology be modified immediately after deposition if it is distorted by shear stress or flowage? The literature contains a number of clues to the answers (Harrison, 1960; Goldthwait, 1971), but again detailed studies at the margins of present day glaciers are urgently required.

Particle-size distributions provide instructive data on the mechanics of till formation, although too often glacial geologists, when faced with widespread variations in the plotted particle-size distributions of specific tills, have found their labours an embarrassment rather than an aid to interpretation (Figure 11.11). As noted by Flint (1971) the curves for most tills are bimodal or even multimodal, suggesting that there is some inherent asymmetry in the process of rock breakdown during glacier transport. Of great interest to the investigation of till matrix evolution are the crushing experiments with ball mills by Gaudin (1926) and Lee (1963). A result from Gaudin's work is given in Figure 11.12a, showing how crushed quartzite has a bimodal distribution, with the first peak coinciding with an optimum rock fragment size, and the second with an optimum mineral grain size. Detailed investigations by Dreimanis and Vagners (1971) have shown that rock components in Pleistocene tills have exactly the same bimodal (Figure 11.12b, c) or multimodal distributions after glacial transport. The precise characteristics of the grain-size curves for particular tills will depend upon the number of rock types making up the till, and the number of minerals making up each rock type. The main conclusions of the study by Dreimanis and Vagners are as follows:

1   At least two modes develop for each lithic component of a till. One is in the clast size, consisting predominantly of rock fragments, and the other is in the till matrix, consisting predominantly of mineral fragments. If the parent rocks consist of minerals with a variety of physical properties, the fine fraction will be poly-modal.
2   Near the source of a till, the coarse fraction is large and mineral fraction small, as in Figure 11.12b. However, the coarse fraction diminishes downglacier, and on the plot of particle-sizes the mineral grain peak will become more and more prominent (Figure 11.12c).
3   In the fine fraction the modes are typical of the minerals present. The modes coincide with certain particle-size grades called the 'terminal grades', because they are the final products of glacial comminution as long as mechanical breakdown is dominant.
4   Predominance of terminal grades over clast-size modes indicates a high degree of maturity in a till. The highest degree of maturity in a till is attained when all of its rocks have become comminuted to the terminal grades of all of their constituent minerals.

Although he plotted his data in quite a different way, Beaumont (1971b) found a similar bimodal distribution of particle sizes in the tills of Durham, and related this to a change in mineral grains.

The results of this work now seem to demonstrate that there are sorting or separating mechanisms which operate in till and that they are completely in accordance with physical laws of crushing. However, one must also remember that certain tills (for example, highly calcareous tills such as the Irish Sea tills of western Britain) are liable to chemical modification during transport and deposition, so tills cannot always be interpreted entirely in terms of Dreimanis's and Vagners's physical rules. In addition, the till in some areas may be composed largely of material derived from an older chemically weathered mantle (Feininger, 1971) or from penecontemporaneous sediments John (1970b).

Structures

It has been known for many years that glacio-tectonic structures (called 'ice-thrust features' by Flint, 1971) are extremely common in present day glaciers. They may be most common in glaciers which are frozen to their beds in the terminal zone, and where glacier movement can only be continued by means of thrusting and other internal deformation (Boulton, 1970a). These structures are easily discernible in glaciers with a high dirt content in layers added by regelation and later disrupted. Both thrusting and folding occur near glacier snouts because deceleration of the ice mass is accompanied by intense longitudinal compression (Figure 11.13). As a response to its own internal stresses, the ice has to deform either by brittle fracture (thrusts and faults) or by bending and folding. The precise proportions of these types of deformation will vary according to the thermal characteristics of the ice and the size of the debris load (Banham, 1975).

Figure 11.13
Thrusting and folding features at the snout of the Thompson glacier, Axel Heiberg Island.

Here we are concerned not so much with structures within a glacier as with structures preserved in till after the glacier has wasted away. Slater (1926; 1927; 1929) long ago argued that some contortions in till were derived from intensely convoluted structures in debris-charged glacial ice. Some of the structures seen in wasting ice margins in other areas also (e.g. Greenland) do not seem far removed from those of contorted Pleistocene sediments in Canada (Kupsch, 1962) and East Anglia. The same is true of the structures of 'stauchänd-moränen' analysed over many years in Denmark and north Germany (Gripp, 1938). It must be concluded that they have all originated through types of stress deformation.

Some of the theoretical aspects of glacio-tectonics considered by Moran (1971), arising out of his studies of various deformation structures in the glaciated parts of the United States. He recognized three types of structures:

1   Simple *in situ* deformation, including disruption of drift and bedrock by simple folds and faults and involving no mass removal of blocks. This type of deformation was explained

simply by the normal stresses exerted on the upglacier side of protuberances, and on the shear stress of the glacier sole moving along its bed (Banham, 1975).

2   Large scale block inclusion. This involves the incorporation of large masses of drift and bedrock into the ice, and transport down glacier. Moran discussed the mechanisms of block removal in terms very similar to those of Boulton (1971), although he believed that shearing in the glacier bed can be achieved in unfrozen sediments if local pore water pressures are high.

3   Transportational stacking within till sheets. This involves the forward transport of till along thrust planes within the debris-laden basal zone of ice, following the ideas of Virk-kala (1952), and Harrison (1957).

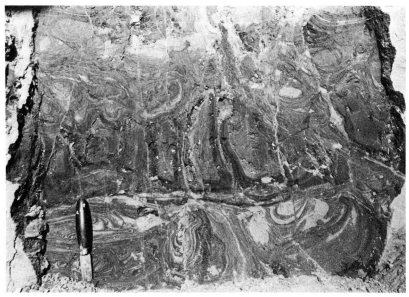

Figure 11.14
Injection features in fine-grained sediments at Noret, Ångermanland, Sweden. Such features may be formed subglacially, but they are difficult to distinguish from normal load-cast phenomena. (*Photograph by Erling Lindström.*)

These theories are interesting because they suggest that deformation from shear stresses in ice can occur even in warm ice and in glacier bed sediments which are unfrozen. They are also relevant for the consideration of end moraine and push moraine formation, topics which are dealt with in the next chapter.

Typical features due to glacial shearing processes (folding and thrusting) where there are low angle axial planes are isolated lenses, folded slump structures, overfolds, and gently dip-ping intercalations of adjacent discontinuous beds. On the other hand 'flame' structures, injection features (Figure 11.14), large scale diapirs and steep folds and faults result from the variable responses of sedimentary layers to differential subglacial loading. In many respects these loading features are similar to those described from northeast England drift sequences by Carruthers (1947; 1948) who explained them as the results of the bottom melting of debris-rich stagnant ice. More recently Marcussen (1973) has given excellent illustrations of such features from Pleistocene flowtills in Denmark also.

The particular relevance of all of these structural features in glacier ice and tills and related deposits is that severe disruption is to be expected in suites of glacial sediments. Too often in the past, fold and thrust structures in till and fluvioglacial outwash sequences have been accepted as indicators of glacial readvances over pre-existing deposits (Penny and Catt, 1967). These ideas should be seriously questioned in view of the wide range of glacio-tectonic and other structures now known to form during till deposition and during the phase of final ice wastage.

Statigraphy

A final fact to be borne in mind in connection with the above paragraphs is that till is seldom deposited in isolation. In subglacial and englacial situations, fluvioglacial materials may accumulate adjacent to dirty ice which is evolving into melt-out or lodgement till. It is not unknown for lodgement till and lacustrine deposits to occur side by side. On downwasting glacier surfaces especially, fluvioglacial and lacustrine materials may be common, and they will have complex spatial and stratigraphic relations with melt-out tills and flowtills. Sequences of deposits in which there may be several tills interbedded with water-laid deposits are common in ice-wastage environments of today; and indeed Figure 11.5 demonstrates how very complex bipartite or tripartite till sequences with interbedded sands and gravels may result from a single phase of glacier wastage.

Boulton (1967) has warned against the facile assumption, unfortunately all too common in some of the older British literature, that every till represents a glacier advance and every layer of fluvioglacial material a glacier retreat. As more workers familiar with modern ice marginal environments write on glacial stratigraphy, much more realistic assessments of the environmental and chronological significance of Pleistocene drift sequences are being made (Price, 1973; Kirby, 1969; Shaw, 1972). In turn, this will lead to more reliable ideas on the thermal characteristics of Pleistocene glaciers (Boulton, 1972c), and to better reconstructions of past glacier behaviour in space and time.

Further reading

BOULTON, G. S. 1968: Flow tills and related deposits on some Vestspitsbergen glaciers. *J. Glaciol.* **7** (51), 391–412.

—— 1975: Processes and patterns of subglacial sedimentation: a theoretical approach. In Wright, A. E. and Moseley, F. (eds) *Ice ages: ancient and modern*, Seel House Press, Liverpool, 7–42.

—— 1978: Boulder shapes and grain size distributions as indicators of transport paths through a glacier, and till genesis. *Sedimentology* **25**, 773–99.

DREWRY, D. J. and COOPER, A. P. R. 1981: Processes and models of Antarctic glaciomarine sedimentation. *Annals of Glaciology* **2**, 117–22.

GOLDTHWAIT, R. P. (ed) 1971: *Till, a Symposium.* Ohio State Univ. Press (402 pp).

LAWSON, D. E. 1979: Sedimentological analysis of the western terminus region of the Matanuska Glacier, Alaska. *U.S. Army Cold Reg. Res. Engng. Lab. Rept.* 79–9 (112 pp).

LEGGETT, R. F. editor 1976: *Glacial till: an interdisciplinary study*, Roy. Soc. Canada Spec. Pub. 12 (412 pp).

POWELL, R. D. 1981: A model for sedimentation by tidewater glaciers. *Annals of Glaciology* **2**, 129–34.

STANKOWSKI, W. editor, 1976: *Till—its genesis and diagenesis*, Poznan: Universytet im Adama Mickiewicza W Poznaniu, *Seria Geografia*, 12 (266 pp).

A Symposium on the genesis of till. Boreas, **6** (2).

12 Landforms of glacial deposition

This chapter is concerned with the various forms assumed by accumulations of till and other deposits related to glaciation. Most of these forms are moraines, which by definition are morphological features. Very few of the morainic landforms described are composed entirely of till, and many well known features such as end moraines and drumlins are often found to contain substantial proportions of stratified drift as well as various types of till.

There are innumerable classifications of morainic forms in the literature (e.g. Okko, 1955; Gravenor and Kupsch, 1959). The broad scheme proposed by Prest (1968) is used here, since it is useful both for describing morainic forms with respect to their situation, and for analysing morainic genesis in terms of glacier dynamics. Table 12.1 (overleaf) shows the classification of moraines adopted here. For the most part groups of features can be related to active ice (controlled deposition) or stagnant ice (uncontrolled or chaotic deposition). Just as till genesis can be related to either throughput or output situations with respect to the glacier system, so can moraines. The following generalizations apply:

1 Moraines formed in throughput situations may sometimes be subglacial steady-state features, reflecting some equilibrium between ice velocity, sediment supply, the thermal characteristics of basal ice layers, and the stress resistance of bed forms and bed materials. Moraines formed parallel to ice flow are streamlined or 'drumlinized', but those transverse to ice flow form as a result of different processes.
2 Marginal moraines created in output situations are often 'dumped', and are hence highly variable in form and composition. They may originate through a large variety of processes, some of them related to active ice movement and others to stagnant ice dissolution. Although many ice edge moraines are easily recognizable as such, there may be a strong random element in their morphology.

Morainic landforms can be extremely shortlived. Medial moraines, lateral moraines and end moraines may be spectacular forms as long as the parent glacier is present (Figure 4.9); but they often depend upon the presence of supporting ice, and may disappear or become so subdued as to be hardly noticeable when the ice has melted out. Particularly complicated inversions of relief may accompany the melting out of ice-cored disintegration moraine (Gravenor and Kupsch, 1959). There are other modifications through contacts with adjacent environmental systems during ice wastage. Periglacial processes modify newly formed moraines in a variety of ways, and fluvioglacial processes are particularly potent either in the destruction of moraines or in the burial of features beneath aggrading outwash material as a glacier snout retreats. During ice edge retreat, glacial processes themselves can modify morainic landforms drastically, for all streamlined forms, like all lodgement deposits, have to pass through the glacier wastage zone before they are exposed to the atmosphere (Figure 11.5). In spite of what is now known about the different ways in which warm-based and cold-based glaciers disintegrate, it is still surprising that some areas of lodgement till can

Table 12.1 Classification of morainic forms (*modified after Prest, 1968*)

Linear features		Non-linear features
Parallel to ice flow (controlled deposition)	Transverse to ice flow (controlled deposition)	Lacking consistent orientation (controlled or uncontrolled deposition)
Subglacial forms with streamlining: (a) Fluted and drumlinized ground-moraine (b) Drumlins and drumlinoid ridges (c) Crag and tail ridges	Subglacial forms: (a) Rogen or ribbed moraine (b) De Geer or washboard moraine (c) Kalixpinnmo hills (d) Subglacial thrust moraines (e) Sublacustrine moraines	Subglacial forms: (a) Low-relief ground moraine (b) Hummocky ground moraine
Ice-pressed forms: longitudinal squeezed ridges	Ice-pressed forms: minor transverse squeezed ridges and corrugated moraine	Ice-pressed forms: random or rectilinear squeezed ridges
Ice marginal forms: lateral and medial moraines, some interlobate and kame moraines	Ice front forms: (a) End moraines (b) Push moraines (c) Ice thrust/shear moraines (d) Some kame and delta moraines	Ice surface forms: (a) Disintegration moraine (controlled) (b) Disintegration moraine (uncontrolled)

be seen at the surface without a cover of ablation till, and that drumlins and other fluted features can be exposed without a chaotic cover of disintegration moraine.

Moraines parallel to ice flow

Subglacial streamlined forms
Where ice is moving, variations in stress on the bed itself often lead to the formation of lineations parallel to the direction of movement. These lineations or streamline features may form

Figure 12.1
Recently formed fluted moraine near the snout of Isfallsglaciären, northern Sweden. *(Photograph by Valter Schytt.)*

in older drift or freshly deposited till, and they represent complex interactions between erosional and depositional processes. Drumlins, fluted ground moraine and other streamlined forms have attracted much attention in the literature. They may or may not be equilibrium forms, but in any case they must be examined in terms of glacial dynamics.

The simplest of this family of features are the flutings with various dimensions which can often be seen emerging from retreating ice edges (Figure 12.1) (Hoppe and Schytt, 1953; Lemke, 1958; Baranowski, 1970). In areas of Pleistocene ground moraine they come in all shapes and sizes. In Montana, for example, there are 'megaflutes' up to 20 km long, 100 m wide and 25 m high, and there are other spectacular flutes in North Dakota, Alberta, Quebec, Ontario, and the Northwest Territories. Generally the individual ridges are less than 1 km long and less than 10 m high. Prest (1968) made a distinction between *fluted ground moraine* (where the ridge crests are at the same level as the adjacent ground moraine surface) and *drumlinized ground moraine* (where the ridges stand above the general level of the ground moraine surface). This implies that the mechanism of fluting is essentially erosional, involving large scale glacial grooving of a till sheet, However, the distinction is not often made by other authors, and the consensus of opinion seems to be that even quite small flutings are positive, constructional features which should be considered in the context of glacial deposition.

Many mechanisms of fluted moraine formation have been proposed. Most commonly it is supposed that the small flutes in front of currently retreating glacier snouts are formed by the squeezing of water-soaked till either into cavities on the lee side of boulders or into basal crevasses (Dyson, 1952; Hoppe and Schytt, 1953; Andersen and Sollid, 1971). Other theories, taking more account of subglacial stress fields and their relations to till fabric characteristics, are proposed by Galloway (1956) and Boulton (1971). The former author considered that ridges beneath glacier ice in Lyngdalen, north Norway, were formed from pre-existing till squeezed or forced to flow under differential overburden pressures into cavities in front of and behind large boulders. If the flowage is considered to be a type of gradual plastic deformation (rather than the more mobile flow of saturated till suggested by Hoppe and Schytt), it may be difficult to explain the evolution of fluted moraine surfaces several square kilometres in extent, for the process would involve the adjustment and redeposition of virtually the whole of the ground moraine mass. That this is not impossible receives some support from the irregular till fabric characteristics of some fluted moraine ridges. Boulton (1971) proposed that fluted lodgement till surfaces in Svalbard have also been formed by the migration of particles of unfrozen till along pressure gradients into cavities which form either on the lee or upglacier sides of large boulders. There will also be some accretion of particles by normal lodgement processes as soon as the sides and crest of the ridge are in contact with the ice. The core of the ridge may be composed of slumped particles such as those portrayed in the lee of the rock knob in Figure 11.3. A general theory of fluted moraine ridge formation is summarized in Figure 12.2 for situations where till particles have migrated in a plastic, rather than semi-liquid, medium. It should be noted that the processes represented are the normal sedimentological processes of till sheets, and that large boulders act in precisely the same way as bedrock knobs in encouraging subglacial melting and cavitation. Hence there need be no fundamental difference between fluted moraine ridges and 'crag and tail' features as far as origin is concerned.

Drumlins are probably the best known and are certainly the most intensively studied of the landforms produced by glacial streamlining. They are widely distributed in glaciated areas, particularly those affected by ice sheets (Flint, 1971) but also occasionally areas affected by broad valley glaciers (Embleton and King, 1975). They occur in a very great variety of shapes and sizes (Gravenor, 1953), ranging from 'ideal' ellipsoidal forms (Reed *et al.*, 1962; Chorley, 1959) to irregular multiple elongated ridges such as those described from Finland by Glückert (1973) and northwest England by Hollingworth in 1931 (Figure 12.3, p. 240). Areas of drumlinized ground moraine imply streamlining by ice but unsuitable basal ice conditions for the formation of discrete drumlins; the same may be true of the various types of 'drumlinization' recognized by Swedish workers, occasionally on top of other types of moraine (Hoppe, 1957). Glückert (1973) argued that drumlins are formed at times of ice advance, but as long as the ice remains active there seems no reason why they should not be formed at times of retreat also. There are almost as many theories of drumlin formation as there are drumlins, and these are adequately summarized by Ebers (1926) and Embleton and King (1975). There have been many studies of drumlin shape (Chorley, 1959; Reed *et al.*, 1962), drumlin spacing and orientation (Vernon, 1966; Doornkamp and King, 1971), drumlin till fabrics (Wright, 1957; Hoppe, 1951; 1959; Andrews and King, 1968), and drumlin lithology and structures (Slater, 1929; Gravenor, 1974).

On the other hand the dynamics of drumlin formation are still inadequately understood, although they have been discussed in a speculative way in many papers. Of particular interest

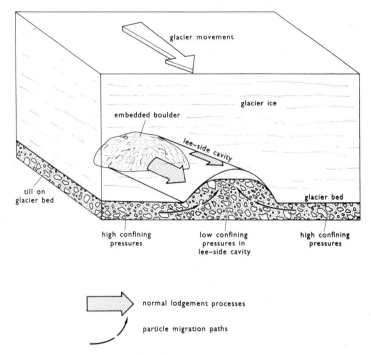

Figure 12.2
A theory of fluted moraine formation. This is probably applicable to most types of fluted moraine ridge.

is the work of Smalley and Unwin (1968), who presented a model of drumlin formation in in which flowing ice, basal load and the glacier bed are considered as separate variables. They imagined a basal till layer being continuously deformed during transport by the ice. This layer has natural variations in stress levels, but if stress locally falls below a certain critical level the till suffers a rapid physical change. From a state of expansion (termed 'dilatancy') it may suddenly fail. The till mass assumes a static stable form on the glacier bed. It becomes highly compact compared with the dilatant material around it, and now there are no local stresses capable of eroding it. The basal dirty ice flows round it, and the compacted mass becomes streamlined, assuming a form which presents the minimum of disturbance to the moving material around it (Chorley, 1959). At the contact between the surface of the drumlin

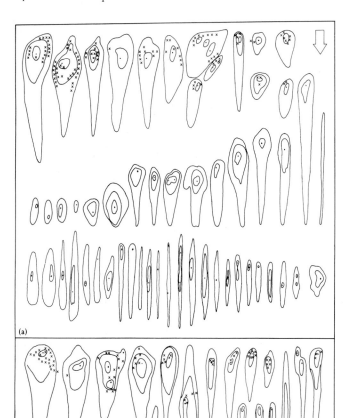

Figure 12.3
Types of drumlins in central Finland. *(Glückert, Fennia 1973.)*

and the enveloping medium, there will still be fluctuations in stress levels; these will either lead to localized erosion of the drumlin, or to growth by accretion. Boulton (1970a; 1971) suggested that especially in areas where ice is at the pressure melting point, a bedrock protrusion or mass of till will be sufficient to encourage pressure melting and particle precipitation from debris-rich ice. Particles will be added to the drumlin flanks by lodgement, and will display long-axis orientations normal to the direction of ice flow (Wright, 1957; Lundqvist, 1948).

Other studies of particle migration in drumlins suggest that the mechanics involved are closely similar to those responsible for fluted moraine ridges. Partly using data from Gravenor and Meneley (1958), Evenson (1971) has analysed particle migration from a theoretical point of view as a response to ice velocity and pressure gradients (Figure 12.4). He considered that as drumlin summits are low pressure zones and intervening troughs high pressure zones,

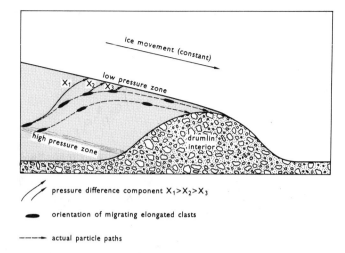

Figure 12.4
Theoretical flow model, showing particle migration paths on the flanks of a growing drumlin. *(Evenson, Ohio State Univ. Press 1971.)*

particles will be attracted to the drumlin flank and will gradually migrate along the gradient to the summit, moving downglacier the whole time as a result of ice movement. Here one may imagine a type of positive feedback operation for, as the relative relief of a drumlin field increases, pressure gradients will also increase, further encouraging drumlin growth at the expense of till deposition in the intervening troughs. Perhaps this explains why some drumlin fields seem to be virtually devoid of ground moraine on the rock surfaces between drumlins.

Ice marginal forms
As mentioned above, there are relatively few morainic features parallel to ice flow which are derived from positions other than the glacier bed. Medial moraines associated with valley glaciers are generally found in throughput situations, often on the moving ice surface. The source of debris supply is usually visible as an upland spur at a glacier confluence (Figures 4.9 and 4.20), but occasionally the source may be hidden beneath an ice fall (Small and Clark, 1974). Medial moraines may protect the ice surface from ablation if they are thick enough, but they have little chance of prolonged survival. As pointed out by Loomis (1970) and Small and Clark (1974) in their models of medial moraine development, supraglacial ridges are eventually lowered to the point of disappearance by such factors as reduced debris supply and lateral sliding of particles down the ridge flanks. By the time they have been carried to the glacier snout, most medial moraines have merged imperceptibly to create a broad cover of surface ablation till, eventually to become incorporated in end moraines or disintegration moraine (Reid, 1970; Tarr and Martin, 1914). One area where ancient medial morainic accumulations are easily recognized is the Faroe Islands. On Streymoy, for example, almost the only glacial depositional landforms are undulating ridges which extend outwards from spur ends, marking the position of former glacier confluences.

Lateral moraines have a better chance of survival, because they may be only partly resting upon the ice surface (Figure 12.5, overleaf). In Figure 12.6 (p. 243) the margin of the Aletsch glacier shows a gradation from actively flowing, relatively clean ice (at the bottom of the picture) to active ice, thinly covered with debris, and then to a zone thickly covered with morainic debris. Line (a) marks the edge of the ice, and line (b) marks the boundary between active and stagnant ice. There are many instances of well preserved lateral moraines in areas of Pleistocene glaciation. In east Greenland there are fine lateral morainic ridges on many valley sides, and in Norway, northwest Iceland, the Alps and New Zealand, ancient lateral moraines often form gently sloping 'shoulders' on the otherwise steep sides of glacial troughs.

Although they are seldom spectacular landscape features, lateral and medial moraines play an important part in the functioning of the glacier system, in that they are the means

Figure 12.5
A ridge of ice-cored lateral moraine at the edge of White glacier, Axel Heiberg Island.

whereby periglacial material is added to the glacier. Above the equilibrium line, frost-shattered material from the valley sides falls onto the glacier flanks, where it is trapped at the break of slope (Price, 1973) and eventually incorporated into the ice. Usually, this material remains at the glacier margin. However, there are important exceptions where the material may be transported in mid glacier, for example when the debris has been derived from the glacier headwall or from a rock avalanche (Figure 2.15), and when glaciers coalesce and convert lateral moraines into medial moraines. The material may remain angular during the whole of its ride on and in the glacier, for it escapes the rounding processes which operate upon basal and englacial clasts (Flint, 1971; Drake, 1971).

Below the equilibrium line on most valley glaciers, lateral morainic zones become the scene of extremely complex particle movements (Figure 11.1). Debris may be temporarily deposited and then reincorporated into the glacier any number of times as the ice margin rises and falls and also expands and contracts laterally over time. Downglacier, as a result of increased meltwater stream activity and increased ice surface downwasting, ridges of ice cored moraine may be 'perched' high above the glacier. Often debris from these ridges slides down into the ice marginal hollow to join the glacier system again.

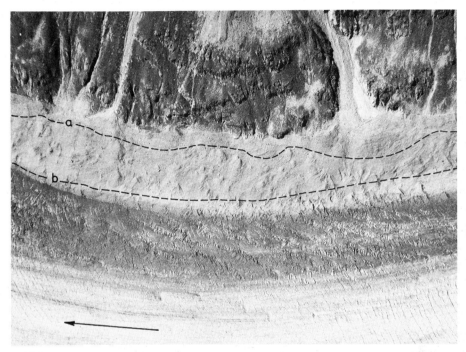

Figure 12.6
The margin of the Aletsch Glacier near Olmen, Switzerland. *(Photograph by the Topographical Survey of Switzerland.)*

Moraines transverse to ice flow

Subglacial forms
Moraines formed in subglacial positions but transverse to the predominant ice flow direction are common in many areas. They have been studied much less intensively than drumlins and other fluted features, and they receive little attention in the standard texts (e.g. Flint, 1971; Embleton and King, 1975). Nevertheless, the academic literature contains lively controversy over their precise modes of formation. In particular, the debate concentrates on the question of whether the ridges are ice front forms (i.e. types of end moraine) or forms developed some way inside the ice margin. The latter interpretation is often favoured by the close associations which exist between the transverse ridges and various types of fluting known to be parallel with the direction of ice movement (Prest, 1968; Wastensson, 1969).

The larger types of transverse moraines are known by a variety of names, although they are generally referred to as 'ribbed moraine' in North America (Hughes, 1964; Cowan, 1968) and 'Rogen moraine' in Scandinavia (G. Lundqvist, 1937; J. Lundqvist, 1969; Kujansuu, 1967). Rogen moraine has fairly large scale transverse lineaments which give an overall, irregular ribbed appearance to the land surface. According to Hughes, individual ridges in northcentral Quebec tend to be 10 to 30 m high, over 1 km long, and with crests 100 to 300 m apart. The depressions between the ridges emphasize the overall ridge pattern (Figure 12.7). Ridges are often gently arcuate, and slightly concave upglacier. They are often linked by irregular cross-ribs, which sometimes link up three or four of the transverse ribs. The moraines are composed of coarse stony till, often with a surface cap of large boulders which is thought to have originated as ablation moraine. As noted as long ago as 1937

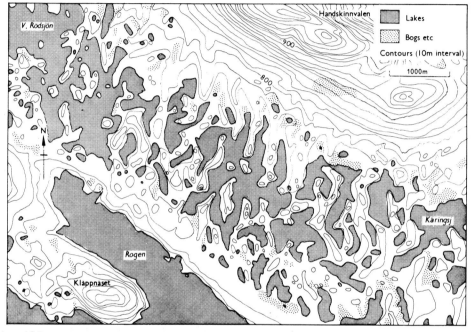

Figure 12.7
Map of part of the classic Rogen moraine area around Lake Rogen, Sweden. The direction of ice movement was from southeast to northwest.

by G. Lundqvist, till fabric in the ridges shows a preferred orientation transverse to the ridge crest. Ridge crests are often fluted or drumlinized. Fields of Rogen moraine ridges are commonly located in slight depressions in till sheets, or in broad bedrock valleys. All of these characteristics are easily recognized in other Rogen moraine areas in Labrador, Newfoundland, Keewatin and the classic area around Lake Rogen, Sweden.

Concerning the origins of Rogen moraine, Cowan (1968) suggested, after detailed studies of till fabric near Schefferville, that the ridges were formed by 'the pushing or overriding of materials by reactivated ice' (p. 1157). He considered that the ridges were formed one by one during a regional ice advance. A subglacial origin has been preferred by Lee (1962), and by Prest (1968), who was more concerned with the explanation of the drumlinized upper surfaces of the transverse ridges. In the most recent of a string of theories in the Swedish literature, Lundqvist (1969) proposed from his studies in Jämtland that Rogen moraine is formed beneath active ice in zones of tension, where basal transverse crevassing causes an irregular precipitation of till from debris-laden ice onto ridges on the bedrock floor. Lundqvist's hypothesis involved no basal shearing of the ice, and the ridges were thought to coincide with the positions of the major transverse crevasses. Because the ice is active, it is quite possible that each ridge, once formed, will be overridden and subjected to surface streamlining during the winter season or during a shortlived climate deterioration. This would explain the drumlinization of Rogen moraine, which has attracted so much attention, and also the presence of cores of 'kalix till' in some Swedish and Finnish Rogen moraine fields (Lundqvist, 1969). However, not all Rogen moraine ridges are made of water-laid or glacio-marine sediments, and not all have drumlinized crests.

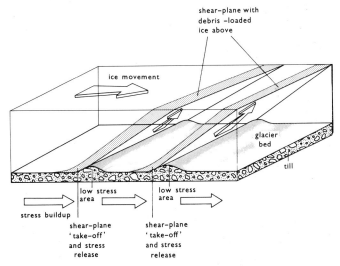

Figure 12.8
A theory of Rogen moraine formation, illustrating debris accumulation at the base of shear-planes in areas of compressive flow. The process may involve particle migration along pressure gradients, and also pressure melting and lodgement at the positions where shear-planes 'take off' from the glacier bed.

Overall, it seems most reasonable to propose, with Prest, that the transverse elements (the Rogen moraine ridges themselves) and the longitudinal elements (the fluted or drumlinized surfaces) have a common origin, possibly reflecting some interaction between debris load, basal ice temperatures, and type of ice movement. As in the case of drumlins, Rogen moraine fields may form where there are irregularly spaced transverse variations in stress fields on the glacier bed. These variations may be the result of longitudinal compression in cold ice which is frozen to its bed. Alternatively they may result from 'normal' compressive flow in warm-based ice, especially near a glacier margin. Again, they may be due to the presence of large scale bedrock obstructions (rock ridges or old river valleys) transverse to the direction of ice flow (Moran, 1971). In any case, consequent upward shearing of ice layers along thrust-planes (Figure 12.8) may accentuate pressure melting, lodgement, and hence gradual ridge construction. The length of the ridge would be controlled by the width of the shear-plane measured across the glacier bed. As in the case of drumlins, increasing ridge size would lead to increased pressure gradients on the glacier bed and to increased particle migration. Also, the presence of the ridge, once established, would encourage continued shearing of ice over the ridge crest, thus assisting in its preservation and allowing further ridge growth by pressure melting, cavitation and lodgement. Where no major shear-plane leaves the glacier bed no transverse ridge will form.

Even where there is shearing, other processes of subglacial moraine construction might operate, including the formation of drumlins, fluted moraine and other streamlined features at a variety of scales. This would explain the irregular cross-ribs and other longitudinal features typical of Rogen moraine fields. Indeed, the common associations of Rogen moraine ridges with drumlins in Sweden has led Lundqvist to propose that there is some type of transition between the streamlined and transverse forms (Figure 12.9). The theory presented above is untested and needs to be examined carefully. It does, however, seem to accord well with the observations of Boulton (1970a) and the theoretical work of Moran (1971).

Another type of subglacial transverse moraine, perhaps related genetically to Rogen moraine, incorporates masses of sheared bedrock and was described by Moran (1971). The

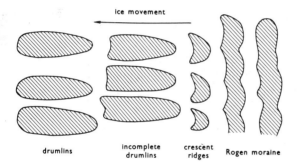

ice movement

drumlins   incomplete   crescent   Rogen moraine
                  drumlins   ridges

Figure 12.9
A suggested sequence of forms related to different ice sheet or ice cap basal conditions, based upon the relations of various morainic features in Hotagsfjällen, Jämtland, Sweden. *(Lundqvist, Sv. geol. Unders. 1969.)*

mechanisms of thrusting and stacking is identical to that mentioned in chapter 8; but the slices or slabs of bedrock may be invisible beneath a mantle of drift, and the resultant ridge could well be mistaken for a true end moraine (Figure 12.10). Kupsch (1962) described spectacular associations of ice-thrust bedrock structures in parts of the interior plains of the USA, where broad arcuate ridges of sheared rock are covered with a thin mangle of drift. Mackay (1959) and Mathews and Mackay (1960) described similar features, formed this time by the shearing of blocks of frozen sediments. The mechanism is presumably the same as that by which large rock erratics are entrained. Prest used the term 'ice-thrust moraine' as a genetic term for all these features.

The other major type of subglacial moraine is De Geer moraine, named after the pioneer of varve chronology. The term is used to describe a succession of discrete, narrow ridges ranging from short and straight to long and undulating. In appearance the ridges are much more delicate than those of Rogen moraine, although they are sometimes similarly connected by cross-ribs. Ridges are seldom more than 15 m high, and they may be regularly spaced anything up to 300 m apart. They are occasionally steep-sided and are made of variable till with a capping of sub-angular and sub-rounded boulders. They are best developed in broad open depressions. Prest (1968) followed the Swedish view (Strömberg, 1965) that De Geer moraines formed some way behind the margin of an ice sheet which calved into the sea or a lake. Lenses of sand and other stratified water-lain deposits may occur in the ridges, and varved sediments occasionally lie in the intervening depressions, supporting the idea that De Geer moraines developed beneath ice which was grounded in deep water.

De Geer himself termed these ridges 'annual moraines', and much critical discussion has centred on the problem of whether such features can have formed by pushing or some other mechanism, annually or at shorter or longer intervals. The problem has been discussed at length for other closely related ridge types such as the cross-valley moraines of Baffin Island (Andrews and Smithson, 1966), 'washboard moraines' (Lawrence and Elson, 1953; Nielson,

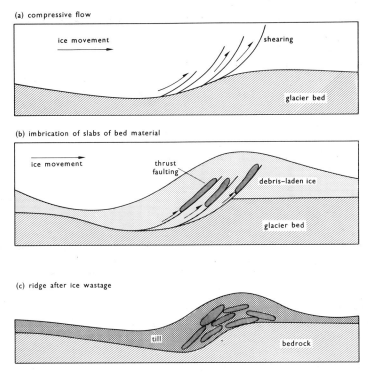

Figure 12.10
Diagram showing the formation of a 'pseudo-end moraine ridge' formed by basal ice shearing, thrust faulting, and the stacking of bedrock slabs. *(Moran, Ohio State Univ. Press 1971.)*

1970) and moraines referred to, for example, as corrugated ground moraine and cyclic moraines (Flint, 1971; Elson, 1969; Prest, 1968). These various types of ridges often appear very similar to one another, but it seems certain that they have a number of different origins. Some, like true De Geer moraines, have probably developed beneath grounded ice. The ridges known as 'kalixpinnmo hills' in northern Sweden are often found close to groups of De Geer moraines but they have a different origin (Hoppe, 1948; 1959). Other small ridges are probably analogous to Rogen moraine ridges, having possibly developed at the base of thrust-planes some distance in from a compressing glacier terminus. And others are probably formed by the squeezing of saturated till upwards into subglacial cavities in the style proposed by Hoppe (1952), Gravenor and Kupsch (1959) and Stalker (1960).

It is no easy matter to distinguish between transverse moraines formed subglacially and those formed at a glacier margin either on land or in water. In particular the elongated moraines referred to by Barnett and Holdsworth (1974) appear to have much in common with both Rogen moraine and De Geer moraine (Figure 12.11). Over the years, several authors have suggested that Rogen moraine is formed in association with pro-glacial or sub-glacial water-bodies, and the mechanisms proposed by Drewes *et al.* (1961) and Holdsworth (1973) for the construction of ridges of 'pseudo-end moraine' deserve consideration. They suggested that large ridges can be formed at the grounding margin of floating glaciers or shelf

Figure 12.11
A sublacustrine moraine ridge, Generator Lake area, Baffin Island. *(Geological Survey of Canada photograph.)*

ice, and they proposed the name 'sublacustrine moraines' for such ridges. The hypothetical mode of formation is summarized in Figure 12.12.

Ice front forms
End moraines are the transverse morainic ridges, or groups of ridges, which delimit former ice-frontal positions. They include recessional moraines and push-moraines. They vary in their precise relationships with the glacier snout, but most of them can be interpreted as the products of some type of glacial output.

There are good discussions of end moraine characteristics and origins in the literature (Gripp, 1938; Flint, 1971; Embleton and King, 1975; Price, 1973), so it will perhaps suffice to emphasize here how very variable they are. Some are 200 m high and stretch for many

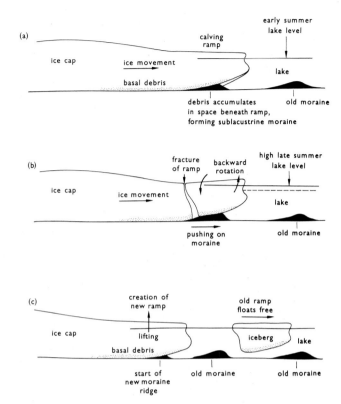

Figure 12.12
A model for the formation of sublacustrine moraines. *(Holds-worth, J. Glaciol. 1973b, by permission of the International Glaciological Society.)*

kilometres to mark the former margins of ice sheet advances or retreat stage positions. Some, like the Holocene terminal moraines of Icelandic and Norwegian glacier troughs, are smaller but no less spectacular in their local context (Figure 12.13, overleaf). Some date from a single glacial stage whereas others may be composite in origin and age. Some are composed entirely of compact tills, whereas others are composed largely of mixed till and fluvioglacial material. Occasionally fluvioglacial processes are so important in the modification of an end moraine complex that the term 'kame moraine' is more accurate. Where glaciers have terminated in standing water, spectacular delta moraines or flat topped moraines may be formed, such as that described from Kjove Land, east Greenland (Sugden and John, 1962). Somewhat similar, but of vast lateral extent and great significance for Scandinavian chronology, are the Salpausselkä ridges of Finland. These are not true end moraines at all, for they are composed almost exclusively of stratified deposits laid down in an ice-contact environment, sometimes well away from the glacier snout (Tanner, 1938; Virkkala, 1963; Hyvärinen, 1973). Potentially of equal importance are the apparent recessional moraine ridges discovered by sounding off the west coast of Norway on the edge of the continental shelf (Holtedahl and Sellevoll, 1972) (Figure 12.14a, p. 251). They occur at depths down to at least 400 m; some of the ridges are more than 1 km long and up to 30 m high, and they seem to be composed of till directly deposited by glacier ice. Similar submerged features are recorded by Winterhalter (1972) from the Gulf of Bothnia (Figure 12.14b) and by Fillon (1975) off Labrador.

(a)

(b)

Figure 12.13
End moraines formed by small outlet glaciers from Norwegian and Icelandic ice caps.
(a) Recent moraine of the Store Supphellebre, Fjaerland, Norway. The glacier and its moraine are in the middle distance, beyond a bedrock ridge.
(b) Large end moraine created by the Kaldalon glacier, northwest Iceland (profile view). The proximal side of the moraine is on the right of the photograph.

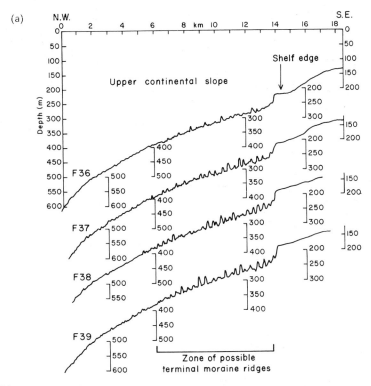

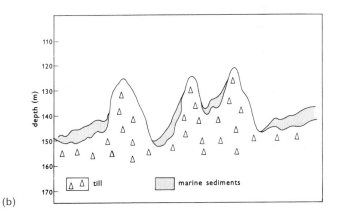

Figure 12.14
(a) Sounding profiles showing possible end moraine ridges just below the edge of the continental shelf off the west coast of Norway. *(Holtedahl and Sellevoll, Ambio Sp. Rept 1972.)*

(b) Profile across possible end moraine ridges in the Gulf of Bothnia in water depth 150 m. Note that the ridges are up to 30 m high. Post-glacial sediments can be seen on the left hand side of the ridges. *(Winterhalter, Fin. Geol. Surv. Bull. 1972.)*

The actual processes of end moraine formation are extremely variable. In addition to the delivery of material to the glacier snout by shearing along thrust-planes (Goldthwait, 1951), much material is delivered by dumping from the ablation moraine cover on the glacier terminus slope. There may be some distortion of material as a result of ice-push, and much detritus will be incorporated at wasting ice sheet edges from meltwater streams and temporary lakes. The processes of till formation referred to in the last chapter (lodgement, melt-out and flowage) will all play their part in the evolution of the proximal side of the moraine ridge, and drastic transformations of the surface of the moraine may occur as the glacier retreats from its inner margin or as buried ice masses melt out (Østrem, 1964). Very often, as in the case of the Kaldalon terminal moraine shown in Figure 12.13b, different processes are responsible for the shaping of the proximal and distal flanks of the ridge. Here the distal flank is composed of massive silt-rich till overlain by fluvioglacial sands and gravels and aeolian deposits, whereas the gentle proximal slope is made only of a stony till with large eratics and incorporated fluvioglacial material derived from the old sandur plain.

Other small end moraines may have much simpler origins. Andrews and Smithson (1966) and Price (1970) proposed that low irregular sequences of ridges adjacent to the Barnes ice cap and Fjällsjökull, Iceland, were formed by the squeezing out of saturated till from beneath the ice margin to the ice edge. This mechanism of moraine formation is probably quite common at the fronts of small glaciers which are warm-based in their terminal zones and which terminate on land.

Push-moraines are a special category of end moraines which may be particularly well represented in high-arctic regions (Dyson, 1952). Flint (1971) claimed that they are not known to exceed 9 m in height, and Price (1973) also suggested that they are of limited importance. There are, however, many examples in the literature of 'Stauchändmoränen' from Spitsbergen, Novaya Zemlya, Holland, Poland and north Germany which attain impressive proportions (Gripp, 1927; Rutten, 1960; Chamberlin, 1893). The Kviarjökull push-moraine in Iceland is more than 100 m high, and on Axel Heiberg Island the Thompson glacier push-moraine has a mean height of 45 m and covers an area 2·1 km long and 0·7 km wide (Figure 12.15). Kalin (1971) showed that it is but one of a considerable group of 35 push-moraines being made by the glaciers of Axel Heiberg Island and Ellesmere Island. The mechanics of formation are complex, but the moraines themselves are fashioned largely as a result of thrusting and faulting in response to shear stresses (Figure 12.16). These stresses are not only exerted horizontally by the advancing glacier snout but also vertically as a result of deep circular shearing in the sediments beneath the glacier, which is frozen to its bed. Hence this is another example of the topographic results of stress relief on the glacier bed, although unlike the cases discussed by Boulton (1972a) and Moran (1971) the stresses beneath the Thompson glacier snout have been exerted in outwash sands and gravels.

Shear-moraine ridges on wasting ice sheet surfaces have attracted much attention (Weertman, 1961; Bishop, 1957; Goldthwait, 1951), and there has been heated debate about their mode of formation and their glaciological and geomorphological significance. Much of the debate has concentrated upon case studies from Baffin Island and the Thule area of west Greenland, where individual ridges may be 15 m high and are continuous over many kilometres more or less parallel to the ice margin. Embleton and King (1975) used the term 'Thule-Baffin moraines' to describe the features in question. The outcropping bands of dirty ice at the glacier surface may represent shear-planes along which there has been differential movement, but it is now known from recent developments in glacial theory that this is not

Figure 12.15
Photograph of part of the Thompson glacier push-moraine, Axel Heiberg Island. The glacier surface is in the foreground.

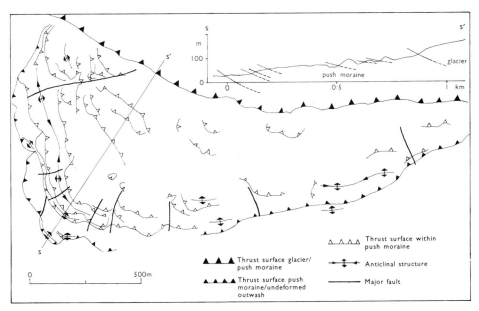

Figure 12.16
Map of the structures discernible in the Thompson glacier push-moraine complex, Axel Heiberg Island. *(Kalin, Axel Heiberg Is. Res. Rept. 1972.)*

necessarily so. Many of the structural planes and dirty layers which eventually outcrop in the glacier snout are simply regelation layers which have been frozen onto the glacier sole with their constituent debris and gradually moved upwards through the ice as a result of surface melting and continued basal freezing. Whatever the pros and cons of the ice thrusting debate, outcropping layers of ice which are heavily charged with debris often give rise to spectacular ridges on the downwasting ice surface. In the South Shetland Islands such ridges often contain striated raised beach cobbles which have been carried up from the glacier bed following an ice edge advance. Like medial and lateral moraines, these features often disappear with the final dissolution of the ice. Where ridges of thrust sediments do survive they can be difficult to differentiate from the subglacial thrust moraines and push-moraines described above.

Moraines lacking consistent orientation

Subglacial forms

In certain circumstances, glacial drift with a constructional surface expression is largely devoid of linear elements, possibly because none of the dynamic conditions referred to earlier (for example, with respect to drumlins or Rogen moraine) could be met at the glacier bed. Nevertheless, there are some forms which deserve a brief mention. Many of them are created in ground moraine, but ground moraine surfaces display a whole continuum of forms from till sheets (where individual morainic landforms may be entirely absent) to areas of violently undulating hummocky moraine where there may be a relief of over 100 m, with steep slopes and deep enclosed depressions. Where the moraine is of low relief it probably formed as a result of broad scale basal deposition. In parts of eastern and northern Canada, the relief is no more than 10 m, and it is often difficult to appreciate except on air photographs. In northern Sweden there is a type of moraine topography created under active ice and referred to as 'Veiki moraine'. It is characterized by irregular hummocks, ridges and plateaux (often with raised rims) with only occasional preferred orientation, and separated by small lakes or hollows similar to those of pitted outwash. The origin of the hummocks was discussed in a series of papers by Hoppe (1952; 1957; 1959). In northern Finland there is a similar type of moraine referred to as 'Pulju moraine' by Kujansuu (1967). Other features which may be related are the prairie mounds, earth mounds and plains plateaux of the interior plains of Canada (Prest, 1968). Some of these may be direct basal depositional features, owing their origin partly to ice pressure and particle migration into basal cavities such as those at the intersections of crevasse systems or in abandoned stream channels. Others may be forms created during ice wastage, for they are often difficult to distinguish from the forms considered below.

Ice surface forms

The relevant glacier margin to be considered here is the surface of contact between the glacier and the atmosphere, particularly in the lower parts of the ablation zone. Most of the features to be described are created largely in a supraglacial position. In areas which have already been deglaciated, high relief hummocky moraine is generally referred to as disintegration moraine, after the well known study of Gravenor and Kupsch (1959). Local relief may be up to 70 m, and the drift topography is characterized by chaotic mounds and pits, generally randomly orientated. Slopes may be steep and unstable, and there will be used and unused stream courses and lake depressions interspersed with the morainic ridges. Hence there will

(a)

(b)

Figure 12.17
(a) Disintegration moraine around the snout of Roslin Gletscher, east Greenland. The glacier has retreated following a surge, leaving a broad terminal zone of chaotic ice cored moraine with many lakes.
(b) Ice cored disintegration moraine at sea-level near the snout of Nathorstbreen, Svalbard. *(Photograph by Gunnar Hoppe.)*

be rapid surface alternations between materials of different lithologies. Perched blocks and stones will be common. In the midst of the morainic mass there may be recognizable fluvio-glacial features such as eskers and kames.

Areas of currently forming disintegration moraine are well known from the literature (Tarr and Martin, 1914; Sharp, 1949). The most essential characteristic is that ablation till and

other deposits must cover a relatively wide area of wasting ice. In other words, the environment is an ice contact one which evolves while the glacier surface downwastes rather than
retreats. In many instances, as in the case of the glaciers shown in Figure 4.14 and 4.15,
debris-covered snouts may still be active, but as wastage reaches an advanced stage the glacier
snout becomes stagnant or dead. Once this happens, the glacier breaks up into detached
portions which melt out one after another. Many authors have remarked on the extremely
confused topography and rapid transformations of the surface which take place while the
underlying dead ice melts out. Clayton (1964) used the effective term 'glacier karst' to describe the appearance of the surface and the processes acting to undermine it, while Flint
(1957) expressed these changes succinctly '. . . in such a place anything can happen, and

Figure 12.18
Sketch of the hummocky dead-ice topography between Sylälvsdalen and Enadalen, Sweden. *(Reproduced,
with permission, from Mannerfelt, Geogr. Annlr 1945.)*

it usually does' (p. 146). Boulton (1967) has described the development of a complex supraglacial moraine in Svalbard, showing how flowtills, melt-out tills and fluvioglacial
materials all coexist on the wasting glacier surface. As mentioned in the last chapter, these
materials become interbedded and mixed as the ice surface is lowered (Figure 11.5). In some
areas known to the authors, such as the dead-ice zones around the snouts of east Greenland
glaciers which have surged, landforms and sediments are even more confused (Figure
12.17a). In Svalbard, chaotic disintegration moraine may occur at sea-level, where marine
processes lead to the rapid modification of forms (Figure 12.17b). In other areas also, the
most chaotic types of dead-ice topography (where thrusts and folds in the wasting ice may
contribute to the confusion) are immediately in front of glaciers which have recently surged
(Johnson, 1972).
    In areas of Pleistocene drifts, spectacular hummocky moraine is occasionally encountered
(Figure 12.18), but it seems to be much less common than the other morainic associations
described above. This may be simply a matter of denudation, and perhaps areas such as

those shown in Figure 12.17 will eventually evolve to form quite subdued drift landscapes. Nevertheless, some wide areas of spectacular features are preserved, such as the moraine plateaux with rim-ridges described by Stalker (1960) and related features described by Parizek (1969). It is not certain, however, whether these are really formed of 'let-down' associations of flowtill and other deposits, for preferred particle orientations suggest that lodgement and ice pressure have been important in their formation.

Very rarely, supraglacial forms such as ridges related to the wastage of high angle debris-bands appear to be locally 'controlled' (Boulton, 1967). They may be linear features when viewed at a high level of resolution, but their orientation is not consistent. Just as difficult to classify are certain reticulated or rectilinear patterns which are referred to as 'disintegration ridges' by Flint (1971). Because they have no *preferred* orientation it is again best to classify them here with non-orientated forms, although they seem to be ice pressed features very similar to the longitudinal and transverse ridges attributed in the literature to the squeezing of till into basal crevasses.

Further reading

AARIO, R. 1977: Classification and terminology of morainic landforms in Finland. *Boreas* **6,** 87-100.

BARNETT, D. M. and HOLDSWORTH, G. 1974: Origin, morphology and chronology of sublacustrine moraines, Generator Lake, Baffin Island, North-west Territories, Canada. *Can. J. Earth Sci.* **11** (3), 380-408.

BIRNIE, R. V. 1977: A snow-bank push mechanism for the formation of some annual moraine ridges. *J. Glaciol.* **18** (78), 77-85.

HOPPE, G. 1952: Hummocky moraine regions with special reference to the interior of Norrbotten. *Geogr. Annlr* **34,** 1-72.

LUNDQVIST, J. 1969: Problems of the so-called Rogen moraine. *Sveriges geol. Unders.* **(C) 648.**

MENZIES, J. 1979: The mechanics of drumlin formation with particular reference to the change in pore-water content of the till. *J. Glaciol.* **22** (87), 373-84.

PREST, 1968: Nomenclature of moraines and ice-flow features as applied to the glacial map of Canada. *Geol. Surv. Pap. Can.* **66-57.**

SMALLEY, I. J. and UNWIN, D. J. 1968: The formation and shape of drumlins and their distribution and orientation in drumlin fields. *J. Glaciol.* **7** (51), 377-90.

SUGDEN, D. E. and CLAPPERTON, C. M. 1981: An ice-shelf moraine, George VI Sound, Antarctica. *Annals of Glaciology* **2,** 135-41.

TARR, R. S. and MARTIN, L. 1914: *Alaskan glacier studies.* Nat. Geog. Soc., Washington DC (498 pp).

13 Landscapes of glacial deposition

The spatial distribution of glacial deposits and depositional landforms at a local scale has been widely discussed in the literature. Many of the standard texts have models showing the relationships between end moraines, lateral moraines, drumlins and other features as they may occur in areas previously affected by valley glaciers. Price (1973) also included a section on the relationships of features which might develop in the vicinity of a wasting ice sheet margin. He stressed the depositional contrasts which occur between areas affected by downwasting and backwasting ice sheet margins, and this aspect has also been discussed by Flint (1971), Mannerfelt (1945), and other authors.

It is most commonly assumed that similar associations of forms will be created at the glacier margin (whatever its type) regardless of where the margin happens to be with respect to the glacier source area. In other words, from a reading of the literature the glacial enthusiast might gain the impression that end moraines, lateral moraines, drumlins, and hummocky moraine have an equal chance of formation and survival virtually anywhere within the peripheries of a glaciated area. This is not the case, and although landform associations in small areas may be highly complex (e.g. Fogelberg, 1970), if a lower level of resolution is employed, the various types of till and of drift landform are seen to possess localized, and almost 'preferential', distributions with respect to the location of the glacier source region. It is the purpose of this chapter to examine some of the links which occur between glacial deposits and depositional forms, and to examine the hypothesis that certain 'zones of deposition' can be recognized beneath glaciers.

Depositional landscape zones in theory

Some of the ideas presented in chapter 3 on the functioning of glacier systems are relevant here. In any discussion of the zones of deposition which might be recognized beneath an ideal valley glacier, we must concentrate on the characteristics of the ablation subsystem. Here, as suggested earlier, ablation is greater than accumulation, and while glacier volume is reduced so is the erosional and transporting capacity of the ice. Deposition becomes more and more the function of the glacier as it has less and less energy available for work of any kind. The same generalizations are true of ice sheet and ice cap systems. In an idealized ice cap it may be possible to recognize two broad zones of glacial deposition:

(a) a zone where deposits are affected by moving ice, some way within the glacier, and
(b) a zone where deposits are laid down near the snout itself, where the ice may be slow moving or even stagnant.

Already, in the last chapter, individual landforms have been classified and discussed with respect to the dynamic characteristics of the ice responsible for their formation, and it is a logical step to treat depositional landscapes in the same way. Each of the two zones has its own characteristic set of depositional processes and its own set of morainic landforms.

The spatial relationships of the zones are portrayed diagrammatically on the simple model in Figure 13.1. In terms of glacial theory, of course, the zones of deposition are much more complicated and numerous than this (Clayton and Moran, 1974). They will have complex relations with erosional zones and with one another. Nevertheless, for a straightforward situation of gradual ice cap dissolution, certain generalizations can be made.

Point (a) on the ground surface, being the outermost point reached by the ice cap margin during a glacial stage, will be affected only by the processes of the 'wastage zone'. These will be the processes of end moraine formation, including dumping, some lodgement, some melt-out and flowage, and perhaps ice-push also. The stratigraphy at the ice edge will be

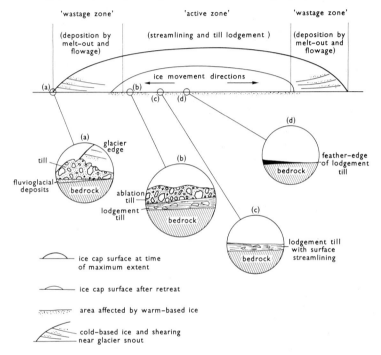

Figure 13.1
Model showing the spatial relations of the two main zones of deposition by the hypothetical ice cap referred to in the text.

relatively simple, consisting of a single till sheet which in places rests on fluvioglacial or periglacial deposits accumulated during the glacier advance (John, 1973). (It may be assumed that in this situation the glacier has not had sufficient energy to remove all pre-existing deposits.)

Further inside the ice margin, the drift stratigraphy and landforms will be much more complicated, as noted from the considerations of till deposition and moraine construction in the last two chapters. Point (b) on the ground surface will be affected successively by the characteristic depositional processes of both the 'active zone' and the 'wastage zone' as the ice margin retreats towards its last centre of outflow. This explains the very frequent and long discussed 'typical' bipartite drift sequence often exposed in glacial deposits, where a basal lodgement till is overlain by a very different ablation till John, 1970a). Here there may be some streamlined depositional forms (if the ice has been able to maintain its forward movement even during retreat), but more probably the land surface will be covered by some types of transverse morainic ridges and a variety of non-orientated forms. The landscape will be one of hummocky drift features or ground moraine. There may be areas of ice disintegration features, and fluvioglacial forms (discussed in chapter 16) will be common.

At point (c), towards the centre of the hypothetical ice cap, the ice may withdraw without ever having experienced a phase of basal freezing. Here there may never have been basal accretion of debris by regelation or a phase in which large amounts of material are carried upwards along shear-planes into englacial positions. Hence, following the deposition of lodgement till, the overlying ice may be relatively clean, and may melt without providing a cap of ablation till. This may explain why some extensive sheets of Pleistocene lodgement till have very clean upper surfaces. If there are any morainic landforms on such till sheets, they will be streamlined features, such as drumlins and drumlinized ground moraine.

Even further towards the centre of the ice cap, there may be a transition from the zone of overall deposition to the zone of overall erosion (Davies, 1969). Here, at point (d), fluted and grooved bedrock will appear, sometimes devoid of any drift cover, and the various features discussed in chapters 9 and 10 will become more and more prominent components in the landscape.

This model is bound to be inadequate in many respects. For instance, it takes no account of the differences between polar and mid latitude ice sheets or ice caps. It takes no account of dynamic differences between the western and eastern margins of ice sheets, well known for example in the case of the Scandinavian Weichselian ice sheet. Further, in upland situations the sequence of events and deposits will be much more complicated than proposed, partly because of the influence of separate ice source areas which may coalesce at stages of intensive glacierization and operate independently at other times. There will also be complications with respect to valley glacier systems, because glacier tributaries can alter discharge characteristics at particular points on the long profile and introduce great differences in basal ice characteristics. Within the boundaries of a cold-based ice cap or ice sheet, there may be sizable areas where the ice is warm-based, particularly in topographic depressions such as outlet glacier troughs. Also, because valley glaciers receive a proportion of their detritus load from the adjacent periglacial system (on the valley sides above the ice surface), input characteristics are highly variable. They appear more complicated still when it is recalled that material may be temporarily deposited, then eroded and entrained in the glacier again any number of times before it comes to its final resting place (Figure 11.1). The diagram, it should be noted, was concerned only with debris which eventually becomes true till and finds its way into morainic landforms of one type or another. In real situations a great deal of material is temporarily affected by fluvioglacial processes on, within, or beneath the glacier before finding its way into the ice again.

Another problem in any discussion of landscapes of glacial deposition is the general lack of long term equilibrium in glacial systems. As demonstrated in chapters 3 and 4, the spatial relationships of the two main glacier subsystems (the accumulation subsystem and ablation subsystem) are different for different types of glacier. In addition, the equilibrium line, and the glacier snout and its margins, are constantly shifting position so that every part of the glacier bed may experience a unique succession of erosional, transporting and depositional events. Although it is occasionally possible for drift successions and sets of morainic landforms to remain constant over wide areas, the task of the glacial geomorphologist is often made extremely hazardous by the variations and discontinuities (not to mention the appearance of fluvioglacial and sometimes periglacial features) which may appear in drift landscapes dating from a single glacial stage (Goldthwait, 1971).

Where drift landscapes are inherited from more than one glacial stage, the problems of interpretation become even more severe. Theoretically, in an area occupied on several occasions by the wastage zone of our imaginary ice cap, there could be several wedges of thick glacial drift (White, 1974). Each of these wedges will thicken towards the ice cap periphery, and there may be examples of overridden terminal moraines, push-moraines and shear structures. Depending on the precise position reached by the ice during each advance, drift stratigraphy will be extremely complex, with tills of all types interbedded with fluvioglacial

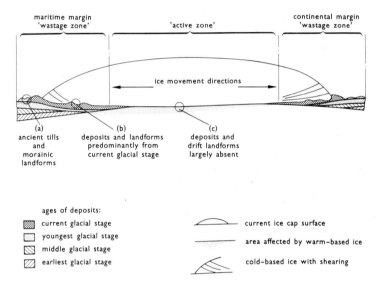

Figure 13.2
A more complex model showing the influence of the *time dimension* on the drift deposits and landforms associated with a hypothetical ice cap.

materials and a variety of interglacial and interstadial sediments also. The wastage zone will certainly be more of a challenge to the geomorphologist and the glacial stratigrapher than indicated in Figure 13.2. Here it is seen to be an area where both surface deposits and features may be a mixture of the ancient and modern, with the proportions depending upon the precise character of glacial activity through the zone during the last occupation by ice. The active zone on this diagram is shown to be an area where fewer ancient sediments have been deposited, and where intermittent erosion by ice may have thinned them even further. Hence the predominant deposits and landforms in this zone will date from the last glaciation. To make the model rather more realistic, a 'peripheral zone' could be added beyond the terminal deposits of the last glaciation, where all of the true moraines and other drift deposits will be ancient features. In reality, as this zone is extended further from the ice cap margin, the ancient tills themselves wedge out one by one, approaching real pro-glacial terrain never affected by glacier activity.

Examples of landscape associations

It is worthwhile examining the extent to which the above generalizations are applicable to the evidence in the field.

In the following paragraphs, a number of examples are cited from northern hemisphere areas affected by the major Pleistocene ice sheets. An attempt is made to relate these examples to the two main zones of depositional landscapes. This is a somewhat simplified framework since landforms shaped by contact with either active or stagnant ice can, and do, occur widely throughout the glaciated area. But the model will serve its purpose. In some respects, particularly concerning relief type and shape, it is similar to the model used earlier in the consideration of erosional landscapes (Figure 10.11). However, because of some important differences (notably the common occurrence of transverse landforms related to glacial deposition), the analogy cannot be carried too far. It is probably true to say that when ice has surplus energy (as when it is eroding), almost all forms are streamlined to some extent. When there is an energy deficit, the ice creates an increasingly large proportion of transverse forms from the equilibrium line to the snout.

There is a great deal of useful information in the literature concerning the location of areas where particular depositional landforms predominate (Bird, 1967; Prest et al., 1968; Clayton and Moran, 1974). The various atlas maps of Pleistocene landforms and geology from various countries provide a wealth of data, although there is seldom any attempt to synthesize the information available.

Landscapes from the zone of active ice

In chapter 12 it was noted that many types of morainic landforms have distinctly gregarious habits, indicating that they are born and thrive in large numbers where basal ice conditions are suitable. This clustering of features is apparent in Figures 13.3 and 13.4, and inevitably many authors have referred to Rogen moraine *landscapes*, washboard moraine *landscapes*, and so on. Occasionally, if one views these landscapes at a low level of resolution, it is possible to generalize about their preferential locations, and even to notice situations where one land-scape type grades into another. Most of the useful examples come from North America, the British Isles, and Scandinavia.

Some of the evidence from Canada has been synthesized by Bird (1967), and he and other authors have emphasized the importance of streamlined depositional forms over wide areas. Using evidence from NWT, Prest (1968) showed that a zone of true drumlins (Figure 13.5, p. 264) gives way on its northern margin to a zone of highly elongated drumlinoid forms and then to drumlinized ground moraine, with ridges standing above the general level of the ground moraine surface. In turn this gives way to fluted ground moraine (Figure 13.6, p. 265) in which the ridge crests are at the same level as the adjacent drift surface, and then to a zone where the drift cover becomes more and more spasmodic and features of ice scouring more and more significant. Clearly this is a transition landscape types across the critical margin between the erosional and depositional zones of the Laurentide ice sheet. In reality the situation is more complex than suggested here (Bird, 1967), since the local details of topography and glacial and deglacial history have inevitably led to widespread anomalies (Prest et al., 1968; Bryson et al., 1969). Nevertheless, over much of the area when viewed at a low level of resolution there seems to be little 'contamination' of the ice sheet

Figure 13.3
Rogen moraine ridges interspersed with lakes, near Lake Rogen, Sweden. The direction of ice movement was towards the viewer. (*Photograph by J. Lundqvist.*)

Figure 13.4
Cross-valley moraines from the upper Isortoq valley, adjacent to the Barnes ice cap, Baffin Island. *(Photograph by J. D. Ives.)*

Figure 13.5
A classic drumlin swarm near Cape Krusenstern, NWT. *(Original photograph supplied by National Air Photo Library, Canadian Department of Energy, Mines and Resources.)*

zones of morainic deposition by fluvioglacial material. There is some interesting data on the depositional landforms of Keewatin, which after 7,500 BP supported one of the last remnants of the collapsing Laurentide ice sheet (Lee, 1959; Bird, 1967). The generalized map (Figure 13.7, p. 266) demonstrates the succession of landscape types northwestwards from the axis of the ice sheet remnant, towards the retreating margin. Beyond the zone of extensive bare rock marking the final ice cap axis, is an irregular zone of drumlins and drumlinoid features. Eskers are found throughout the area, but the densest concentrations are, for the most part, restricted to their expected position in a broad zone outside the zone of drumlins (Bird, 1967). Again, in keeping with the idea that the inner parts of an ice sheet may be

Figure 13.6
Strongly fluted ground moraine west of Beverly Lake, NWT. *(Original photograph supplied by National Air Photo Library, Canadian Department of Energy, Mines and Resources.)*

composed of relatively clean ice above the basal layers, wide areas are largely devoid of ablation moraine or ice disintegration features.

Turning attention to the British Isles, it is difficult to restrict discussion to the zone affected predominantly by highly active ice, largely because nobody knows precisely where this zone was (Sparks and West, 1972). The configuration of the area affected during the Weichselian glaciation is so complex, and the pattern of ice cap retreat so irregular, that the models represented in Figures 13.1 and 13.2 are difficult to apply. Nevertheless, it is instructive to examine how some depositional features are distributed.

Drumlin landscapes are quite common within the Weichselian glacial limit (Figure 13.8,

p. 267). They occur for the most part in lowland situations, although examples are known from upland valleys also. The orientations of drumlins in particular swarms are always thought to reflect the directions of regional ice movement at some relatively late stage of the last glaciation (Hill, 1970; Embleton and King, 1975). In the Eden Valley and the Solway area, the ice is thought to have achieved some spectacular acrobatics in order to shape the local drumlin fields (Hollingworth, 1931). It did, however, create more rounded drumlin forms beneath the local ice shed and more elongated forms where ice was more active. The main swarms are located between the areas of greatest glacial scouring and the areas of thickest drift deposits (Sissons, 1964b; 1967). Occasionally swarms are found close to the drift limit of the last glaciation (as in the Welsh borders and part of Yorkshire), but more normally they are located at some distance from the margin where glaciological conditions favoured streamlining processes.

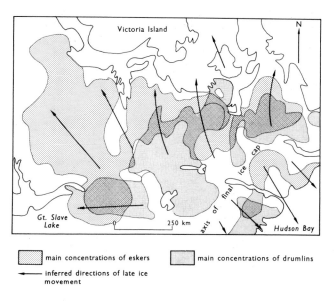

Figure 13.7
Spatial relations of the drumlin zone and esker zone on the northwest flank of the Keewatin ice cap, a remnant of the Laurentide ice sheet. *(Bird, Johns Hopkins Press 1967.)*

The area which is geographically closest to the 'ideal' situation portrayed in Figure 13.1 is Ireland, where ice movement towards the southeast was largely unimpeded by upland massifs. The vast drumlin field (one of the largest in Europe) which stretches across the northern half of the island consists of several different components, and Synge and Stephens (1960), Vernon (1966), and Hill (1971) suggested that these components represent such different directions of ice movement that they were not all formed contemporaneously. Indeed, since the ice cap axis is thought to have migrated during the last glaciation across the drumlin field, and since rapid ice evacuation is not possible beneath an ice shed, the drumlin belt was probably formed in several different phases as the ice shed migrated parallel with the direction of drumlin orientation. It is unproved that the drumlin field was formed entirely during a late 'drumlin readvance' which terminated just to the south of the belt, as suggested by Synge (1970). Perhaps more important for an interpretation of regional landscapes is the very clear manner in which the drumlin field gives way on its southern margin to a wide zone of kames, eskers and mixed glacial deposits. In turn this zone is succeeded in south

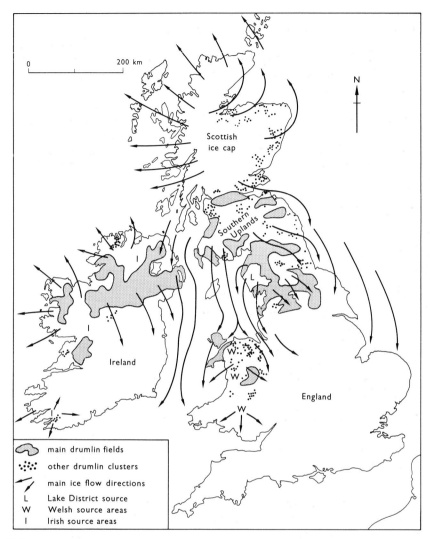

Figure 13.8
Map showing the distribution of the main drumlin fields of the British Isles with respect to the main centres of ice outflow.

central Ireland by a zone of hummocky morainic deposits, kame moraines and other fluvio-glacial landforms supposed by Mitchell (1960), Synge and Stephens (1960) and Synge (1970), to represent the limit of the Irish ice cap during the last glaciation.

   Another lowland area in which there has been very little impediment to ice sheet advance and retreat is Finland, and again it is useful to examine the succession of large scale deposi-tional landscapes encountered on a north–south transect of the country. In Finnish Lap-land, there is an undulating upland affected by complex but weak ice movements beneath and near the ice shed; there are, nevertheless, many depositional features including areas of Rogen moraine (Kujansuu, 1967), fluted till surfaces and drumlins. Much more spectacular drumlin fields are seen further south. Two of these fields have been described by Glückert (1973), and there are other discussions of the distribution of Finnish drumlins by Sauramo (1929), Tanner (1938), Penttilä (1963), and Virkkala (1952). Some idea of the dimensions, density, forms and orientations of drumlins from the Pieksämäki field can be obtained from Figure 13.9. The drumlins themselves are by no means the only drift features located within the drumlin fields, and it can be seen from the map that esker ridges are interspersed among the drumlins (and sometimes riding over them). The eskers follow orientations which are

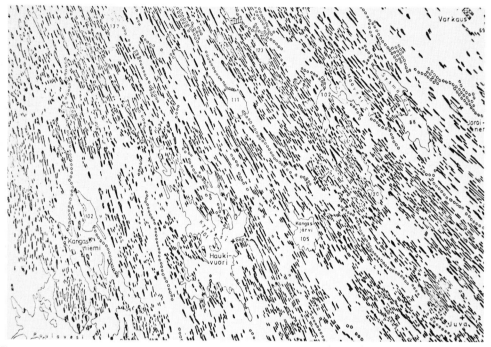

Figure 13.9
Map of part of the Pieksämäki drumlin field, central Finland. *(Glückert, Fennia 1973, copied by permission.)*

generally consistent with the directions of ice streaming inferred from drumlin long axis orientations. Farther to the west, Glückert (1973) has shown that there is occasionally some divergence between drumlin and esker orientations, suggesting that the direction of ice movement during retreat shifted slightly between the time of drumlin streamlining and the time of esker formation. In Finland, as in Ireland, there is a considerable mixing of morainic and fluvioglacial landforms, but it is nevertheless still possible to differentiate between a southern zone of predominantly fluvioglacial deposition (in which eskers are the main landforms), and a northern zone of predominantly streamlined till with drumlins.

Since much of the ice sheet retreat in Finland and Sweden was achieved while the ice margin was floating or grounded in deep water, the overall distribution of depositional landscapes is complicated by the appearance of certain of the morainic types attributed to

grounded ice. The situation of the highest regional shoreline becomes critical in any explanation of the distribution of various drift landscapes. De Geer moraines are commonly
encountered on both shores of the Gulf of Bothnia (Figure 13.10), and often they extend
below present sea-level (Aartolahti, 1972). In many areas they are juxtaposed with eskers,
kames and delta moraines, but very seldom' with drumlins (Fromm, 1949). From a number
of detailed studies in Norrbotten, Hoppe (1948; 1959) has described the groupings of certain
landscapes which include De Geer moraines, kalixpinnmo hills and 'kalottberg' hills which
have a cap of unwashed till and washed lower slopes beneath the level of the highest shoreline.
These various elements seem to have consistent relationships with one another. Although
they are not strictly restricted to areas washed by the sea, the kalixpinnmo hills are generally
located in valley bottoms. In this respect they are similar to the cross-valley moraines and
the 'sublacustrine moraines' referred to in the last chapter (Barnett and Holdsworth, 1974).

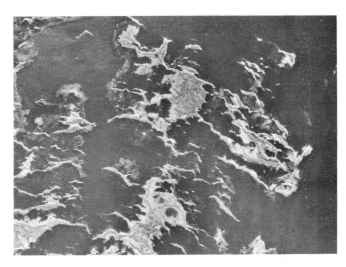

Figure 13.10
De Geer moraines near Årås,
Sweden. *(Reproduced by permis-
sion of Statens Lantmäteriverk,
Sweden.)*

On the flanks of the valleys, the kalixpinnmo hills give way to De Geer moraines, as in the
example shown in Figure 13.11 (overleaf).

Farther inland in Sweden, other types of depositional landscapes appear. Between the
highest shoreline and the Scandinavian mountains there are many more forms which were
attributed in the last chapter to the action of streaming ice; in particular, drumlins and
Rogen moraine. Drumlins occur very commonly in small concentrations, but there are no
extensive fields to match those of Finland. In Norrbotten, drumlin fields occur adjacent to
areas of Veiki moraine, while in other parts of Sweden, just to the east of the uplands, they
occur adjacent to Rogen moraine landscapes (Lundqvist, 1969).

All these forms may be part of a continuum. Using the ideas put forward in the last chapter,
it may be that at one extreme drumlins represent 'ideal' streamlining conditions beneath
active ice. Where shearing of the ice mass occurs, as in conditions of compressive flow in
topographic concavities, Rogen moraine ridges are formed (Figure 13.12, overleaf). Above
the highest shoreline, where ice flow is sluggish and local ice wastage well advanced, various
types of dead-ice moraine are formed. In valley bottoms, ice disintegration features interspersed with fluvioglacial landforms represent the other extreme of the continuum. Below

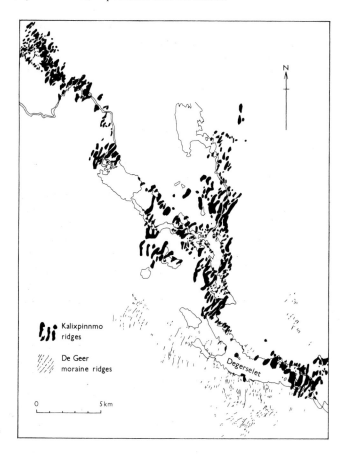

Figure 13.11
Relationships between kalix-pinnmo ridges and De Geer moraines in part of Norrbotten. *(Hoppe, Geogr. Annlr 1948.)*

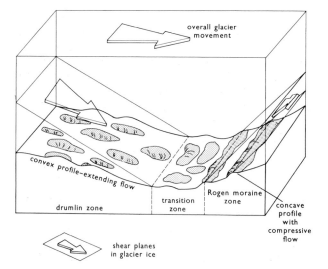

Figure 13.12
Illustration of the topographic relations of drumlins and Rogen moraine ridges in Jämtland. Note the correlation of the drumlin zone, transition zone and Rogen moraine zone with different dynamic conditions at the glacier bed. See also Figure 12.8.

the highest shoreline, there are significantly different landform associations, with De Geer moraine and kalixpinnmo hills more characteristic, and with well developed deltas, strand-lines, eskers and spreads of varied glacio-marine sediments in favourable locations (Lind-ström, 1973). Clearly, over much of Scandinavia it is extremely difficult to differentiate between the active zone and the wastage zone of the last ice sheet.

Landscapes from the zone of wasting ice

The drift landscapes close to the peripheries of large glaciers are highly confused, since the range of processes involved in till deposition and in the creation of moraines are highly vari-able (chapters 11 and 12). In the British Isles, for example, the outer zone of glaciation is, for the most part, a zone of subdued drift landforms and thick drifts. The most complete sequence of drifts is known from East Anglia, where there may be glacial deposits dating from three or more distinct glaciations (Mitchell *et al.*, 1973). Two of the tills are differenti-ated on the basis of their fabrics (West and Donner, 1956), weathering characteristics and degree of denudation by subaerial processes, and by their relations with interglacial deposits. However, it is fair to say that the margins of these drift sheets are not established with any certainty, and there is doubt about the real stratigraphic significance of several nonglacial horizons also (Bristow and Cox, 1973). Nevertheless, the drifts are undoubtedly thick across most of this zone, and coastal cliff exposures and borings in Yorkshire, Lincolnshire and Norfolk show that the drift sequence is often more than 100 m thick and extremely variable in its lithology and stratigraphy. Morainic landforms older than the Weichselian are not common, but there are occasionally traces of ancient end moraines, and the Saalian end moraine near Buckingham is a spectacular feature in spite of its age. On the western flank of Great Britain the largest of all the British glaciers, the Irish Sea glacier, extended as far as the Celtic Sea on at least two occasions. However, most of its terminal zone deposits are now submerged beneath sea-level, and on land the few discernible morainic features can all be related to the Weichselian stage. On both sides of St George's Channel, in north Pembroke-shire and southeast County Wexford, there are areas of well developed dead ice topography. Both areas have stratified sands and gravels and a mixed ablation till overlying a character-istic calcareous lodgement till referred to as the 'Irish Sea till'. These deposits all date from the same glacial stage, but overall it is the fluvioglacial landforms, rather than the true morainic forms, which are the most impressive features of the landscape (John, 1970a).

End moraine zones which are much more spectacular occur at the mouths of certain valleys leading from the Alps. Here glacial discharge from the Alpine ice cap at its maximum stage was achieved largely through confined troughs. The glaciers were fast flowing and heavily charged with detritus, and some of them created vast end moraine complexes which have impounded elongated lakes. Lake Garda, with its 'dam' of morainic materials looping out onto the Po plain, is a famous example (Figure 13.13, overleaf).

The southern marginal zone of the Scandinavian ice sheet is characterized by a much broader belt of spectacular drift landscapes. In this zone, there are four well known terminal moraines indicating the terminal positions of successive Scandinavian ice sheets all within 200 km or so of one another. As in North America, the reasons for this very close coincidence are not entirely clear, but repeated invasions by ice and repeated stagnation in the same narrow strip of land in Denmark and along the north coasts of Germany and Poland have resulted in a highly compressed band of morainic ridges, fluvioglacial landforms, urström-taler, old lake depressions and other landforms of various ages (Woldstedt, 1969; Hansen,

1965). In places, the drift sequence is over 400 m thick, and the mean drift thickness in north Germany is estimated to be 48 m. Within the sequence there are push structures and a great variety of other glacio-tectonic features related to active ice movement, and also extensive areas of dead ice topography. Nieriarowski (1963) has shown some of these zones in Poland to have been 250 km wide.

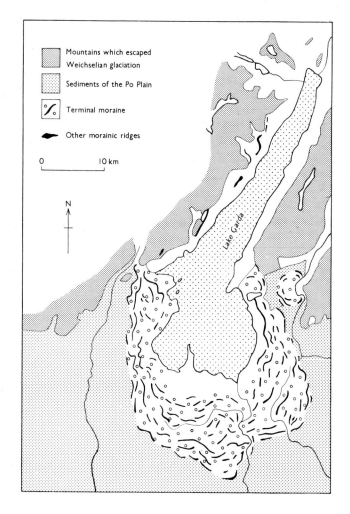

Figure 13.13
The end moraine belt at the southern end of Lake Garda, Italy.

Unfortunately much of the zone immediately inside the southern morainic belt of the Scandinavian ice sheet is submerged beneath the waters of the Baltic. It is, therefore, difficult to discern any 'characteristic' associations of depositional landforms. Furthermore, the whole history of Weichselian deglaciation in southern Scandinavia was so profoundly influenced by the various Baltic water-bodies (Lundqvist, 1965; Donner, 1965) that the mode of glacial deposition was, since at least 12,000 BP, largely glacio-marine and glacio-lacustrine. Nevertheless, in parts of south central Sweden there is a mixture of lodgement till features, eskers and other fluvioglacial landforms, and ice disintegration forms (Nilsson, 1960). It is difficult to see any relationship between these assemblages and the model outlined above.

The situation with respect to the south marginal zone of ice sheet glaciation in North America is much clearer, since the extent of ice dammed water-bodies during deglaciation was considerably less than in Scandinavia. The distribution of the main end moraines and drift sheets is adequately discussed by Flint (1971). However, the close coincidence of the end moraines of the various glaciations (mentioned in chapter 7) is again worthy of note,

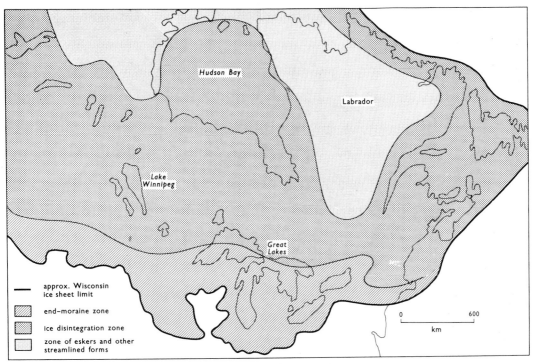

Figure 13.14
Generalized map of the three depositional zones near the southern margin of the Laurentide ice sheet. The outermost zones are the end moraine/till zone and the ice disintegration zone. *(Flint, 1971 by permission of John Wiley and Prest et al. 1968.)*

as is the thickness of drift in this zone. In parts of Idaho, Washington and New York, drifts are more than 300 m thick, whereas the average thickness in Iowa is estimated to be 45–60 m, and the average for Illinois 35 m. These figures compare with average thicknesses of only 2–3 m for central Quebec–Labrador, an area representative of the active zone in Figure 13.1. The landscapes represented through this zone are highly variable. Occasionally, as in the area between Chicago and St Louis, there are prominent end moraines both of Wisconsin and earlier age. Elsewhere the drift landscape is subdued, with drift sheets ending unspectacularly in 'attenuated drift borders'. There are some areas of dead-ice topography, and meltwater channels cut through moraine are locally important also. Fluvioglacial deposits are widespread, and shorelines and other features relating to the late Wisconsin

Figure 13.15
A good example of a landscape of disintegration features in NWT, Canada. Most of the moraines here are devoid of any clear orientation. *(Original photograph supplied by National Air Photo Library, Canadian Department of Energy, Mines and Resources.)*

pro-glacial lakes (including Lake Agassiz, Lake Chicago and Lake Whittlesey) occasionally noticeable. Further to the east there are prominent end moraines, including spectacular push structures at Martha's Vineyard.

Flint (1971) makes some important observations concerning the arrangement of drifts and landforms in and within the Laurentide ice sheet's terminal zone (Figure 13.14). He suggests that there is an irregular outer belt 300–500 km wide, where drift is predominantly clay-rich till in the form of ground moraine and large numbers of concentrically arranged end moraines. These moraines are for the most part recessional moraines, although as noted in chapter 7 some of them date from the readvances which occurred intermittently on the southern periphery of the ice sheet. Inside this belt lies a wide zone with further, less numerous, end moraines and subdued ground moraine features but more common ice disintegration features (Figure 13.15). Some of the areas where these features are best developed were described by Flint (1929) in Connecticut, Gravenor and Kupsch (1955) in western Canada, Elson (1957) in Manitoba, Stalker (1960) in Alberta, and Winters (1961) in North Dakota. Ice disintegration forms, including some of the ice pressed and transverse linear features referred to in the last chapter, are thus common across the northern Great Plains, in much of New England, and in the Appalachian region. The distribution of these forms is partly related to conditions of sediment supply and local topography, but Flint also suggested that increased climatic amelioration after the formation of the lake border moraines played an important role. Whatever the real reasons, the two broad zones of (a) end moraine creation, and (b) ice disintegration seem to be well established. Inside the latter of these zones, as mentioned by Flint, 'great systems of eskers' become more and more prominent; these gradually give way to the real streamlined forms such as drumlins (Craig and Fyles, 1960).

Towards a model of drift landscape types

Having looked at some examples of landform associations and drift distributions at a variety of scales, it is possible to look again at the simple models presented in Figures 13.1 and 13.2. Although these models are highly idealized, they are by no means far removed from reality, and it is now possible to add the further model shown in Figure 13.16, using material published by Prest (1968), Flint (1971) and other authors. It has been shown from Ireland,

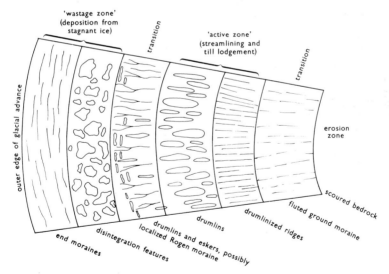

Figure 13.16
A model of the types of depositional landscapes expected to develop beneath part of the periphery of a mid-latitude Pleistocene ice sheet at, or just after, the phase of maximum glaciation. The same sequence may also be expected beneath an ice cap periphery. Not all of these landscapes will develop contemporaneously, and the landscape type is determined above all by the character of the ice above a site when rapid dissolution sets in.

North America and Finland, for example, that in spite of the climatic fluctuations which interrupted the last deglaciation, and in spite of the constraints of local topography, there are areas where the models appear to be more or less directly applicable. In other areas, where ice sheet, ice cap and valley glaciations have been juxtaposed, there are complications.

But from the evidence in Norrland, it seems that the models could be adapted to explain local depositional landscapes, especially if the variable of sea-level or lake-level control over deglaciation is introduced. Also, it is surprising that parts of Canada, affected so drastically by the vast water-bodies which followed in the wake of the retreating Laurentide ice sheet margin, still allow broad landscape subdivisions to be made on the basis of glacier dynamics. For example, the subdivision of the drift landscapes of the Canadian North-West Territories based on the active ice/stagnant ice criteria still holds, in spite of the influence which various ice edge stillstands and readvances may have had in the destruction of old landforms and the construction of new. It seems to be a point of fundamental importance, therefore, that at a large scale the distribution of landscapes of glacial deposition is determined by

(a) the glaciological conditions which prevailed at or near the maximum of a glaciation, and

(b) the location of a region with respect to the ice sheet axis or centre.

This seems to hold true for medium scale and local scale depositional landscapes also, but here the controlling ice body may be a local ice cap of later date (perhaps representing a late-glacial readvance) rather than the full-glacial ice sheet. This is supported by the fact that in some regions affected by shortlived readvances, drumlin and other morainic orientations are controlled by the direction of the last powerful ice flow (Bird, 1967).

These hypotheses of landscape zonation will continue to be tested by the geographical techniques of spatial analysis. With improved maps, air photographs and ERTS imagery available, the discernment of large scale depositional zones has become easier. Also, it is now possible to apply glaciological criteria to the analysis of depositional landscapes, providing valuable supplementary evidence to that already collected in the fields of glacial stratigraphy and geomorphology.

Further reading

BIRD, J. B. 1967: *The physiography of Arctic Canada*. Johns Hopkins Press, Baltimore (336 pp).

GRAVENOR, C. P. and KUPSCH, W. O. 1959: Ice disintegration features in western Canada, *J. Geol.* **67,** 48–64.

HOPPE, G. 1959: Glacial morphology and inland ice recession in north Sweden. *Geogr. Annlr* **41,** 193–212.

MANNERFELT, C. M. 1945: Några glacialmorfologiska formelement. *Geogr. Annlr* **27,** 1–239.

PREST, V. K. 1970: Quaternary geology of Canada. In *Geology and Economic Minerals of Canada*. Economic Geology Report No. 1. Geological Survey of Canada. Department of Energy, Mines and Resources, Ottawa, 676–764.

WOLDSTEDT, P. 1969: Quartär. *Handbuch der stratigraphischen Geologie* **2,** Stuttgart (263 pp).

Part V
Meltwater: a glacial subsystem

14 Meltwater as part of a glacier

Meltwater can be envisaged simply as that part of a glacier above a certain temperature threshold. The difference between the solid and liquid states of water above and below its freezing point is reflected in the dramatic contrast between the process/form systems associated with glacier ice on the one hand and glacier meltwater on the other. Meltwater is an integral part of most glacier systems and fulfills several roles. It is the main ablation product of most glaciers. It is intimately involved in the movement of glacier ice, through its influence on creep, regelation at a glacier base and drag between glacier ice and bedrock. Also it is responsible for the throughput of quantities of rock debris.

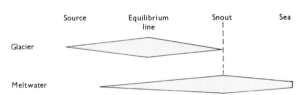

Figure 14.1
A comparison of the relative importance of glacier and meltwater activity along the length of a glacier.

Figure 14.1 is an attempt to show the relative significance of meltwater along the length of a glacier. Meltwater increases in quantity from above the equilibrium line, where it may originate both from surface and bottom melting, as far as the glacier snout where it is at a maximum. Thus, as the glacier system loses its capacity for geomorphological work towards the snout, so the strength of the overlapping fluvioglacial system is built up, and a greater and greater proportion of erosive work may be achieved by fluvioglacial, rather than glacial, processes. Beyond the snout there is loss of volume by evaporation and ground water absorption as the meltwater makes its way to the sea.

Input

Two main sources of meltwater can be recognized – (1) surface and (2) basal and internal. Each main source can be subdivided further into a number of distinctive components. Before investigating these in any detail, it is helpful to highlight some fundamental contrasts between meltwater of surface origin and that of mainly basal origin. It should be borne in mind that surface supplies exceed basal supplies by at least an order of magnitude on most glaciers (Shreve, 1972). Furthermore, surface supplies are seasonal and reach a peak usually in the later part of the summer, whereas basal supplies fluctuate less markedly and commonly exist throughout the year.

Surface sources
Surface melting is the most important meltwater source and accounts for by far the greatest amount of total glacier ablation. The well known decline in meltwater production with

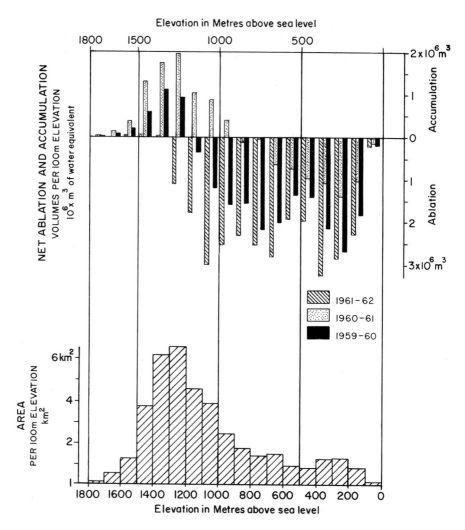

Figure 14.2
The decrease of net ablation with altitude on White glacier, Axel Heiberg Island for three years. Amounts of meltwater produced on the glacier are closely influenced by the altitudinal distribution of the glacier. *(Müller, Axel Heiberg Is. Prelim. Rept. 1963.)*

increasing altitude is illustrated by ablation data for three years obtained for White glacier, Axel Heiberg Island (Figure 14.2). This reduction of melting with increased altitude, called the summer balance gradient or the ablation gradient, is steeper in maritime environments and gentler in continental environments (Schytt, 1967). For example, the maritime Blue glacier, Washington has an ablation gradient of 10 mm per m (LaChapelle, 1965) whereas Decade glacier, Baffin Island has a gradient of only 1·1 mm per m (Schytt, 1967). Although many other factors are involved, the relative importance of the supply of meltwater along the length of a glacier and the total quantity depends greatly on the altitudinal distribution of the glacier, as can be seen from Figure 14.2. Suppose that there are two adjacent glaciers of comparable size and altitudinal range, but one has most of its area at low altitude and the other at high altitude; not only will the former produce more meltwater than the latter but the amount will increase more rapidly towards the snout than in the case of the latter.

Another source of glacial meltwater is rain which falls on most glacier ablation areas, especially in the summer months. The rain itself has little effect in actually melting ice which is already at 0°C, but it may drain off the glacier along surface and marginal meltwater routes and thus contribute to overall meltwater flow. It is an especially important component in maritime environments (Pytte, 1970; Gudmundsson and Sigbjarnarson, 1972). Rain early in the summer may freeze on the glacier surface or be retained in the pores of the firn in an unfrozen state and thus have its runoff delayed. Late in the summer when the glacier surface consists of much bare ice, the rain runs off almost instantly.

On all valley glaciers, glacial meltwater incorporates water draining into the glacier from adjacent valley sides. Such water flow will vary according to the characteristics of the periglacial environment, such as the amount and nature of snow melt, the extent and effect of frozen ground, summer rainfall, bedrock permeability, vegetation cover and slope steepness.

Basal and internal sources

When a glacier base is at the pressure melting point, geothermal heat is used to melt a layer of ice – about 6 mm in thickness beneath the glacier each year (chapter 2). Confirmation of the reality of the process has come from measurements beneath alpine glaciers where bedrock temperatures are above freezing point (1·5°C) and a film of meltwater underlies the glacier (Vivian, 1970). It is to be expected that there will be variations in the amount melted depending on spatial variations in the geothermal heat flux (Table 2.2).

Internal movement of the glacier and glacier sliding together produce heat which will contribute to the production of meltwater if the glacier base is at the pressure melting point. For the reasons explained in chapter 2, most of this meltwater will be produced at or near the glacier base. Since movement of 20 m per year at the stresses common in glaciers releases approximately the same amount of heat as obtained from the geothermal heat flux, one can suggest that most warm-based glaciers may produce enough heat by movement to melt the equivalent of a layer of ice approximately 6–60 mm thick each year.

Meltwater flowing as a stream within or beneath a glacier at the pressure melting point can melt ice surfaces with which it is in contact. The melting may be due to the frictional heat released by the flow of meltwater itself (Röthlisberger, 1972; Shreve, 1972) or to the heat carried down from the surface by relatively warm surface water (Vivian, 1970; Röthlisberger, 1972). The total melting achieved within a glacier in these ways is probably limited. However, as will be seen later, it assumes great significance in the creation of internal drainage channels.

Just as ground water contributes to the flow of a river, it may flow to the bedrock surface beneath a glacier and contribute to the overall meltwater flow. The amount will vary from glacier to glacier depending on local topographic and geological conditions. It may be a major or even dominant component of meltwater flow in winter in numbers of alpine glaciers when overall discharge rates are low (Stenborg, 1965).

Contrasts between glaciers

After considering the characteristics of the various sources of water contributing to glacial meltwater streams, it is clear that one can expect marked contrasts between glaciers. One of the most important causes of variation is the climatic environment. Total amounts of meltwater production are highest in temperate maritime environments, and they decline towards high and low latitudes and continental interiors. As mentioned in chapter 3, this is largely a reflection of the magnitude of the turnover of ice and the need for melting to balance accumulation. However, it is also a reflection of the increasing importance of evaporation which accounts for some ablation, especially at high altitudes and in high latitudes. The seasonality of meltwater production also varies with latitude. In equatorial latitudes, meltwater is produced almost daily throughout the year. From here, seasonal variations become progressively more marked until in polar environments meltwater may be produced for only a few weeks each year. Important spatial variations in meltwater supply may occur depending on the thermal characteristics of the ice. Whereas surface melting dominates in ablation areas, basal melting can only take place where the basal ice is at the pressure melting point. In polar latitudes this means that basal meltwater tends to occur only beneath thick ice or ice nourished in a maritime climate.

The physical characteristics of a glacier can induce variations in meltwater supply. There are obvious differences between ice sheets and valley glaciers. Whereas water from surrounding unglaciated slopes is an important component of valley glacier meltwater, it is unimportant in association with ice sheets. Also rates of ablation tend to be higher on valley glaciers where rock walls enhance melting and where surface moraines may decrease the albedo of the glacier. Perhaps the most important contrast is between glaciers terminating on the land and those calving into the sea. In the case of the former, most of the output is achieved by conversion of ice to meltwater. Thus in those parts of west Greenland where the ice sheet terminates on land, there is massive meltwater activity on and beneath the ice sheet margin In the Antarctic on the other hand, meltwater activity seems restricted to ice in the vicinity of rock outcrops, especially at low altitude such as the Molodezhnaya oasis (Klokov, 1973) and Victoria Land (Davis and Nichols, 1968). Figure 5.13 is another example which shows meltwater activity on the ice sheet surface near the Schirmacher Ponds oasis.

Routes

A fundamental distinction exists between discharge routes which run along the surface of a glacier and those which flow within or beneath the glacier. Few field observations exist about either type. This reflects the relative inaccessibility of the sub-surface routes, as well as the fact that the impermanence of all routes makes it difficult to document their characteristics with any precision. Recently, however, understanding has vastly improved. In part, this results from physical contact with the ice bed in association with HEP tunnels in Europe, and also drill holes penetrating glaciers to their beds (Kirchner, 1963; Mathews, 1964; Weertman, 1970); in part it arises from continuing developments in the theory of meltwater flow (Weertman, 1972; Röthlisberger, 1972; Shreve, 1972). The time now seems ripe for linking these results with the vast amount of information gathered over the years by glacial geomorphologists.

Surface streams

Surface streams are common in the ablation areas of glaciers (Figure 14.3). They form when surface melting produces more water than can be absorbed into the glacier. Typically they begin to flow when there is a good deal of slush on the glacier. After a phase of rapid discharge early in the season as water temporarily stored in the snow runs off the glacier, the discharge responds to climatic conditions influencing ablation and precipitation.

Figure 14.3
Surface meltwater patterns near the snout of Iceberg glacier, Axel Heiberg Island. *(Photograph by F. Müller, Axel Heiberg Research Expedition and Swiss Federal Institute of Technology.)*

Typical surface channels vary from a few millimetres to a few metres in depth (Figure 14.4, overleaf). The sides of the channel are smooth and offer less resistance to water flow than normal river channels, as anyone who has slipped into such a stream and been carried swiftly down-glacier will testify. The channel may or may not be in a valley in the surface of the glacier. When walking across the Sukkertoppen ice cap in west Greenland, it is often impossible to pick out the location of a stream until one stumbles on to it, for it is incised into the glacier surface like an artificial ditch. Elsewhere, as for example on the Lyngbrae outlet glacier of the Sukkertopen ice cap, the stream channels may occupy distinct river valleys in the ice. Presumably in the former case, ablation of the surrounding ice surface

Figure 14.4
Above, a shallow surface melt-
    water stream on Britannia
    glacier, northeast Greenland.
    *(Scott Polar Research Institute,
    British North Greenland Col-
    lection.)*
Below, surface meltwater stream
    flowing in a steep gorge down
    the edge of the Greenland ice
    sheet near Camp Tuto. *(Photo-
    graph by Valter Schytt.)*

proceeds as fast as any melting achieved in the stream channel. In the second case melting in the stream channel exceeds the ablation rates on the surrounding ice.

As has long been realized, surface meltwater streams frequently meander (Figure 14.5). Knighton (1972) found that on Østerdalsisen the meanders on one stream were similar in general form to those developed in alluvial valleys even though they carry virtually no sediment. As is to be expected in a uniform medium like ice, there was greater symmetry in terms of the relationship of wavelength to curve amplitude. Features related to irregularities in the substrate, such as cut-offs, were not observed. However, not all surface streams meander, and structural weaknesses such as former cracks or closed crevasses are often selectively used by streams. In places of strong structural interference, rectilinear stream channels dominate.

Figure 14.5
A small channel, *c.* 0·5 m wide and 2 m deep, cut by a meandering meltwater stream on the surface of Kaldalon glacier, northwest Iceland.

Stream patterns may evolve over areas of the order of several hundred square kilometres and include individual streams several kilometres in length, for example those on the peripheries of the Greenland ice sheet (Figure 14.6). As in other parts of the world the streams tend to assume a dendritic network and, indeed, they have been used as general models of drainage basin evolution (Leopold *et al.*, 1964; Haggett and Chorley, 1969). However, any analogy must be treated with care since there are a number of differences which seem apparent, at least qualitatively. (1) The network is often dense and rill-like. In part this may reflect the impossibility of cutting a major valley system into a moving dynamic mass like ice which tends to adjust its flow so as to obliterate major irregularities. This means that large integrated drainage patterns are rare. Another factor concerns the rapid rates of runoff which demand a dense network in order to drain the water provided. (2) There is a strong sub-parallel element in the drainage pattern. Probably this arises from the fact that in most ablation areas the glacier surface is relatively steeply sloping towards the margin, and the sub-parallelism reflects this control. (3) There is a decrease in stream density upglacier. Two factors seem relevant here. In the first place and in contrast to most other rivers, the amount of water to be discharged decreases with elevation since ablation rates are lower at higher altitudes. Also, unlike most river valleys, overall slope angles decrease upglacier away from the glacier margin. (4) The channel pattern is highly changeable. Not only may moulins truncate a stream, but the channels move with the ice (often differentially), and superimposed ice may block existing channels at the end of an ablation season.

Individual stream networks, like their 'normal' counterparts, are influenced by the overall relief, structure and lithology of the surface over which they flow. On an ice sheet, the overall relief favours a regular but slightly radial outflow downglacier towards the margin. On a

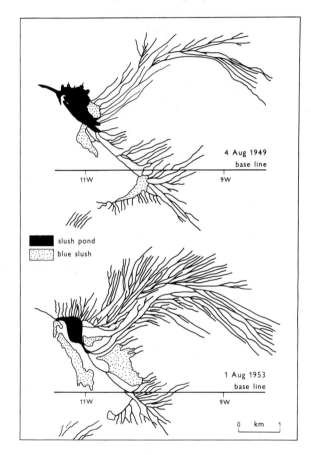

Figure 14.6
Surface meltwater stream patterns on the Greenland ice sheet. (Holmes, U.S. Air Univ. Pub. 1955.)

valley glacier there is also divergent pattern towards the lateral margins which are lower than the centre of the glacier.

Ice structure is important especially in truncating surface streams by allowing access into the glacier (Figure 14.7). Stenborg (1968) notes how tensional crevasses created by the friction at a glacier's valley side provide not only surface routes tending to divert surface streams laterally but also the sites of moulins. On Mikkaglaciären, the crevasses are suitable for use by moulins for about 6 years before ice movement carries them past the area of crevassing. Compressive ice flow conditions are suitable for extensive stream development since crevasses are unlikely to occur in such circumstances. Indeed, mountaineers can find large streams

Figure 14.7
Surface meltwater stream disappearing into a moulin, Britannia Glacier, northeast Greenland. *(Scott Polar Research Institute, British North Greenland Collection.)*

useful as navigation aids on glaciers since they normally indicate the absence of crevasses for some kilometres upglacier.

Ice temperature is analogous to lithology, in that cold ice can be equated with impermeable rock and warm ice with permeable rock. Cold ice is highly favourable for surface flow and inhibits moulin development since meltwater tends to freeze if it descends into the glacier. If a glacier is wholly comprised of cold ice, it is unlikely to have any streams flowing into and through the ice. However, under certain conditions crevasses may allow surface melt-water to penetrate a surface layer of cold ice and link up with meltwater flowing in warmer basal layers of the glacier, as recorded on White glacier by Iken (1972). On almost all glaciers there is a cold surface layer at the surface for part of the year and at least for the early part of the ablation season. This is one reason why surface drainage is more common early in the season than later on when internal passages are exploited.

Internal and basal routes: karst analogy
There seem good grounds for regarding meltwater in a glacier as analogous to water flow in karst (Shreve, 1972). The analogy seems especially helpful in glaciers at the pressure melt-ing point. There is an analogy with primary and secondary permeability of karst, the former reflecting the porosity of the rock and the latter the permeability of the rock in terms of its joints. Nye and Frank (1973) have argued that ice at the pressure melting point is per-meable to water which moves through a three dimensional network of veins surrounding in-dividual ice grains. They argued that water can move through the ice sufficiently effectively to flush out impurities in the ice. Secondary permeability is provided by tunnels which, as in limestone, are likely to exploit structural weakness in the ice (Stenborg, 1968). The tunnels in ice are developed by melting which is analogous to solution in limestone. Tunnel cross-

sections in both mediums vary according to whether or not they form under hydrostatic pressure. Surface streams frequently drain into sink holes or moulins, which like their karstic counterparts exploit vertical weaknesses (Stenborg, 1968; Iken, 1972; Dewart, 1966).

In addition, there are now different lines of evidence which suggest that water tables can exist in glaciers, as in karst areas. Fluctuations in water pressure in a regular and predictable way, often in a wide variety of locations, are difficult to explain except in terms of a fluctuating water table (Vivian, 1970). Regular fluctuations in the water surface in a wide variety of different moulins, as noted for example by Iken (1972), are most easily interpreted in a similar way. Furthermore the sedimentary and morphological characteristics of glacial deposits have caused glacial geomorphologists to postulate water tables in glaciers, as for example in Scotland (Sissons, 1967). The water table in a glacier is thought to fluctuate sharply according to the supply of meltwater and the efficiency of drainage routeways, but generally lies some way below the surface (Röthlisberger, 1972). This means that, as in karst, there is an upper

Figure 14.8
Water under pressure issuing from beneath the snout of Hoffelsjö-kull, Iceland. *(Photograph by Sigurður Thorarinsson.)*

vadose zone where water movement is essentially downward under the influence of gravity, and a lower phreatic zone where meltwater flows under hydrostatic pressure (Figure 14.8). The energy supplied to the hydrostatic system derives essentially from the slope of the water table. Subglacial water pressures have been measured beneath the Glacier d'Argentière. In winter values attain relatively steady values of 9–10 bars, while in summer they fluctuate diurnally with a range of 3–6 bars and occasionally reach values of 11–12 bars in the afternoon (Vivian, 1970). On occasions the water table may rise to the glacier surface. Under such circumstances, water may issue from surface holes as springs or more spectacularly as water spouts, as for example described by Wyllie (1965), Rucklidge (1956) and Baranowski (1973).

The most important arteries of water movement within and beneath a glacier are tunnels (Figure 14.9). At present, the reasons for their development and the pattern they assume are imperfectly understood. However, there are several good theoretical reasons why streams of meltwater are able to maintain channels in ice. One, sometimes called the Glen mechanism (Glen, 1954), reflects the greater density of water than ice and the deformability of the latter under stress. Glen notes that a column of water extending *c.* 150 m down from the glacier surface exerts sufficient pressure on the ice at the bottom physically to deform the ice and expand the hole, although rates of deformation are slow. It is likely that under certain hydrostatic conditions, Glen's mechanism may be important in maintaining and developing tunnels. Another more rapid process has been mentioned earlier where heat, carried from the surface and generated by the friction of water flow, may melt the tunnel walls. Calculations and observations of the catastrophic drainage of glacier-dammed lakes suggest that this mechanism can increase the size of a tunnel materially in a matter of hours.

Until recently it has been difficult to see how either of these mechanisms could create a long tunnel. The Glen mechanism is too slow to be effective and melting can only take place when water is already flowing in a tunnel. However, Shreve (1972) suggests that the tunnels form by water flowing through the ice mass along the grain intersections. Bigger passages will tend to get bigger at the expense of the smaller ones, largely because more heat relative to wall area is carried by water in the larger passages than in the smaller ones.

Figure 14.9
The exit of a subglacial stream tunnel in glacier ice in Spitsbergen. *(Photograph by A. Kosiba.)*

Eventually this process of selective development leads to the development of a dendritic pattern of tunnels. Eventually it is likely that water in tunnels will find its way to the bottom of a glacier. The operation of Glen's mechanism below the water table will tend to cause water to move downwards, other variables being equal. Such a view is accepted by Röthlisberger (1972) who also makes the observation that most streams issue from the bottom of a glacier and carry large quantities of debris likely to have been derived from flow along the glacier bed higher up the glacier.

Tunnels forming in the upper vadose part of a glacier above the water table reflect the downward passage of water under the direct influence of gravity. Moulins are vertical or steeply sloping tubes usually 0·5 to 1 m in diameter (Dewart, 1966; Iken, 1972) but occasionally larger. On Mikkaglaciären and Storglaciären, Stenborg (1968) found that most were 25–30 m deep (Figure 14.10, overleaf) and had abrupt lower limits where they led into presumed horizontal stretches of tunnel. The vertical component of the moulin is often slightly stepped. As in the vadose zone in karst, air is free to circulate in moulins and is probably partly responsible for some melting of passages above the water table. Vivian and Bocquet (1973) recorded how air circulates freely at depth beneath the Glacier d'Argentière and they considered that some is brought down by meltwater.

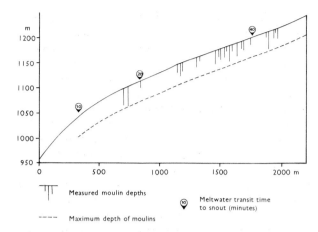

Figure 14.10
Longitudinal section on the tongue of Mikkaglaciären showing position and depths of moulins. Meltwater transit times were calculated by measuring the time taken for salts dropped in a moulin to reach the glacier snout. *(Stenborg, Geogr. Annlr 1969.)*

⊓⊓ Measured moulin depths

⊽ (10) Meltwater transit time to snout (minutes)

- - - - Maximum depth of moulins

In the phreatic zone of a glacier below the water table, meltwater flows under hydrostatic pressure. It is likely that the tunnels form some equilibrium between the rate of melting and the rate of closure of the tunnel by ice deformation. If the meltwater flow decreases, allowing water pressure to fall below that of the ice, the tunnel will close. If it increases, the higher velocities will produce more frictional heat and tend to enlarge the passage. There is never complete equilibrium, since closure probably adjusts over periods of weeks or months (depending on ice thickness) whereas melting can adjust tunnel dimensions in a matter of hours. However, it is probably fair to assume some tendency towards equilibrium at least over time spans of weeks. This is particularly likely to occur at depth where closure will occur most rapidly.

There are few records of the cross-sectional shapes of stream tunnels. Figure 14.11 is a

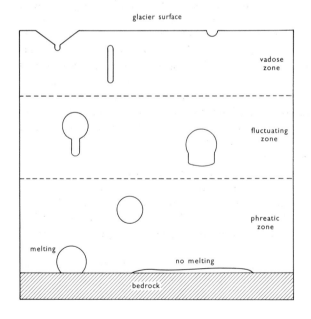

Figure 14.11
A model showing the relationship of contrasting tunnel cross-sections to different zones in a warm glacier.

speculative model based on the analogy with tunnels in karst. In the vadose zone, cross-sections tend to be narrow and deep because the stream occupies and melts only the floor of the tunnel. In ice in the phreatic zone below the water table, the cross-sectional shape is probably a circle, which is the equilibrium shape in the hydrostatic medium of ice (Nye, 1965c). Certainly this view is confirmed by the tunnel cross-sections which can sometimes be seen in freshly carved icebergs in the fjords of east Greenland. In a subglacial position in the phreatic zone there may be two cross-sectional shapes depending on whether or not there is melting of the tunnel walls (Shreve, 1972). In the zone of fluctuating water table one might expect keyhole cross-sections analogous to those in karst. In this situation periods of flow under vadose conditions have the effect of deepening the tunnel cross profile.

It has been suggested that sub- and englacial drainage patterns will tend to form three dimensional dendritic tunnel patterns with the master tunnel at the base of the glacier (Shreve, 1972). Assuming that there is equilibrium between ice and water pressures below the water table, the movement of water will be controlled by the ambient pressures in the ice. It is important to emphasize that it is not hydrostatic pressure which causes the water to flow, but the gradient of the excess of pressure over hydrostatic. The gradient is of the potential Φ which is given by the equation

$$\Phi = \Phi_0 + p_w + \rho_w g z$$

where Φ_0 is an arbitrary constant, p_w is the water pressure in the tunnel, ρ_w is the density of water, g the acceleration due to gravity and z is the elevation of the point considered (Shreve, 1972). In a glacier this means that the direction of flow will be influenced primarily by the direction of slope of the ice surface and secondarily by the shape of the underlying topography. Under an ice sheet or ice cap the subglacial flow is approximately parallel to ice flow but there are deviations in sympathy with the underlying topography as the water avoids high pressure areas associated with bumps and hills and flows by means of cols and valleys. This agrees with the conclusions of glacial geomorphologists and geologists (Flint, 1971). Sollid (1963/64), for example, introduced the term 'ice-directed' to describe features related to such meltwater flow. In a valley glacier, patterns of meltwater flow are more variable. The fact that any valley bed is lower in the middle than at the sides tends to favour the development of a central stream in mid valley at the deepest point. Tributaries will tend to move towards the trunk stream especially if the glacier cross-profile is concave upward (Figure 14.12, overleaf). However, in the snout area the situation may differ. Here the convex cross profile of the glacier surface tends to exert a lateral pressure on subglacial meltwater and, if the valley floor is relatively wide and gentle, may favour subglacial streams in marginal situations, as confirmed by Stenborg (1969). The steeper the slope of the glacier sides and the gentler the slope of the valley sides the stronger this tendency is.

Pressure differences beneath the ice may induce an anastomosing flow of meltwater resembling a braided stream. Meltwater may be diverted either side of a rock obstacle by the diverging flow of ice. It is notable that anastomosing rock channels attributed to subglacial flow have long been recognized in glacial geomorphology, for example by Sissons (1963), as well as being observed beneath glaciers by adventurous speleological expeditions (Figure 14.13, overleaf).

Considerable problems remain concerning the size and permanence of subglacial tunnels. Weertman (1972) suggested that any rock channel slightly oblique to the direction of ice flow will tend to be an area of lower pressure and thus will be a permanent and efficient

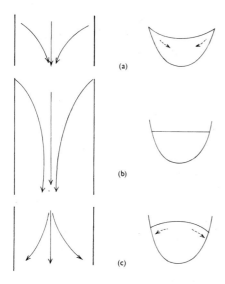

Figure 14.12
Theoretical subglacial meltwater routes in a valley glacier. The bedrock topography favours flow in mid valley **(b)**. Varying transverse ice surface profiles fortify **(a)**, or counteract this tendency **(c)**. *(Weertman, Reviews Geophys. and Space Phys. 1972.)*

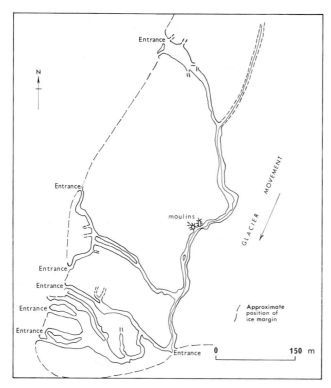

Figure 14.13
A subglacial tunnel pattern beneath Stevens glacier, Mount Rainier, Washington. The tunnel occupied by the stream is the trunk passage. *(Halliday and Andersen, Studies in Speleology 1970.)*

means of drainage. A distinctive tunnel in the ice will be obviously less permanent but may survive the rigours of passage across an irregular rocky bed and survive for many years. On the other hand if it is small (of the order of 0·5 m diameter or smaller) it might be obliterated as the ice passes over and round an obstacle by enhanced basal creep and/or pressure melting.

Sheets, lakes and pockets

Subglacial sheets or films of water have been postulated beneath glaciers in a series of papers by Weertman (1964; 1966) and observed in reality (Vivian, 1970; Gow, 1970). Weertman (1972) argued that the subglacial sheet varies in thickness, depending on ice pressures, and may coexist with channels. When water pressures locally exceed ice pressures, as for example may occur in bedrock hollows, the sheet may thicken and form a subglacial lake (Shreve, 1972). The existence of such subglacial lakes has long been postulated by workers in mid latitudes in order to explain the characteristics of certain glacial deposits (chapters 12 and 16). Also, the existence of subglacial lakes is sometimes suspected from the evidence of radio-echo sounding of ice sheets (Oswald and Robin, 1973). For example, Morgan and Budd (1975) suggested that a subglacial lake c. 15 km across occupies a rock depression beneath part of the huge Lambert glacier, Antarctica.

At a smaller scale, pockets of water are a common feature both within and beneath warm glaciers. These provide a temporary means of water storage which may be tapped and drained suddenly by main meltwater routes. Some are thought to form subglacially in cavities (Lliboutry, 1968a) and have been discovered in such situations (Miller, 1952). Others occur englacially at a variety of depths in the ice (Paterson and Savage, 1970) or at the surface in the form of closed moulins (Stenborg, 1968). The englacial pockets are likely to be the remnants of partly closed crevasses or meltwater tunnels.

Output

Water

A critical problem concerns the length of time required for water to traverse a stretch of glacier. Although there is wide variability, transit times in tunnels are in the order of 1–2 km per hour. Stenborg (1969) mentioned an average figure of 40 minutes for 1·7 km based on measurements on Mikkaglaciären (Figure 14.10). Transit times tend to decrease during the summer melt season. Vivian (1970) commented that the velocities represented by such transit times are less than those of comparable subaerial rivers. In some situations, flow is very much slower. Fisher (1973) noted a transit time of between 39 and 52 hours for 11·65 km through Salmon glacier in British Columbia, a figure consistent with flow through tubes with diameters as small as 2 mm. Such rates probably reflect blockages in a tunnel system or sheet flow of water beneath the glacier. These transit times are consistent with general estimates for glaciers as a whole. In a report edited by Pytte (1969) it was considered that on average a 2 day lag exists on the larger Norwegian glaciers between the formation of meltwater and its departure from the glacier. Golubev (1973) suggested that on certain Russian glaciers the lag varies from 1 day for glaciers of about 1 km² in area to 8 days for glaciers of about 300 km² in area.

It is a well known fact that the discharge of glacial meltwater streams fluctuates greatly from time to time, reflecting a wide range of environmental oscillations. On a short time

scale, continuous records of meltwater discharge reveal the occurrence of small scale fluctuations lasting a matter of seconds or minutes. Such fluctuations have been measured in the
Caucasus by Golubev, and are thought to be related to the tapping and draining of pockets
of water in the glacier. Perhaps the best known fluctuation is diurnal, one which is responsible
for marooning innumerable expeditions at the end of a day's fieldwork. Typically the meltwater stream is low in the early morning and rises rather abruptly towards the late afternoon
and early evening as the discharge increases dramatically (Figure 14.14). The diurnal
fluctuation is virtually non-existent in winter and begins to appear in early summer. Its
amplitude tends to increase towards late summer when ablation rates are highest. An important point made by Elliston (1973) is that, however characteristic the fluctuation, it is

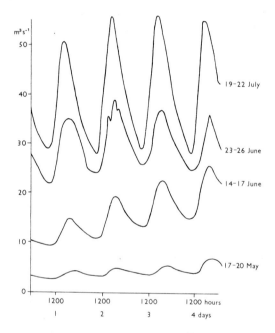

Figure 14.14
The diurnal fluctuation in discharge of the Matter-Vispa river
over selected four day periods in
1959, showing the increase in
amplitude as the summer pro-
gresses. (Elliston, Int. Ass. scient.
Hydrol. 1973.)

superimposed on a base flow which accounts for the bulk of the flow. Elliston suggested that
the peaks are due to an increase in discharge caused by a rise in the water table within
the glacier. The daily ablation has the effect of topping up the glacier's meltwater reservoir,
thus increasing the rate of outflow from the snout. This is confirmed by the correlation
between diurnal variations in discharge and pressure variations on Glacier d'Argentière
(Vivian and Zumstein, 1973). The suggested mechanism explains how a diurnal fluctuation
can occur in a glacier where the transit time for meltwater flow is measurable in days. The
time of the daily peak tends to become progressively earlier during the summer and autumn.
On the Glacier d'Argentière, it occurs at 23–24.00 hours in early summer; 18–20.00 hours
in mid-summer and 16–17.00 hours in September (Vivian and Zumstein, 1973). A similar
pattern has been recorded elsewhere (Elliston, 1973; Østrem et al., 1967). It is thought to
reflect the gradual improvement and widening of the internal drainage network throughout
the ablation season, which has the effect of speeding transit times.

Fluctuations of the order of days occur in response to changes in weather conditions (Fig. 6.6). A brief summer snowfall may reduce ablation and cause a fall in overall discharge lasting several days (Elliston, 1973). Similarly a peak of base flow may follow a sunny spell. Summer precipitation is important at this time scale, and one depression may increase discharge rates dramatically for several days. A point made by several workers is that rain-induced peaks are more effective in the autumn when much of the glacier surface is bare ice and the discharge routes are open (Ostrem *et al.*, 1967; Meier and Tangborn, 1961).

Seasonal fluctuations in meltwater flow are marked. When compared to ice-free river basins, glaciers have the effect of storing winter precipitation and discharging it in the summer. For example Boulton and Vivian (1973) noted that the meltwater flow beneath the Glacier d'Argentière increases from $0 \cdot 1$ to $1 \cdot 5$ m^3 s^{-1} in winter to 10–11 m^3 s^{-1} in summer. In winter most flow is due to basal sources while in summer it is mainly due to surface sources (Stenborg, 1965). Minimum flow is usually in late winter. There is a delay in early summer

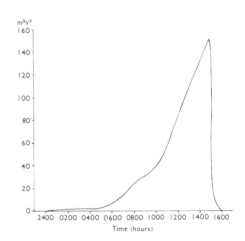

Figure 14.15
Rating curve of the discharge accompanying the subglacial drainage of Strupvatnet, a marginal lake in Norway. *(Whalley, Norsk geogr. Tidsskr. 1971.)*

runoff for a variety of reasons. The internal tunnel network has decayed by closure and by freezing and must be reactivated before efficient discharge can take place. Also the first rain and snow melt is retained in the firn either as superimposed ice or as slush. Stenborg (1970) estimated that this phase stores about 25 per cent of the melted run-off before releasing it later in the summer.

A special category of meltwater fluctuation occurring on an annual or longer basis concerns the periodic catastrophic draining of glacier-dammed lakes through or over a glacier dam. Typical of such outbursts is the 1 km long marginal Strupvatn discussed by Whalley (1971) (Figure 14.15). The 1969 outburst began at 23.30 on 19 July and reached a peak discharge of c. 150 m^3 s^{-1} at about 15.00 hours on the following day. It was fully drained by 16.00 hours. A number of theories have been put forward to explain such outbursts, and these are discussed by Whalley. It may simply be that ice dammed lakes drain when the internal drainage network developing during the ablation season taps the lake (Röthlisberger, 1972). Starting with a small trickle the lake outflow melts open a suitably big passage in a matter of hours. The speed of this process partly reflects the head of water involved. A number of factors such as the rate of filling of the lake and the annual development of the internal drainage net will influence the periodicity of such outbursts and may explain why some lakes drain on an irregular basis or every few years.

In Iceland such outbursts are sufficiently common to have been given the expressive name *jökulhlaup*. A spectacular example with a periodicity of about ten years is Grímsvötn, Vatnajökull (Figure 14.16). Here water builds up subglacially above a volcano and then most drains

Figure 14.16
A pit formed by subglacial melting in the northwestern part of the Grímsvötn caldera, Vatnajökull, Iceland. *(Photograph by S. Thorarinsson, August 1955.)*

catastrophically in the space of a few hours. In 1934 at its peak 'forty to fifty thousand cubic metres of muddy, grey water plunged forth every second from under the glacier border bringing with it icebergs as big as three storeyed houses. Almost the whole of the sandur or outwash plain, some 1,000 km² in area, was flooded' (Thorarinsson, 1953; p. 268). Grímsvötn is especially spectacular because the stresses in the crater associated with the unloading of the water may sometimes trigger a volcanic eruption.

Sometimes marginal lakes may drain over the surface of a glacier rather than underneath. Such events are common on sub-polar glaciers in Axel Heiberg Island and it seems that, catastrophic though they are, their drainage peaks are lower and more prolonged than those of subglacial drainage (Maag, 1969).

There are other types of fluctuation with periodicities of over a few years. One concerns the variation in meltwater produced by melting. Meaningful measurements covering a large number of years are not available. However, some indication of variability can be quoted. Østrem et al. (1967) presented an impressive account of meltwater flow for Decade glacier for the year 1965. However, they noted that the peak discharge in the following year was twice the amount recorded in 1965. Glacier surges introduce an often cyclic variation in meltwater produced on time scales of tens of years. As noted in chapter 3, surges are often associated with high rates of meltwater output. Finally, one might turn to time scales of hundreds of years and consider the excessively high rates of meltwater output produced during the melting of the mid latitude ice sheets. The probable rapid rates of ice sheet wastage have been discussed in chapter 7. If, for the moment, one extrapolates a far from extreme ablation rate of 10 m of ice per year and applies it to the whole surface of a completely stagnant ice mass, then this is sufficient to release the stored precipitation of many seasons in one summer. Using average mid latitude precipitation figures, the amount released mainly in one ablation season could be the equivalent of 5–15 years' precipitation. Perhaps when envisaging meltwater conditions during the decay of such Pleistocene ice sheets, it is fair to be thinking in terms of Thorarinsson's description of the Grímsvötn jökulhlaup.

Sediment load

As is clear to even the most casual observer who looks at the muddy colour of glacial meltwater issuing from a glacier, sediment loads are high. The ready availability of debris and turbulence associated with swiftly flowing streams contribute to these high loads. Suspended sediment loads of as high as 3,800 mg per litre were measured on the Evdalsbreen meltwater stream in Norway (Pytte, 1969). At this order of concentration, one can take a glass of meltwater from the stream and observe a layer of sediment which settles on the bottom after a few minutes. Even then one is not advised to drink the water. Bed loads of glacial meltwater streams are notoriously difficult to measure. However, a quantitative insight into the importance of the bed load can be gained by anyone attempting to cross major streams in Greenland. Not only can one stand on the bank and hear the rumble and feel the vibration of boulders rolling along the bed but, when crossing relatively small streams, it is common to collect bruises on one's legs and two boots full of pebbles. Probably the bed load is in excess of one third the total suspended sediment load.

Østrem et al. (1967) noted that there is no simple relationship between discharge and sediment load on Decade glacier, Baffin Island. As is normal on rivers, there is a peak in suspended sediment load some 2–3 hours before the peak of discharge, presumably reflecting the entrainment of the bed load as the flood picks up. In addition, there is a tendency for greater concentrations of sediment early in the season compared to the later part of the season. Probably this reflects the flushing out from beneath the glacier of the debris eroded by the glacier during the previous winter, a process which is accomplished in the first part of the season. The periodicity of sediment transport is very marked. For example, the Evdalsbreen meltwater stream carried 70 per cent of its suspended sediment load in 1968 in a flood lasting from 2–4 July (Pytte, 1969).

There are indications that meltwater streams also carry important quantities of material in solution. Ca and SiO_2 are found in subglacial meltwater beneath Glacier d'Argentière,

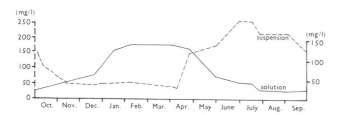

Figure 14.17
The suspension and solution load of the subglacial stream beneath Glacier d'Argentière in the balance year 1967–8. The left hand scale refers to suspension and the right hand scale to solution. *(Vivian and Zumstein, Int. Ass. scient. Hydrol. 1973.)*

and in winter during the period of low discharge the dissolved load far exceeds the suspended sediment load (Figure 14.17). In summer during the period of high discharge, the chemical load is still significant although *in toto* it is exceeded by the suspended sediment load. The winter meltwater is derived from basal melting and any groundwater flowing into the glacier. It is reasonable to attribute the solution load to the results of chemical action taking place at least partly beneath the glacier.

Meltwater models

By way of concluding this chapter it is perhaps worth trying to highlight some features of meltwater within model glaciers (Figure 14.18). A fundamental contrast exists between cold glaciers and warm glaciers. In the former (Figure 14.18a), basal ice is below the pressure melting point and frozen to the bed. No meltwater can exist within the glacier in liquid

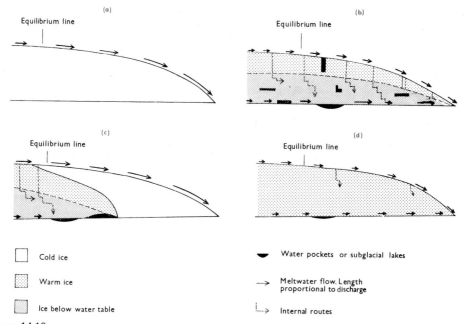

Figure 14.18
Meltwater models — **(a)** cold ice, **(b)** warm ice, **(c)** combination of warm ice in the accumulation zone and cold ice in the ablation zone, and **(d)** high altitude equatorial glacier.

form because of sub-freezing temperatures. Thus all meltwater activity is concentrated at the surface or at the glacier margins. Since ablation increases towards the snout, so also will the amount of surface meltwater. Warm glaciers may contain water throughout their mass (Figure 14.18b). Surface meltwater tends to be relatively uncommon because sooner or later it penetrates the ice. Within the ice there is a zone of moulins above the water table, and below it a mass of tunnels generally leading down to the glacier base. At the glacier base the amount of meltwater flow increases towards the snout as it receives increments from above. Pockets of water may exist anywhere in the glacier and at the base in depressions. Figure 14.18c is an intermediate type of glacier common in sub-polar environments, with cold ice at the surface and snout of the glacier and warm ice in the firn collecting grounds and on much of the bed. Here the ablation zone resembles that of a cold glacier, with surface meltwater increasing towards the snout and no basal meltwater. Above the equilibrium line, however, the glacier resembles a warm glacier with englacial and subglacial meltwater. Unless the basal meltwater is refrozen on to the base of the glacier, it may accumulate as a subglacial lake. Perhaps this is one of the situations discussed in chapter 3 which is conducive to the triggering of a surge, especially on glaciers like those of Spitsbergen. Finally, Figure 14.18d shows a warm glacier in a high altitude tropical environment. Relatively little melt-water is produced on the surface and within the glacier because a significant proportion of ablation is due to evaporation. In such situations there is less likely to be a water table than, for example, in glaciers in mid-latitudes.

Further reading

BJÖRNSSON, H. 1976: Marginal and supraglacial lakes in Iceland. *Jökull* **26,** 40–51.

COLLINS, D. N. 1979: Quantitative determination of the subglacial hydrology of two Alpine glaciers. *J. Glaciol.* **23** (89), 347–62.

ELLISTON, G. R. 1973: Water movement through Gornergletscher. *Symposium on the hydrology of glaciers. Cambridge, 9–13 Sept. 1969, Int. Ass. scient. Hydrol.* **95,** 79–84.

OSTREM, G., BRIDGE, C. W. and RANNIE, W. F. 1967: Glacio-hydrology, discharge and sediment transport in the Decade glacier area, Baffin Island, NWT. *Geogr. Annlr* **49**A, 268–82.

RÖTHLISBERGER, H. 1972: Water pressure in intra- and subglacial channels. *J. Glaciol.* **11** (62), 177–203.

SHREVE, R. L. 1972: Movement of water in glaciers. *J. Glaciol.* **11** (62), 205–14.

STENBORG, T. 1969: Studies of the internal drainage of glaciers. *Geogr. Annlr* **51**A (1–2), 13–41.

VIVIAN, R. and ZUMSTEIN, J. 1973: Hydrologie sous-glaciaire au glacier d'Argentière (Mont Blanc, France). *Symposium on the hydrology of glaciers, Cambridge, 9–13 Sept. 1969, Int. Ass. scient. Hydrol.* **95,** 53–64.

15 Meltwater erosion and its effects

In the light of the discussion in the previous chapter, it can be seen that glacial meltwater has large amounts of energy available for geomorphological work. Not only does much of this meltwater find its way into large subglacial channels, but velocities are often high. Furthermore, the extreme variability of flow ensures that there are brief periods of extremely high velocity associated with high discharge peaks. Under certain circumstances, the energy available can perform major feats of erosion. Vivian (1970) mentioned alpine examples where resistant rocks were eroded dramatically within a few years. Potholes 2·3 m wide and 2 m deep were eroded in a channel cut in ophiolites near Zermatt within the space of three years, while in an experiment 0·1 m of vertical erosion in quartzite blocks was accomplished by meltwater in five years. It is the purpose of this chapter to discuss first the basic processes of erosion, and then the landforms and landscapes associated with such activity.

Processes of meltwater erosion

Whereas the erosion of coherent bedrock by meltwater is widely accepted, the various mechanisms and their relative importance are far from clear. Observations and experiments suggest that corrasion is an important mechanism of erosion by a glacial meltwater. Hydro-electric engineers have long been aware of the destruction of pipes, pumps and turbine installations by the coarser fraction of the suspended sediment load in meltwater streams. For example, Bezinge and Schafer (1968) have shown that, if particles greater than 0·2 mm in diameter are removed by settling in a lake, this greatly reduces the erosion of surfaces in contact with the meltwater. The load of finer rock flour which does not settle out has little erosional effect. Likewise Vivian (1970) stressed the erosional importance of the coarse fraction of the suspended sediment load in the meltwater stream beneath Glacier d'Argentière, where over 50 per cent of the suspended sediment load is between 0·63 mm and 2 mm in size. Examination of the sand grains shows fractures reflecting the violence of transport. Corrasion by particles of this size can perhaps be envisaged as a form of wet sand-blasting.

There seems no reason to doubt that the hydraulic forces that occur in subaerial rivers occur also in subglacial meltwater streams and are capable of uplifting rock fragments if they are suitably shaped or weakly attached to bedrock. In view of the high peak discharges and associated high velocities in subglacial streams, the process is likely to be more effective beneath a glacier than in subaerial rivers. In addition, there may be hydraulic forces more or less unique to meltwater erosion. Cavitation is a process favoured by high water velocities in a channel with rough or irregular sides. Under certain circumstances of localized high velocity, as for example at a narrow point in the channel, the pressure in the stream drops sufficiently for airless bubbles to form. A slight increase in pressure, such as accompanies a subsequent decrease in velocity in response to a broadening of the channel, causes the collapse of these bubbles. The resultant implosion can effectively deliver a hammer-blow to any solid surface in the immediate vicinity. The process has proved highly destructive in certain conditions and there is a record of erosion of concrete in a dam spillway to a depth of 0·45 m in 23 hours (Barnes, 1956). Barnes suggested that stream velocities of the order of 8–15 m per second, might be sufficient for the initiation of cavitation. However there are many unknowns and the figures are only very approximate. Hjulström (1935) and Barnes (1956) have agreed that, because of the high velocities, cavitation may be an important process in subglacial environments and may initiate the erosion of potholes and bowls. It is notable that subglacial water velocities as high as 50 m per second have been recorded by Vivian (1975) and these would seem to be adequate for the process to occur.

There are several lines of evidence to suggest that mineral solution is an important means whereby meltwater can attack bedrock. Vivian (1970) emphasized that it is important not only because of the load removed by direct solution of minerals but also because in crystalline rocks it selectively attacks feldspars and weakens the crystal matrix of the rock. The latter process then makes the rock more susceptible to physical disintegration by direct ice action.

Figure 17 in the previous chapter gives some hint of the importance of mineral solution by glacial meltwater. Further possible evidence comes from the discussion in the literature concerning the relative efficiency of the solution of limestone by glacier meltwater. Page (1971) argued for greater than average quantities of carbon dioxide in meltwater and implied that solution of calcium carbonate may proceed rapidly beneath glaciers. High concentrations of carbon dioxide compared to the free atmosphere were found in ice near the equilibrium line of Storbreen (Coachman et al., 1956), while Vivian and Zumstein (1973) attributed the high concentration of carbon dioxide in the summer meltwater beneath Glacier d'Argentière at least partly to meltwater percolating through snow. Although there is still some doubt as to the relative importance of the solution of calcareous rocks by subglacial meltwater (Smith, 1972), it seems likely that the process plays a significant role beneath warm-based glaciers.

A further group of processes involving chemical agencies contributes to the deterioration of clay-rich rocks beneath wet-based glaciers (Falconer, 1969). The importance of the role of clay minerals in rock decomposition has been stressed by Yatsu (1966), and it seems that there is ample scope to extend and refine these ideas in relation to subglacial meltwater.

Landforms of meltwater erosion

There is a good deal of uncertainty about the nature of some landforms of meltwater erosion. Although meltwater channels have long been recognized especially in formerly glaciated mid latitude areas, attention has tended to focus on the more obvious features generally more than 100 m in length. Smaller forms have posed many problems of recognition. For this reason, any classification is highly speculative and major doubts persist as to the relative size, shape and importance of many forms. Indeed, it is likely that some forms are still unrecognized. In this chapter it is proposed to use the same major subdivisions as in the discussion of landforms of glacial erosion, namely areal and linear features.

Areal forms
It seems that at scales of less than approximately 20 m a number of different types of depression and smooth rock surfaces are related to meltwater flow. Frequently such forms have been classed as *p-forms* or plastically sculptured forms (Dahl, 1965). As explained in chapter 9, it is likely that some of the elongated and striated p-forms are the result of abrasion by

debris held in ice. On the other hand there are other depressions not easily explained by ice action,

Sichelwannen are crescent-shaped depressions sculpted on surfaces of hard, generally crystalline rocks (Figure 15.1). There are particularly good descriptions of the forms in Scandinavia (Ljunger, 1930; Dahl, 1965; Holtedahl, 1967) and on rocks of part of the Canadian Shield (Bernard, 1971a; 1971b). Sichelwannen occur on vertical or horizontal surfaces with the horns of the crescents pointing down glacier. Generally they are 1–10 m in length and 5–

Figure 15.1
Sichelwannen on a steep rock face, Skorpo Island Hardangerfjord, Norway. Ice movement was from right to left. *(Photograph by H. Holtedahl.)*

6 m in width. *Potholes* and a variety of shallower forms range in size from depressions a few tens of millimetres across to giant potholes 15–20 m deep and 16 m in diameter (Figure 15.2). They are commonly found in association with sichelwannen and have been described by Ljunger (1930), Dahl (1965), Gjessing (1965; 1967) and Holtedahl (1967). Typically, potholes are round rimmed shafts, deeper than they are wide, with internal spiral grooves. Often when a pothole occurs on a steep rock face, it has only one side and resembles an upright test tube cut along its length.

Although Boulton (1974) argued that abrasion can create sichelwannen, and Gjessing (1965; 1967) argued in favour of saturated till as the main agent responsible for both types of depression, there is wide support for the view that many p-forms are due to the action

Figure 15.2
Perkils Kättil, a giant pothole in
Sweden. *(Photograph by H.
Hansson.)*

of subglacial meltwater (Ljunger, 1930; Dahl, 1965; Holtedahl, 1967; Allen, 1971). Allen,
for example, considered that sichelwannen are strikingly similar to bedforms sculpted by
water on less resistant rocks and that they may be explained by differential corrosion caused
by separated flow within a subglacial stream. Also, following Alexander (1932), Allen
accepted that potholes are the result of corrosion in swift jetlike streams of water involving
flow separation. In addition, Hjulström (1935) drew attention to the possible effect of cavitation in initiating some of the depressions, while Vivian (1970) argued that the 'polished'
surfaces often associated with such forms are the result of corrosion.

Linear forms: meltwater channels
Meltwater channels have long attracted attention, especially in Europe. In part this may
reflect the often spectacular nature of their canyons, well known to tourists in many Alpine
and Scandinavian countries, and in part to their often unusual locations breaching con-
spicuous watersheds. Near Aberdeen there is a hill with a meltwater channel and a tor (Clach
na Ben) conspicuously silhouetted against the sky. A long-standing explanation is that the
devil took a bite out of the hill, found it too heavy and dropped it nearby (to form the tor).
Cynics might be tempted to ask how far understanding has subsequently progressed, other
than substituting an ice sheet for the devil.

There is no attempt to present a complete classification of meltwater channels in this section. In 1961, Sissons argued that such a classification was not advisable or practicable owing
to our present inadequate knowledge and this view seems equally appropriate today. Instead
the channels are discussed under two broad headings in terms of their probable association
with ice conditions. *Ice-directed channels* are those subglacial forms thought to reflect the primary control of active ice movement in their formation. *Marginal and sub-marginal channels* are

those which, as the name implies, are related to ice conditions near a glacier margin. Although such a discussion raises problems of overlap, it serves as a temporary framework for discussion. No attempt is made here to describe meltwater channels in detail. Such information may be obtained from the many detailed descriptions of channel morphology and characteristic locations, especially from former ice sheet areas, for example in Scandinavia (Mannerfelt, 1945; 1949; Holtedahl, 1967), in Britain (Derbyshire, 1961; Clapperton, 1968; 1971; Sissons, 1958; 1961a; Price, 1973), and in North America (Ives, 1958; 1960; Derbyshire, 1962; Sissons, 1960b).

Ice-directed channels   The distinguishing feature of such channels is that they tend to be aligned in a direction parallel to ice movement. In former ice sheet situations, where areal ice flow obtained, series of short channel segments comprise the rock components of systems extending for many tens of kilometres. A typical situation occurs in the Cairngorm mountains, Scotland (Figure 15.3). The individual channels tend to be most marked on convexities in the underlying topography and to be sharply incised into cols, spurs and hills. Two main types of channel can be distinguished. Some breach drainage divides from one side to the other while others occur on the lee side, beginning at the crest and running downslope for some distance. Both types are usually incised clearly into the divide. Often, there is no trace of a channel

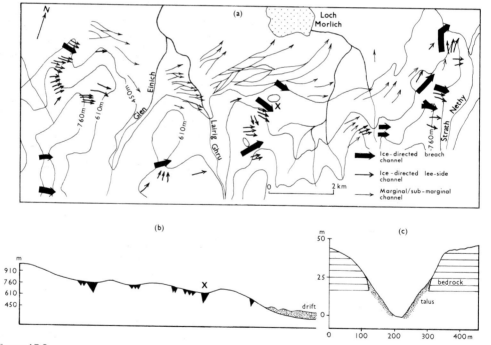

Figure 15.3
Meltwater channels on the northern slopes of the Cairngorm mountains, Scotland. **(a)** The pattern showing ice-directed subglacial forms and marginal/sub-marginal forms. **(b)** The relationship of ice-directed channels to cols on the spur east of the Lairig Ghru. **(c)** Cross-section of channel x. Ice moved from west-southwest to east-northeast.

in the depression between two divides although channels may be linked by linear fluvioglacial deposits. Although the primary control on channel alignment can be demonstrated to be the direction of ice movement, an important secondary control is the shape of the bedrock topography. For example, there is a tendency for channels to seek out the low points in divides. Where several channels cross one col the largest is usually at the bottom of the col.

In situations where former ice masses flowed as ice streams, either in outlet glaciers or valley glaciers, a slightly different pattern occurs. Here the channels tend to be aligned down valley with traces of a main channel along the axis, perhaps with tributaries joining the trunk channel from either side. The channels are most clearly developed at convex irregularities in the trough floor, for example on riegels, trough heads or at the mouths of hanging

Figure 15.4
A meltwater channel with sides over 200 m deep incised into a trough head southeast of Sukkertoppen ice cap, west Greenland. The channel has exploited a fault.

tributary troughs where they join the main trough. Holtedahl (1967) described these associations in relation to several Norwegian troughs and includes a study of the spectacular gorge near Fossli at the head of Hardangerfjord, known to many tourists. Many other examples could be quoted and Figure 15.4 is a west Greenland example cut in a trough head near the Sukkertoppen ice cap. In Alpine valleys there are many comparable examples, such as the Gorge du Guil above La Maison du Roy (Tricart, 1963).

With regard to channel pattern, there seems to be a continuum from individual straight channels, through channels with double intakes or one bifurcation to highly complex anastomosing patterns resembling braided channels (Figure 15.5, overleaf). In typical anastomosing channels, the junctions are often discordant, with one channel taking off from the side of another and joining yet another anywhere between the floor and the top of the side. Sometimes the pattern is so dense that the main feature is a series of elongated ice moulded hillocks separating the channels. Sometimes it can be demonstrated that bifurcating channels were used contemporaneously (Sissons, 1963). Straight and angular segments of channels

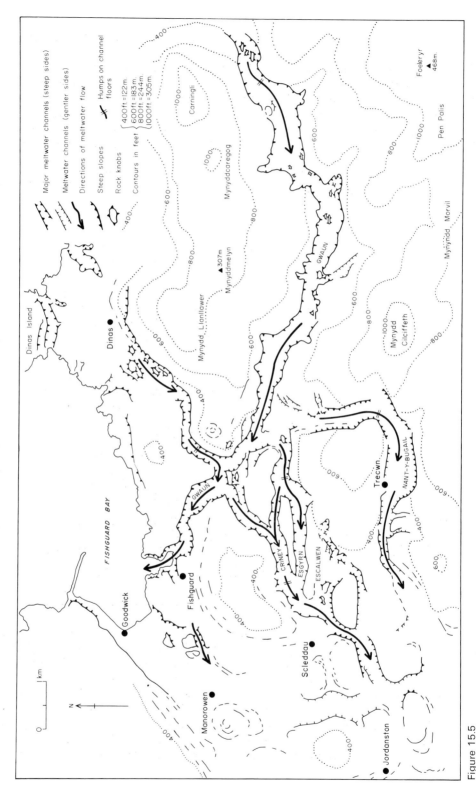

Figure 15.5
The Gwaun-Jordanston system of meltwater channels in north Pembrokeshire, Wales *(John, Longmans 1970a.)*

or channel patterns are commonly found to be associated with lines of structural weakness. Often such channels are much larger than others in the vicinity. Single channels appear to be most common on the up-ice sides and crests of divides while anastomosing patterns seem common on the lee sides of divides. In Figure 15.6, the two patterns are shown together, with a simple pattern of isolated channels on the up-ice side of the divide of a peninsula and a complex anastomosing pattern on the lee side (John and Sugden, 1971; John, 1972a).

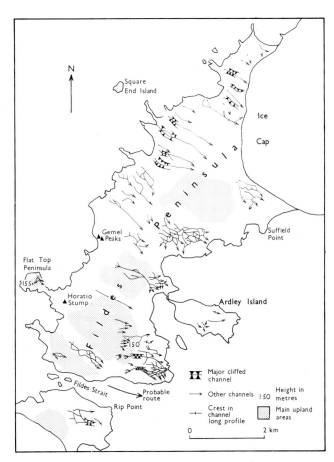

Figure 15.6
The pattern of meltwater channels on Fildes Peninsula, South Shetland Islands, Antarctica. Ice moved from northwest to south-east *(By permission of British Antarctic Survey, Brit. Antarct. Surv. Bull.)*

The morphology of individual channels is highly variable. In many areas channels tend to be flat-floored with steep sides, especially where gradients are gentle (Figure 15.7), but sometimes slope processes have attacked the steeper sides and bequeathed a V-shaped cross-profile (Figure 15.3c). This is particularly common in the case of smaller channels associated with former mid latitude ice sheets. Although it is most usual for channels to be several times wider than they are deep, in certain favourable situations their depth may exceed their width. For example, Corrieshalloch gorge in northwest Scotland is 10–20 m wide and 60 m deep, while Holtedahl (1967) noted that a gorge at Berekvam in Norway is 2 m wide and 30 m deep at its narrowest point. In many instances it has been recorded that the channel sides

are striated and moulded by glacier action. Channel lengths vary from a matter of metres
to the 75 km represented by some tunnel valleys of Denmark (Schou, 1949).

It appears that some channels have relatively smooth gradients falling in the direction
of ice and water flow. However, others, particularly long channels, have irregular long profiles with a series of ridges and basins, while still others have one main crest. The latter have
been termed 'humped' or 'up-and-down' channels. Although many uphill segments have
an amplitude of a few metres there are much greater recorded instances. Sissons (1961b)
recorded an uphill section which rises 80 m in Lanarkshire, Scotland, while an uphill segment
rising 142 m occurs at Naqerdloq kangigdleq near the Sukkertoppen ice cap, Greenland.

Many writers have produced morphological evidence to suggest that the channels described
above are subglacial in origin. In former ice sheet situations and particularly in Britain,

Figure 15.7
Oblique air photograph of the
Gwaun meltwater channel, Pembrokeshire, showing the flat-
floored cross-section (see Figure
15.5 for location) *(J.K. St.
Joseph, Cambridge University
collection; copyright reserved.)*

it had previously been argued that such channels were cut by the overflows of lakes dammed
by glaciers. As the ice sheet retreated, such lakes were thought to form wherever the relief
sloped towards the ice margin. As can so easily happen, a hypothesis tentatively put forward
with full acknowledgement of its limitations (Kendall, 1902), was developed uncritically
into a ruling hypothesis which survived into the 1950s (Charlesworth, 1955). In a number
of closely argued papers, Sissons (1958; 1960a; 1961a) built on ideas developed somewhat
earlier in Scandinavia by Mannerfelt, and argued in favour of a subglacial origin. Not only
were other traces of lakes such as deltas, shorelines and bottom deposits often lacking, but
the arrangement of channels could not be reconciled with lake overflows, and it could indeed
be demonstrated that ice must have occupied the site of the suspected lake at the time a
particular channel was used. In addition, the morphology of the channels suggests a subglacial origin. For example, up-and-down long profiles are most easily explained by water
flowing under hydrostatic pressure in enclosed conduits. The exceptions to this rule
mentioned by Schumm and Shepherd (1973) are on a much smaller scale. The widespread

occurrence of ice-moulded sides is easily explained if the channel formed subglacially. Finally, bifurcations in rock are difficult to explain under subaerial conditions, but are likely under a glacier.

A comparable change of interpretation can be recognized for meltwater channels in valleys formerly occupied by glaciers. In the Alps and in Norway, the presence of sharply incised river gorges cut into riegels and the lips of hanging valleys has often been interpreted in terms of post-glacial or inter-glacial adjustment by subaerial rivers, as for example argued by de Martonne (1957) and Gjessing (1965–66). In the Alps, however, the relationships of the gorges to glacial deposits contained within them demonstrate that at least some must have been cut beneath a glacier (Tricart, 1963). Furthermore, observations beneath the Mer de Glace have revealed the presence of gorges incised into riegels (Tricart, 1963), a feature also found beneath the Glacier de'Argentière (Vivian, 1970). In Norway, Holtedahl (1967) used channel patterns and small scale features of meltwater erosion to argue that canyons such as those around Fossli and Flåmsdal are subglacial in origin.

The theory of meltwater flow within a glacier summarized in the previous chapter lends much support to the subglacial interpretations, and offers some hint of the processes involved. Shreve (1972) showed why, in the hydrostatic medium of ice, basal meltwater tends to flow in the direction of the slope of the ice surface, and also how it is influenced by ambient pressures at the base of the ice caused by the flow of ice across irregular bedrock topography. Meltwater tends to avoid areas of relative high pressure associated with ice flow round upstanding obstacles and, where possible, flows in intervening topographic depressions. At a small scale, this process favours the development of contemporaneous bifurcating channels. At a large scale, it helps to explain the preference for channels to occur in cols in the topo-graphy underlying a former ice sheet. It also means that the irregularity of the relief will determine the complexity of the pattern of meltwater channels. The smoother the relief, the more closely will channels run parallel to the direction of ice flow. Where the relief is more varied, channels will diverge more readily from the overall direction of ice flow. This is apparently confirmed by the contrast between the relatively simple and straight tunnel valleys of low lying Denmark and the complex patterns characteristic of areas of higher relief, as shown in Figure 15.3.

One problem concerning the origin of meltwater channels is why they form in hard rock when it would seem easier for a subglacial channel to be excavated in the less resistant medium of ice. Nye (1973b) suggested that the answer to the problem concerns the relative permanence of channels cut in rock and ice. Channels cut in ice are in constant danger of being closed, not only by closure induced by the overlying weight of ice but also by the forward movement of the ice which might pinch off whole sections of a channel as the glacier reaches an obstacle. On the other hand a channel in bedrock, once formed, is much more permanent and in little danger of being closed off. This allows a greater length of time for a channel to develop. Also, the very presence of a rock channel favours its continued use by meltwater. This is because the hollow represented by the channel forms an area of locally low basal ice pressure which in turn tends to attract meltwater flow.

A second problem concerns the large size of many meltwater channels. It is sometimes assumed that the rock channel is the subglacial counterpart of a subaerial river valley. Whereas a subaerial river valley is widened vastly in relation to the river channel due to slope processes, a meltwater channel may represent the full dimensions of the former channel simply because active slope processes are unlikely to have operated. However, many channel cross-sections would represent, if filled, vast rivers with improbably high discharges. One possibility is that only the lower parts of the channel were utilized by water and that the remainder was occupied by ice. This view agrees with the widespread evidence of ice mould-ing on channel walls. Another possibility is that the dimensions of some rock channels are related to infrequent discharge events, such as might accompany the sudden draining of a subglacial lake.

A third problem concerns the discontinuous nature of meltwater channels and their

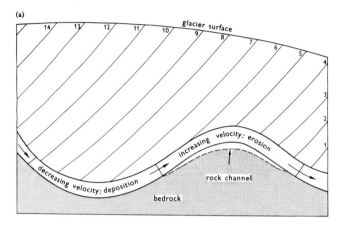

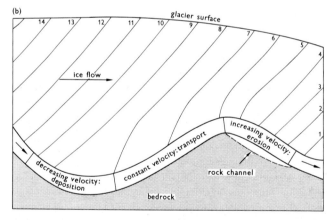

Figure 15.8
The erosion of ice-directed rock channels across a divide. See text for explanation.

apparent relationship to bedrock convexities. An answer is provided by Shreve (1972) and illustrated in Figure 15.8. The diagram portrays water flowing from left to right at the base of the glacier as well as the equipotentials (contours of equal potential), which are in essence the contours of the pressure gradient responsible for en- and subglacial flow (chapter 14). Meltwater velocity will be highest where the equipotential lines are closest, and lowest where they are widely spaced. More important, the velocity of the stream and thus its transporting capacity increases downstream where the spacing of the equipotentials becomes closer in the downstream direction. It is this latter situation which favours erosion. Figure 15.8a, which for simplicity ignores the effect of forward ice flow, shows that the spacing of the equipotentials is widest on the upstream side of the divide and progressively decreases across the crest of the divide. Other variables being equal, the resultant downstream increase in transporting capacity will cause erosion of a rock channel across the divide. Figure 15.8b allows for the effect of ice movement across the divide. Pressure will be elevated on the upglacier side of the divide and therefore the equipotentials will be depressed there. This reduces any variation in the pressure gradient on the upglacier side of the divide, and has the effect of shifting the zone of erosion towards the lee side. This would create a channel on the lee side of the divide. The magnitude of this effect depends on several factors, for example the shape of the bed and the depth and velocity of the ice.

The interpretation of subglacial meltwater channels associated with active ice as presented above differs in some respects from earlier interpretations, which have been influenced by a belief that meltwater is unlikely to flow beneath great thicknesses of ice. For example, a view widely accepted in Britain is that many channels are the result of superimposition of englacial meltwater routes onto the underlying topography as the ice thinned during deglaciation (Price, 1960; 1973; Derbyshire, 1961; Embleton, 1961; Sissons, 1963; Clapperton, 1968). Whereas this may have occurred under certain conditions, it seems that the concept of gently sloping levels of meltwater flow was influenced by the difficulty of envisaging flow beneath great thicknesses of ice over the depressions between channels. Now that theoretical studies have removed the latter difficulty, it may be that many channels are simply the result of normal meltwater flow beneath a warm-based glacier. Furthermore, some channels may relate to the flow of water during glacial maxima rather than during deglaciation.

Marginal and sub-marginal channels   The distinguishing feature of such channels is that they tend to run approximately parallel to a lateral or frontal glacier margin. A common situation

Figure 15.9
Marginal   and   sub-marginal
channels, Klövsjöfjällen, Sweden.
(Photograph by J. Lundqvist.)

in formerly glaciated areas is to find channels running along a valley or mountain slope at a relatively shallow angle to the contours (Figures 15.9 and 10). The channels cover a range of types, including those that are strictly marginal and ran between ice and bedrock, those that ran in or on ice near the margin, those cut wholly in bedrock or moraine close to the glacier, and a whole series of sub-marginal forms thought to have been formed under the ice close to the glacier margin. Indeed the word sub-marginal is deceptively simple and includes a continuum of forms from subglacial forms influenced mainly by basal ice pressures to near marginal forms.

Understanding of marginal and sub-marginal channels has come from work both in formerly glaciated mid latitude areas and in currently glacierized areas. In two classic articles, Mannerfelt (1945; 1949) gave descriptions of series of channels in parts of Scandinavia. The channels formed during deglaciation were thought to reflect ice gradients of between 1 : 33 and 1 : 100, though individual channels usually slope more steeply. It is difficult to distinguish a truly marginal channel from a sub-marginal channel. In the case of a channel near Tregarth, north Wales, which slopes at a gradient of 1 : 300 for a distance of 2 km, it is reasonable to infer that it was cut by water flowing at the glacier margin (Embleton, 1964). From

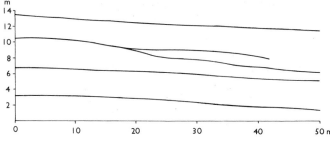

Figure 15.10
Profiles along sub-marginal melt-water channels at Högfjället. *(Mannerfelt, Geogr. Annlr 1945.)*

observations in Scotland, Sissons (1961a) suggested that as a rule gradients less than 1 : 50 may be marginal. Gradients of around 1 : 30, which are commonly observed, presumably represent flow which cut channels obliquely beneath the ice margin (Figure 15.10). It is common to find an anastomosing pattern with channel segments running at relatively gentle gradients along the hill slope separated by segments running more directly downslope. Whereas this pattern may sometimes represent a blend of marginal and sub-marginal channels, Sissons (1961a) showed that it can be developed entirely beneath the ice margin. In other situations, the downslope pattern may dominate and form a series of sub-parallel channels running directly downslope. These are termed subglacial chutes (Mannerfelt, 1945; Sissons, 1961a). Frequently they can be shown to originate at a former ice margin.

Channel forms are highly variable and reflect the influence of such factors as the nature of the ice margin, ice temperature, structural features in the ice, the slope and nature of the bedrock and the supply of meltwater and debris. Channels may be two-sided gullies or simply flat benches cut into the hillside. Occasionally more complex forms such as old meander scars may be preserved.

Valuable observations of drainage at the sides of active glaciers come from a series of classic studies on North American glaciers. There are many early and vivid descriptions of conditions at the margins of 'warm' glaciers in southeast Alaska (Russell, 1893; Tarr, 1909; Tarr and Butler, 1909; von Engeln, 1912). Surface meltwater from the glacier and valley side streams contributed large quantities of water to the glacier margin. Von Engeln drew attention to the role of valley side streams in bringing water at relatively high temperatures (6–7°C) into contact with the ice. One of the main conclusions was that true marginal drainage was less common than systems of sub-marginal drainage where water flowed below the margin of the ice either parallel to or oblique to the margin. Such a conclusion has been confirmed by Stenborg's observations in northern Sweden (Stenborg, 1973). On cold glaciers the situation is quite different. Maag (1969) described drainage associated with sub-polar glaciers on Axel Heiberg Island. Most channels occur at the margin although they may be incised slightly into the glacier (Figure 15.11). Maag noted how most erosion of marginal channels

Figure 15.11
Marginal drainage beside Crusoe glacier, Axel Heiberg Island. In the photograph taken in August, 1972, Accident river flows towards the glacier and then beneath the ice margin. The cross-section shows the form of the channel at the same ice margin on 8 July 1961, the day before it collapsed. Excavation of the channel beneath the glacier margin took place in three days during the sudden emptying of a marginal lake. This sequence of events takes place every year. *(Maag, Axel Heiberg Is. Res. Rpt 1969.)*

is accomplished during floods. These may be due to the rapid draining of a multitude of marginal lakes or to summer rain storms. Rapid and sudden runoff is favoured in such environments where the infiltration capacity of the glaciers and permafrost is low, vegetation is scanty or non-existent and where there is little debris in glacier surface meltwater channels to counteract a sudden increase in discharge.

When trying to explain the pattern and form of marginal meltwater and sub-marginal channels, it is helpful to recall the fundamental contrast between marginal meltwater flow beside warm and cold ice. In warm ice, sub-marginal flow of meltwater is favoured. In part, this is because water can penetrate ice at the pressure melting point, especially if it is several degrees above 0°C. In part, it is a reflection of the tendency of water in the ablation zone of a glacier to be driven out towards the valley sides. As explained in the previous chapter, this tendency reflects the englacial pressures induced by the slope of the ice surface towards the ice margin (Weertman, 1972) as well as the influence of marginal structural features

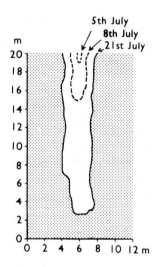

Figure 15.12
Rapid channel erosion in permafrost, Accident river, Crusoe
glacier. *(Maag, Axel Heiberg Is.
Res. Rpt 1969.)*

(Stenborg, 1973). In cold ice on the other hand, truly marginal flow is favoured. This is because no water can penetrate into the cold ice without freezing.

From this it can be suggested that long, continuous marginal channels extending over several kilometres are likely to reflect marginal flow beside cold ice, while the complex sub-marginal channels noted in formerly glaciated mid latitude areas are likely to reflect the presence of warm ice. In the latter situation, the channel pattern and orientation reflect both the movement of the ice and the shape of the topography. If there is a constant valley slope and the ice is moving fast, there will be a strong down-valley component in response to basal ice pressures and channels will run obliquely across the valley side slope. If the ice is moving slowly, the down-valley component will be less marked, and in stagnant ice channels will run directly downslope. This latter situation may be especially favourable for the formation of sub-glacial chutes. Sometimes there is an abrupt lower limit to marginal and sub-marginal channel erosion on a hill slope and, indeed, in some situations it may suddenly give way to the deposition of eskers (Sissons, 1958; Sugden, 1970). Sissons suggested that this change may represent the former existence of a water table and that erosion ceased at this level in sub-marginal situations. This, of course, does not exclude the occurrence of erosion well below the water table where hydrostatic pressures are important. A further contrast between cold and warm glaciers concerns the mechanisms of channel erosion. Beside cold glaciers the ground may consist of permafrost with a high ice content, as for example in Axel Heiberg Island. Maag (1969) noted how rates of erosion in ice-cemented gravels are enhanced by thermal erosion in addition to other normal mechanisms of fluvial erosion (Figure 15.12).

Landscapes of meltwater erosion

There are considerable spatial variations in the role of meltwater erosion from one glaciated area to another. These reflect (a) the characteristics of past or present ice masses, (b) the glacial history of an area and (c) the shape of the underlying topography.

(a)   Since subglacial and sub-marginal flow is favoured where the ice is at the pressure melting point, it is reasonable to expect sub-marginal and ice-directed channels to be associated with glaciers such as those in the Alps and southeast Alaska, and with the warm-based parts of former mid latitude ice sheets, such as near their maritime peripheries. This would include Norway, Britain, west Greenland, the Atlantic fringe of the Laurentide ice and the warmer parts of the Antarctic coastal areas. Reference to many mid latitude examples has been made above, while polar examples have been described in west Greenland (Sugden, 1972), Arctic Canada (Ives and Kirby, 1964), the South Shetland Islands (John and Sugden, 1971) and in Victoria Land (Calkin, 1973). True marginal channels are particularly well developed in areas where cold glaciers experience considerable summer melting, such as in north Greenland and northern Canada. In extremely dry continental environments, there may be little melting and thus little meltwater of any type available for erosion.

(b)   The glacial history of an area is clearly important in influencing the type of meltwater erosional landform. A common landscape association can be related to the disappearance of the mid latitude ice sheets as patches of ice became isolated in depressions, and progressively became stagnant as they downwasted. Frequently, marginal channels reflect the gradual decrease in the ice gradient as the glacier thinned. At the same time sub-marginal channels show a progressive change as ice activity decreased with thinning and they tend to run more directly downhill nearer the bottom of the depression. These associations are apparent in many regional studies of deglaciation, for example in Scotland (Soons, 1960; Sissons, 1967) (Figure 15.3), Scandinavia (Gjessing, 1960) and North America (Ives, 1960).
    Although in areal terms, marginal and sub-marginal channels are best displayed in areas of former deglaciation, their overall erosive effect is diffused over a wide area. The erosional effects are much more concentrated where a glacier margin is stable for a long period of time. In such cases one might expect to find that marginal meltwater erosion can affect the cross-profile of a glacial valley by eroding benches and causing some widening. Maag (1969) claimed to have recognized such features on Axel Heiberg Island.
    One difficulty in assessing the influence of glacial history on a landscape of meltwater erosion is uncertainty about which phase of glaciation was responsible for meltwater erosion. There are two facets to the problem. One, hinted at earlier, is that many subglacial channels may have been initiated during an early or maximum stage of a glaciation. They may then

have been utilized during later stages of deglaciation simply because they were there. A second problem common in mid latitudes concerns the number of glaciations. Which glaciation initiated the channels? In parts of eastern Scotland channels are cut into interfluves which retain their pre-glacial form. If such interfluves have survived several glaciations, then it is likely that channels incised into the interfluves could easily survive from earlier glaciations and simply be re-used subsequently. It could be argued that the most likely phase of erosion was during the first glaciation of an area, and that the channels, once formed, were merely re-used by meltwater of subsequent glaciation. Sometimes, as in Pembrokeshire, there are indications that meltwater channels were wrongly aligned for excavation during the last glaciation and that they must relate to an earlier glaciation (John, 1970a). In other cases where ice conditions have varied through a glaciation, then channels may reflect earlier phases of meltwater activity. A good example of this is the complex pattern of channels known as the Labyrynth in Victoria Land (Calkin, 1973).

(c)   The nature of the underlying topography is a particularly powerful factor influencing landscapes of meltwater erosion at a local scale. The effect is probably most obvious with regard to marginal channels, where the angle of slope and the attitude of the slope in relation to the glacier affect not only the type of channel, but also help to determine whether or not meltwater will flow on the ground beside the glacier or on the glacier itself. Thus Sissons (1961a) and Price (1973) have noted that marginal and sub-marginal channels are most common on gentle slopes, especially where they are drift covered. Concerning ice-directed subglacial channels, the favoured location on divides carries an interesting implication, namely that the clearest channel systems occur where the direction of ice movement is across the grain of the underlying topography. Where pre-existing valleys run in the direction of ice movement, there may be no recognizable channels since the meltwater tends to follow the valley. There are abundant examples of this phenomenon and it is likely that meltwater flow down suitably orientated valleys has been underestimated. In the eastern Highlands of Scotland, for example, the major meltwater discharge routes are likely to have been down valleys such as that of the Dee, Don and Spey, although there is little obvious erosional sign of this in the landscape.

The interpretation of landforms of meltwater erosion has, until recently, leaned heavily on evidence obtained in the formerly glaciated areas of the mid latitudes. A fundamental problem of any such landform interpretation is the difficulty of distinguishing features associated with deglaciation from those formed during glacial maxima. Most workers have tended to assume that they are seeing the work of deglaciation and erosional meltwater features are interpreted in the light of this. However, the recent advances in the theory of meltwater activity within a glacier help to explain many existing field observations, and it seems that many features may prove to reflect earlier glacial conditions. Under these circumstances one can expect important changes in interpretations over the next few years.

Further reading

CALKIN, P. E. 1973: Glacial processes in the ice-free valleys of southern Victoria Land, Antarctica. In FAHEY, B. D. and THOMPSON, R. D. (eds) *Research in polar and alpine geomorphology*, Geo Abstracts, Norwich, 167–86.

DAHL, R. 1965: Plastically sculptured detail forms on rock surfaces in northern Nordland, Norway. *Geogr. Annlr* **47,** 83–140.

HALLETT, B. 1976a: Deposits formed by subglacial precipitation of CaCO₃. *Bull. Geol. Soc. Amer.* **87,** 1003–15.

MAAG, H. 1969: Ice dammed lakes and marginal drainage on Axel Heiberg Island. *Axel Heiberg Island Res. Rept.*, McGill Univ., Montreal (147 pp).

SISSONS, J. B. 1960: Some aspects of glacial drainage channels in Britain, part I. *Scott. geogr. Mag.* **76** (3), 131–46.

1961: Some aspects of glacial drainage channels in Britain, part II. *Scott. geogr. Mag.* **77** (1), 15–36.

SOUCHEZ, R. A. and TISON, J. L. 1981: Basal freezing of squeezed water: its influence on glacier erosion. *Annals of Glaciology* **2,** 63–6.

VIVIAN, R. 1970: Hydrologie et érosion sous-glaciaires. *Revue Géogr. alp.* **58,** 241–64.

16 Meltwater deposition

This chapter is concerned with the formation and surface expression of stratified deposits laid down by glacial meltwater. By their internal composition, these deposits are generally easy to recognize, but the landforms built of fluvioglacial sands and gravels are often similar to those made of till, particularly if they have originated in ice marginal environments where settling, collapse or flowage have modified both the arrangement of the sediments and the original ice-contact forms (Hartshorn, 1958). As noted in chapter 12, dead-ice topography in which fluvioglacial materials predominate may be completely inverted as a result of the gradual melting out of buried masses of dead ice. There are, nevertheless, distinctive forms produced by meltwater deposition; notable among these are eskers, kames and kame terraces, and these are discussed in the following paragraphs. Since this book is concerned only with the effects of glaciation under and at the margins of ice masses, pro-glacial depositional land-forms are not considered in this chapter. The most important of these are pro-glacial deltas, outwash plains (sandar) and valley trains, which are treated in some detail by Church (1972) Mickelson (1971), and Hjulström *et al.* (1954).

As in the previous two chapters, it is helpful to consider meltwater deposition in the context of glacial, fluvioglacial, and fluvial systems. The fluvioglacial system provides a link between the other two, overlapping with each (see chapter 14). Within and under the glacier there are a number of situations in which meltwater deposition can occur (Figure 16.1, a–e). Often the deposition is of a strictly temporary nature, for there are such violent fluctuations of stream-flow on the part of glacial streams, both in the short term and in the long term, that their transporting capacity is subject to constant variation. This explains the discontinuous nature of many fluvioglacial landforms, and also the rapid alternations of coarse and fine-grained sediments which have attracted so much attention from sedimentologists.

Beyond the glacier snout, the transporting capacity of meltwater is suddenly reduced, and deposition of stratified materials is the result. Often this deposition is caused by localized reductions of stream velocity, perhaps because of rapid shifts of stream courses (thus creating local areas of 'stagnant' water ideal for deposition) or because the provision of detritus in the vicinity of the glacier snout becomes too great for meltwater to transport. Hence part of the detritus load must be released (Figure 16.1 f).

As in the case of glacial sedimentation, the input, throughput and output relationships of fluvioglacial deposition are often extremely complex. As shown in the simplified model in Figure 16.2 input of material into the fluvioglacial system may be from four sources:
a   detritus derived directly from fluvioglacial erosion of bedrock;
b   detritus derived from previously deposited materials;
c   detritus released by glacier ice;
d   detritus derived from the periglacial system (particularly in supraglacial and marginal situations).

The path followed by these materials may be complicated if they are frozen into the glacier, or else incorporated into existing deposits englacially or subglacially. In particular,

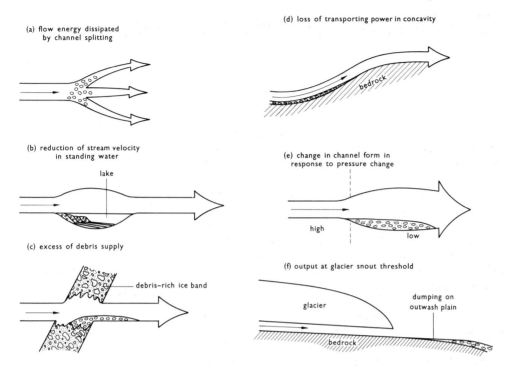

(a) flow energy dissipated
by channel splitting

(b) reduction of stream velocity
in standing water

lake

(c) excess of debris supply

debris-rich ice band

(d) loss of transporting power in concavity

bedrock

(e) change in channel form in
response to pressure change

high

low

(f) output at glacier snout threshold

glacier

dumping on
outwash plain

bedrock

Figure 16.1
Common situations which may lead to fluvioglacial sedimentation. All except (f) are in tunnels in ice-contact situations, and the sediments and landforms will probably be modified during ice wastage. At (f) there is a real output situation, where sediments and landforms have a better chance of survival.

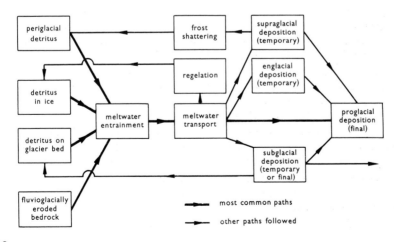

periglacial
detritus

frost
shattering

supraglacial
deposition
(temporary)

regelation

englacial
deposition
(temporary)

detritus
in ice

meltwater
entrainment

meltwater
transport

proglacial
deposition
(final)

detritus on
glacier bed

subglacial
deposition
(temporary
or final)

fluvioglacially
eroded
bedrock

most common paths

other paths followed

Figure 16.2
Flow diagram showing in simplified fashion some of the paths followed by detrital particles in the fluvioglacial system.

fluvio-glacial materials may be mixed with till, lacustrine sediments and frost-shattered materials in ice-wastage environments during the process of flow-till formation. True output from the fluvioglacial system is not generally achieved until the stratified sediments are in a pro-glacial situation, either carried by meltwater and laid down well beyond the ice edge, or abandoned by the glacier as a result of ice edge retreat or downwasting.

It is of interest to note that in spite of recent advances in understanding of sedimentation processes (Allen, 1971) and in the analysis of fluvioglacial landforms (Price, 1969), there is still great confusion over the interpretation and significance of ice-contact features built largely of bedded sands and gravels. For example, research into the deglaciation of Scotland has been marked by the reinterpretation of certain groups of landforms once called moraines and now accepted as accumulations of eskers, kames and other fluvioglacial forms (Sissons, 1967; Clapperton and Sugden, 1972). In Finland, the Salpausselkä ridges continue to attract attention (Aartolahti, 1972: Hyvärinen, 1973; Fogelberg, 1970), and there is still no general agreement over their precise geomorphological or chronological significance. It is no simple matter to differentiate between the roles of glacial and fluvioglacial processes in the complex zone of system overlap occupied by glacial meltwater.

Processes of deposition

The processes of deposition by meltwater are very similar to those of normal streams, except that contact with glacier ice imparts certain peculiarities in flow regime, stream velocities and channel characteristics. Some of these peculiarities have been discussed already in chapter 14, but they have seldom been considered in detail in so far as they affect sedimentation. In the following paragraphs there is some discussion of meltwater sedimentation, particularly with respect to situations where deposits are built up either englacially in ice channels or against an ice margin.

Stream power and bed forms

Many structures in channel sediments can be correlated with variations in stream transport rate and stream power. Where bed material is of fine sand with a fall diameter of less than 0.65 mm, ascending stream power will give first ripples with cross-lamination (Figure 16.3), then dunes with cross-bedding, and then a plane bed with even lamination (Allen, 1970). Coarser sand whose median fall diameter is more than 0·65 mm will not be moved until stream power reaches about 300 ergs/cm²/sec, but then ascending stream power will give plane beds, then dunes with cross bedding, and then plane beds again.[1] As shown in Figure 16.4 the commonest structures in fine sands in fluvioglacial sediments will be ripples and cross-lamination (generally formed when stream power is between 100 and 1000 ergs/cm²/sec), whereas the most common structures in coarser sands will be dunes and cross-bedding, formed when stream power is between 600 and 10,000 ergs/cm²/sec.

There is a much more efficient sorting of material in fluvioglacial sediments than in till, although there may be sharp transitions of particle-size between one bed and another. For example, exposures in the flanks of eskers and kames often reveal beds of coarse boulders and cobbles, gravel horizons, and also sandy sediments. Silt and clay layers are less common, for stream flow velocities are generally sufficient to transport the finer fractions of sediments in transport to englacial or marginal water-bodies or to pro-glacial situations. The sorting

[1] The unit of stream power referred to here is not an SI or otherwise approved unit.

Figure 16.3
Fine grained sediments showing annual varves and ripple cross-lamination at Bageråsklack, Ångermanland, Sweden. *(Photograph by kind permission of Erling Lindström.)*

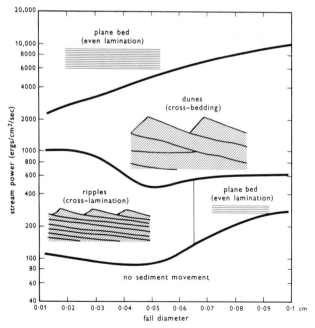

Figure 16.4
Bed form in relation to stream power and the fall diameters of particles in transport. *(Allen, Unwin 1970.)* The units used here are not SI-units.

of fluvioglacial sediments is caused both by the great variations of detritus input which occur with respect to meltwater streams inside the glacier margin (due, for example, to the sudden collapse of an englacial channel wall releasing debris-rich ice into the stream), and also by day-to-day, seasonal and annual variations in stream velocity. In chapter 14 it was pointed out that winter and spring accumulations of sediment are often 'flushed out' during the early summer flood. Accordingly, in some situations coarse boulder and cobble beds may be interpreted as annual layers related not only to high stream velocities but also to the release of accumulated coarse subglacial and englacial sediment. If a sudden input of coarse till occurs at a time of low meltwater flow, a highly localized bed of boulders or coarse gravels may

Figure 16.5
Sands and gravels in a fluvioglacial sequence at Sjöbo, southern Sweden. Cross-bedding is prominent in the middle part of the sequence. *(Photograph by Sven Stridsberg.)*

be formed, to be buried by cross-laminated sands and gravels which are more truly representative of local sedimentological conditions (Figure 16.5). It is extremely common for individual beds to be discontinuous laterally, and for beds (or whole bed successions) to be removed by erosion if stream velocity increases or if the meltwater stream migrates laterally across the surface of a sediment sequence.

In an interesting study, Allen (1971b) noted several erosional breaks in the sequence of sands which he analysed in the Uppsala esker. From his analyses of climbing-ripple cross-laminations, he was able to relate variations in the angle of climb of ripples to local variations in meltwater flow. He concluded that flow was unsteady and non-uniform, with the threshold between plane bed deposition and ripple formation being crossed at times rhythmically. These fluctuations were related to the short term variations of discharge typical for glacial streams (Figures 6.6 and 14.14). Individual beds were thought by Allen to have formed in a matter of hours or days; he suggested that one bed 0·6 m thick formed in only 2·8 hours, while another bed more than 2 m thick formed in 26·4 hours. The implications of this work are great, for it holds the possibility of calculating possible rates of esker or kame terrace construction. The 22·5 m of sands and gravels examined by Allen in the Uppsala esker could easily have accumulated within the space of a single melt season, but because most of the deposits are cross-bedded sands (indicating higher stream velocities), and because of the interference of gravel beds and erosional contacts, it is impossible to know how much of the sedimentary history is represented in the presently preserved sequence. In particular, the erosional contacts may represent erosional phases which were more prolonged than the depositional phases. However, it seems extremely unlikely that the sequence represents deposition over centuries or even decades. It is most probable that the whole 24 m of deposits accumulated over a few seasons or even during a single season. This would be consistent with observations on the speed of esker formation in present day ice marginal environments (Price, 1973; Stokes, 1958), and with other rates of sedimentation from glacial meltwater streams (Maag, 1969; Ziegler, 1972).

Sedimentation in ice channels
Allen has shown that the processes of englacial and subglacial deposition are essentially 'normal'; but there have as yet been few attempts to quantify the depositional environment of a channel which is completely filled with meltwater flowing under high hydrostatic pressure (Röthlisberger, 1972; Lister, 1973). For example, many of the irregularities which occur in ordinary fluvial sedimentary sequences are explained by lateral shifts in stream channels. This explanation can seldom be used in the case of englacial or subglacial sedimentation, because as long as a tunnel is filled with meltwater, stream flow must be strictly confined to one vertical plane. Stream beds can be lowered or raised by erosion or deposition, but they seldom migrate laterally unless discharge is so low that only the tunnel bed is affected. In this case a normal meandering route might be followed by the stream, with deposition and erosion alternating on both flanks of the channel. On the other hand, for times of high meltwater discharge there is a possibility of approximate equilibrium between channel dimensions and rate of deposition; as the floor of the channel is raised by a buildup of deposits, the channel roof may be raised by thermal erosion in order to maintain the dimensions required for a specific discharge.

So far there have been few attempts to assess the transporting capacity or depositing characteristics of streams flowing in ice channels of various shapes, which have much lower coefficients of friction than bedrock channels (Shreve, 1972). Shaw (1971) has suggested that in channels where deposition is taking place, there is a gradual melting of the ice walls by heat generated through friction, leading to channel widening and a gradual reduction of stream power. However, in view of what has been said in chapter 14, it seems that this will only occur beneath thin or stagnant ice. In active ice, where channels survive the tendency for closure by plastic deformation, this may indicate a delicate equilibrium with the thermal properties of the ice mass, and the flowing water, and with frictional heat generated. This latter will depend upon the roughness of the ice-water interface, the turbulence of flowing water and also the contacts which transported detrital particles have with the floor and walls (and perhaps the roof also) of the ice channel. These interactions have not been adequately investigated, but they are certainly of fundamental importance for an understanding of the processes of meltwater deposition.

Lacustrine and sheet-flow sedimentation

The foregoing discussion has been entirely concerned with the processes of deposition which characterize channels or other constrained situations in which meltwater flow is maintained for most of the melting season. Fluctuations in the character of the sediment load and in stream power will lead to the crossing of several depositional thresholds, giving rise to the various types of structures discussed by Allen (1970) and Shaw (1971). In other ice-contact situations there may be much more abrupt losses of transporting capacity by meltwater streams, particularly if they flow into ice marginal lakes or subglacial water-bodies. If deposition takes place in a marginal lake, there will be close analogies with deltaic sedimentation, with bottom-set beds representing the distal settling-out of fines some distance from the point of stream discharge, and with fore-set and top-set beds representing the buildup of the proximal coarser material (Sollid, 1963/4; Maag, 1972). If the lake receives sediment from a number of different discharge points then the sedimentary sequence will be more complex.

Meltwater sedimentation in subglacial standing water is still imperfectly understood, and the processes are likely to be complex. Following on the ideas of Flint (1930) for central

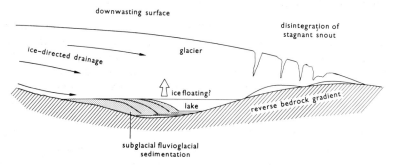

Figure 16.6
Subglacial fluvioglacial sedimentation in a large water-body, beneath a warm-based glacier. *(After the ideas of Gjessing, Ad Novas 1960.)*

Ireland and Andersen (1931) for Denmark, Sissons (1958b) has argued for subglacial sedimentation in the Eddleston valley, Scotland, which was controlled by water-bodies beneath stagnant ice. Gjessing (1960) has taken these ideas much further, in particular referring extensive terraces and deltas of stratified sediments in Atnedalen, Norway, to deposition in large subglacial water-bodies. As argued in chapter 14, such water-bodies are likely to have been ponded up against reverse bedrock gradients where the rock is impermeable (Figure 16.6). The extent of the meltwater buildup was, according to Gjessing, sometimes sufficient to lift the glacier off its bed, thus allowing 'free' sedimentation in a subglacial lacustrine environment. While this idea has been the subject of some criticism (Hoppe, 1960; Lundqvist, 1972), its fundamentals seem sound enough (Shreve, 1972), especially when applied to areas of wasting warm ice. There is, furthermore, no reason why deposition should not also occur in the subglacial water-bodies known to occur beneath sub-polar glaciers (Baranowski, 1973) and even beneath the warm-based parts of the Antarctic ice sheet (Oswald and Robin, 1973; Budd, 1975).

As in so many other cases, little is known about the present day processes of sedimentation which occur in subglacial lakes, where the pressures exerted by the overlying ice or by the hydrostatic head of water may severely affect meltwater stream discharge rates. In particular, drastic post-depositional distortion of sediments may be expected as a result of ice loading. Indeed, such distortion is known from thick sequences of fluvioglacial sediments which have puzzled glacial geologists in northeast England and other areas (Francis, 1970).

Subglacial sedimentation of fluvioglacial materials near retreating ice margins, which are grounded in the sea or in pro-glacial lakes, is perhaps easier to understand, and since the classic work of De Geer (1897) the processes concerned have received much attention in the Scandinavian literature (Hoppe, 1948; Granö, 1958; Wisniewski, 1973). As noted in chapter 11, many of the deposits from areas below the highest shoreline in Norrbotten, Sweden, are not true tills at all. Hoppe (1948) has called some of them 'pseudo-moraine', and the Kalixpinnmo hills are formed predominantly of fine sands and silts which point to a slow flow of subglacial meltwater in sheets close to the ice edge. It is apparent that fluvioglacial sedimentation is possible beneath a glacier even where ice channels are absent, although Shreve (1972) points out that meltwater flow without channels is rather unstable.

Characteristics of fluvioglacial deposits

Stratified deposits laid down by meltwater can be recognized by a variety of criteria, although often it is difficult to differentiate between them and true fluvial deposits laid down in upland environments or during times of flood discharge. Cailleux and Tricart (1959) decided that fluvioglacial clasts could be distinguished from fluvial clasts on the basis of a substantially lower degree of rounding, and this has been confirmed by a variety of other studies (King, 1969). It is often possible to differentiate between the roundness of clasts in till and those affected by fluvioglacial processes, although it should be borne in mind that till often contains a proportion of fluvioglacial materials and that stratified sands and gravels often contain a proportion of fragments derived from till and then redeposited only a short distance from their point of input. King and Buckley (1968) decided that there were significant differences in the roundness values for materials from Baffin Island moraines, kames and eskers, and current studies by D. K. Chester in northeast Scotland reveal that irregular kames have less well rounded cobbles than eskers, which in turn contain less well rounded cobbles than outwash deposits. It remains to be seen whether these distinctions can be sustained in other areas.

Most pebbles in fluvioglacial deposits are not particularly far travelled, and this explains their relatively poor degree of rounding. Lee (1965) found that most pebbles in a Canadian esker had travelled less than 10 km from their source, whereas sand and gravel particles had travelled much further. In Finland, Hellaakoski (1930) discovered rather longer distances of travel for pebbles of granite in an esker at Laitila, but concluded that most pebbles were probably derived from till rather than the bedrock outcrop itself. In southwest Wales, most of the rock types represented in fluvioglacial deposits are of strictly local origin, and there is seldom more than 5 per cent of erratic pebbles which have travelled more than 5 km from their source.

Particle size distributions for fluvioglacial deposits are clearly different from those of tills, and plotted curves show a less obvious (or completely absent) bimodal distribution. This is because most of the particles in sand and gravel sequences are *rock fragments*; the mineral particles (of silt and clay size) which are so characteristic of till have been largely flushed out. Plotted as cumulative distribution curves on semi-log paper, many curves for fluvioglacial sands and gravel samples are lacking in sharp breaks of slope although differing widely in the precise positions occupied on the graph (Figure 16.7). It should be borne in mind that there is effectively no upper size limit for material contained in fluvioglacial deposits. Boulders or even huge blocks of bedrock may find their way from the ice into meltwater streams (Lister, 1973). Maag (1972) recorded an instance of one boulder weighing 1000 kg being moved 200 m in one flood season in a meltwater stream on Axel Heiberg Island. While even larger clasts may be only intermittently transported by the stream, they may be subject to rounding by attrition before being buried by sands and gravels.

The stratigraphy of ice-contact fluvioglacial deposits can be complex, although as noted above many 'normal' bedding features are developed. Often, beds are sharply truncated. Eskers are frequently characterized by arched bedding, indicating to Granö (1958) and Shreve (1972) that deposition of individual layers may take place on the crest and flanks of a ridge at the same time. Again, esker stratigraphy as clear as that displayed in Mannerfelt's

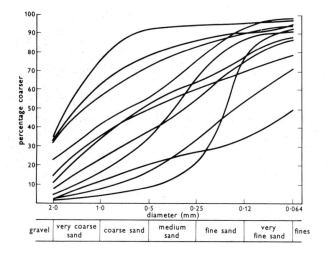

Figure 16.7
Particle-size distributions for eleven samples of fluvioglacial deposits from Pembrokeshire, west Wales.

(1945) classic diagrams suggests that ice-contact fluvioglacial features can retain most of their sedimentary structures even after the removal of the containing ice walls (Figure 16.8). However, in some situations it is possible to interpret arch structures as let-down arching, particularly if there is evidence that an esker has formed subglacially.

One interesting characteristic of kames and kame terraces is the 'coarsening-upwards' commented upon by many investigators (e.g. Shaw, 1971; Francis, 1970; John, 1972). Sometimes this sequence of basal silts and clays, middle sands, and upper coarse gravels can be explained by the buildup and progression of a fluvioglacial sequence across an ice marginal or proglacial area, with the finer sediments representing the early distal end of the sediment suite and the coarser sediments the late proximal end. Sometimes, it is possible to interpret the sequence as a deltaic or partly deltaic succession, and sometimes it may be due to a gradual increase in stream discharge carrying and then depositing increasingly coarse detritus. Alternatively, a sedimentary sequence capped by a cobble or boulder bed might be the result of a decrease in stream discharge (perhaps due to the onset of winter or the closure of a tunnel section further upstream) leading to the abandonment of the migrating bedload. On occasion, as in the case of the kame terrace at Mullock Bridge, Southwest Wales (John, 1972b) and in parts of County Durham (Carruthers, 1953), the sequence grades upwards into till. If such a sequence is in reality related to one glacial phase, the till cap may be interpreted as a flow till derived from adjacent downwasting ice (Hartshorn, 1958; Boulton, 1967). If on the other hand, the till cover is thick and continuous there may be

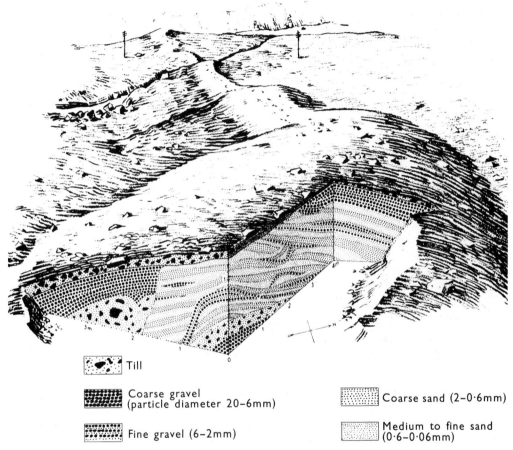

Till

Coarse gravel
(particle diameter 20–6mm)

Coarse sand (2–0·6mm)

Fine gravel (6–2mm)

Medium to fine sand
(0·6–0·06mm)

Figure 16.8
Cross-profile and longitudinal section through a subglacially engorged esker in Sylälvsdalen, Sweden. *(Reproduced by kind permission of C. M. Mannerfelt, Geogr. Annlr 1945.)*

no alternative but to interpret the underlying stratified sequence as deposited subglacially as suggested by Carruthers, perhaps in a subglacial water-body.

Much has been written about the secondary structures which characterize meltwater deposits. Most of these are the direct result of sediment collapse or settling after the melting of buried or containing ice. It has often been noticed that there are violent ice-contact structures on the flanks of eskers (Embleton and King, 1975; Flint, 1971; Synge, 1950), including faults, recumbent folds and steeply dipping, occasionally vertical, bedding. Similarly, complex contortions of bedding are known from the ice-contact slopes of many kames and kame terraces (Figure 16.9), and some examples are given by Shaw (1971) from Shropshire. Not all of the structures are related to faulting and folding; diapiric structures, dyke injections, collapse structures and convolutions are common also. Often these are related to the overloading of saturated fine sediments, and it is possible that some diapirism, for example, may be due to liquefaction effected by the shocks of faulting. In addition to the arched bedding referred to above as characteristic of subglacial eskers, it should not be forgotten that many eskers and kames are formed englacially or supraglacially, which means that they receive their final form only after being let down by the melting of underlying ice (Tanner, 1932; 1934; 1937). Ice cored eskers in process of formation have been described by Lewis (1949) and Stokes (1958) in Norway, Price (1966; 1969) in Alaska and Iceland,

Figure 16.9
Faulting in sands and gravels on the ice-contact flank of a kame terrace at Mullock Bridge, southwest Wales.

and Meier (1951) in Wyoming. The environments described are so complex, and slumping and settling on the flanks of the ice core so rapid and widespread, that it is surprising that any sedimentary beds and structures can survive at all. On the other hand, where the rate of ice dissolution is very slow even quite delicate sedimentary structures may be preserved as pointed out in the section on melt-out processes in chapter 11. Nevertheless, the fact that bedding does survive reasonably well in the vast majority of meltwater deposits from widely differing environments suggests strongly that most of them were laid down in subglacial, rather than englacial or supraglacial, situations.

Fluvioglacial landforms

Already there has been some mention in the foregoing paragraphs of the main types of landform resulting from meltwater deposition – eskers, kames and kame terraces. All of these

Table 16.1   A classification of fluvioglacial ice-contact forms

Linear features				Non-linear features
Parallel to ice flow		Transverse to ice flow		(no consistent orientation)
Subglacial	Marginal	Subglacial	Marginal	
Most types of eskers	Some eskers Kame terraces	Subglacially engorged eskers Some kames	Crevasse fillings Delta-moraines Delta-kames	Most kames Collapse features Kettles and pitted outwash

are formed in ice contact situations and all of them may be modified more or less during the process of deglaciation. Some features, particularly eskers such as those described by Price (1973) from glacier snout situations, may disappear entirely as the glacier retreats and as they are affected by the processes of the pro-glacial zone. But very many features do survive, and they are described and analysed in a vast literature. However, it is worth referring here to the main landforms associated with meltwater deposition in ice-contact situations. These are classified in Table 16.1. As in the classification of morainic landforms (Table 12.1) the subdivisions are based both upon morphology and location with respect to the direction of ice movement. It is worth noting, as with all the other glacial landforms discussed in this book, that the selected features are but parts of a whole continuum of forms.

Linear features parallel to ice flow
Eskers are the classic linear features which generally show a close correspondence with the most recent direction of regional ice movement (Embleton, 1964; Sissons, 1967). Figure 16.10 gives some indication of the main characteristics of esker ridges in a part of Sweden. The rather sinuous appearance of the main esker is typical, as is the occasional 'beaded' form

Figure 16.10
Oblique aerial photograph of an esker system in Lake Rörströms-sjön, Ångermandland, Sweden. *(Photograph by Erling Lind-ström.)*

and the appearance of confluent and diffluent eskers joining the main ridge. There is little purpose to be served here by a detailed description of esker forms or by a review of the extensive literature on esker formation. The topic has been covered in great detail by Tanner (1938), Embleton and King (1975) and Price (1973), and the reader is referred to these books for enlightenment.

There seems little doubt that eskers can form beneath ice which is still active, for their orientation is generally in close sympathy with the most recent set of striations on glaciated bedrock (Strömberg, 1962; Lundqvist, 1960). In addition, eskers can partially disregard the details of local relief (Shreve, 1972), crossing depressions and hills over a height range of 250 m or more as they follow an overall orientation controlled by the ice surface gradient (Flint, 1971). It can be argued that the factors controlling the orientation of long tunnels in active ice (see p. 291) are also relevant here. Subglacial meltwater discharge must be sufficient for channels to remain open in spite of the natural tendency of ice to close them, but water velocity must decline in order to allow sedimentation to take place. Although Shreve (1972) has argued persuasively for esker formation beneath active ice, it should be added that esker preservation is most likely beneath dying or inactive ice.

Another interpretation of the relationship between esker orientation and ice surface gradient is that of Tanner (1937). From his studies of eskers above the highest shoreline in Finland, he concluded that they must have been formed in open channels on the ice surface and then let down or superimposed upon the relief. In Finland and Sweden, many eskers were formed beneath the highest shoreline in association with grounded ice which was generally active right up to the snout. Such eskers, like that in Figure 16.10, follow the low points in the relief, supporting De Geer's hypothesis of formation as a series of connected marginal or sub-marginal accumulations (Granö, 1958; Donner, 1965; Fogelberg, 1970).

On the problem of esker genesis, the most critical points may be summarized as follows:

1   The longest eskers are ice-directed features which are even more accurate indicators than drumlins of the last direction of regional ice movement.
2   Eskers indicate the position of the major routes by which meltwater and fluvioglacial debris is transported from within the confines of the glacier system to achieve output at the ice margin.
3   Eskers are probably not true equilibrium forms, although the channels in which they form may have been at the time of channel formation. Eskers form only when stream velocity is falling, for example following the early summer meltwater discharge peak. Some eskers may form in tunnels in the late summer and autumn and be destroyed during the following early summer flood, as long as the containing tunnel remains in use. If a tunnel is abandoned then the esker will be able to survive as long as the glacier does not remove it.
4   The mere presence of eskers indicates free meltwater drainage towards the glacier margin. If meltwater flow is inhibited, the ice becomes saturated and channels (if they form at all) are filled with standing water. Again this may be a seasonal phenomenon; in the spring existing englacial tunnels may experience free meltwater flow in the vadose zone, but later on, during and after the discharge peak, the tunnels may find themselves well below the water table, in the zone of phreatic flow. Under these circumstances sediments are less likely to be deposited in tunnels.

Apart from eskers, other ice-directed subglacial or englacial fluvioglacial forms are rare.

Very close to a glacier snout, individual eskers may merge to give a group of elongated kames which may grade into a delta-moraine or delta-kame. Examples are some of the feeders for the Galtrim moraine in Ireland (Synge, 1960) and for the Salpausselkä ridges in Finland (Fogelberg, 1970). These feeders may be barely discernible in a mass of hummocky dead-ice features, but they may retain traces of orientation in the general direction of meltwater flow (De Geer, 1897). In parts of Sweden, there are esker nets composed of many inter-connected esker ridges together with sediment plateaux and mounds, transverse ridges and kettle holes (Lindström, 1973). Although these nets probably formed beneath stagnant ice, there is a preferred orientation parallel with the last direction of ice movement.

The second major type of landform parallel to ice flow is the kame terrace, normally formed by meltwater deposition between a valley side and a downwasting ice margin (Goldthwait, 1968). Like lateral moraines, they may be considered as ice-contact forms which may be modified drastically by the melting out of the ice walls which support them on their down-slope sides, or of ice cores (Mickelson, 1971; McKenzie, 1969). Price (1973) considered that they are probably formed as a result of fluvioglacial deposition in standing water, but this is unnecessarily restrictive, and many kame terraces are composed largely of coarse bedded gravels and sands with no trace of true deltaic bedding. As noted by Flint (1971), kame terraces often slope downvalley with a gradient approximately similar to that of the ice surface. Terraces may be 'paired' on opposite valley walls, and sometimes terrace series on a slope indicate successive positions of the downwasting ice margin. Sissons (1958) provided an excellent example from East Lothian, where four distinct kame terraces can be distinguished on slopes between altitudes of 244 and 320 m. Often in Scotland the highest terraces are sloping and the lowest almost horizontal.

Linear features transverse to ice flow

A category of esker which has received inadequate attention in most textbooks is the 'subglacially engorged esker' common in many parts of Scandinavia. Eskers of this type were described in detail by Mannerfelt (1945) and they also occur in parts of Arctic Canada (Bird, 1967). Bird has called them 'valley eskers' because they are generally formed at an advanced stage of ice disintegration when their orientation is controlled above all by local topography. They are located on valley sides where meltwater streams from lateral situations enter the ice; they are orientated diagonally down-valley, partly influenced by the direction of ice movement (see chapter 14). Often they are associated with kame terraces or lateral drainage channels at their upper ends and with kame and kettle topography at their lower ends. They are shorter and straighter than normal eskers, and because of their situations they are seldom destroyed by pro-glacial meltwater erosion or buried under sandurs or outwash plains. Also because of their situations they may be easily mistaken for fragments of end moraines (Sugden, 1970).

Other related linear forms composed largely of stratified drift are the crevasse fillings described by Flint (1928). These are only occasionally subglacial, being formed normally by the filling of crevasses open to the atmosphere. If the filled crevasses were of the radial type (Sharp, 1960) and thus orientated perpendicular to the glacier margin, the crevasse fillings could be aligned with local eskers and easily confused. But crevasses of this type are unusual in ice cap and ice sheet margins, and most commonly crevasse fillings are orientated apparently randomly or transverse to the direction of the ice movement.

The other major meltwater features controlled by ice margin orientation are the delta-moraines or delta-kames referred to above (Figure 16.11). These are the only real meltwater features consistently formed transverse to the direction of ice movement. They are almost always marginal, although rarely they may be sub-marginal if a glacier snout is located in deep water. An excellent example of a delta-moraine is the impressive Holger Dansker Briller moraine in Kjove Land, east Greenland, formed at the culmination of a glacier re-advance when sea-level was 101 m higher than at present (Sugden and John, 1965). This moraine has many of the classic attributes of both deltas and moraines. Delta-moraines in northwest Iceland commonly show a gradation from till with a litter of erratic boulders on the proximal flank to bedded gravels, sands and silts on the distal flank.

Figure 16.11
A fine example of a delta-moraine from **Angujartorfik** Søndre Strømfjord, Greenland. Note the very flat upper surface and the steep proximal slope. *(Photograph by Valerie Haynes.)*

Non-linear features

There is a large group of fluvioglacial landforms which seem to be devoid of any consistent orientation with respect to the direction of ice movement. Best known are the features broadly classified as kames (Holmes, 1947; Cook, 1946a). They are made of stratified sediments, and have a variety of surface expressions such as mounds, hummocks, sediment plateaux and discontinuous terraces. They are really ice disintegration features, formed in ice marginal, subglacial, englacial or supraglacial situations wherever cavities happened to be available for the receipt of water-borne sediment (Clayton, 1964). Depending upon the manner of ice disintegration, kames may be relatively minor or predominant features of the landscape. If sedimentation has taken place in widely spaced cavities in a more or less continuous ice mass, groups of kames may stand above the ground surface as positive features often referred to as 'kame complexes' (Cook, 1946b). Alternatively, closer to the wasting glacier snout where meltwater deposition may have been much more extensive, the greater part of the ground surface may be composed of kames and outwash plateaux. There is an excellent example in the 'Valle härad' kame complex to the west of Billingen, Sweden.

Figure 16.12
An example of pitted outwash
from an old sandur in Gröndalen,
Sweden. (*Photograph by G. Lundqvist.*)

In many low lying areas subjected to rapid ice wastage, detached dead ice masses may be buried beneath outwash sands and gravels. When these masses melt out they produce negative relief features such as pits and depressions (Rich, 1943; Price, 1969). This type of topography is referred to as pitted outwash or kame and kettle topography (Figure 16.12), and it is obviously closely related to the disintegration moraine referred to in chapter 12. Often there is so much mixing of till and fluvioglacial materials, and of morainic and melt-water landforms, that differentiation becomes impossible. Hence the term 'dead-ice topography' is in some respect more suitable since it can be used to encompass both morainic and meltwater features.

The various types of kame and kettle topography are adequately discussed in the texts, and they will not be considered further here in view of the overlap with what has been said in chapter 12. The reader is referred to the papers by Holmes (1947), Cook (1946a; b), McKenzie (1969) and Clayton (1964) for further interesting data on kame and kettle characteristics.

Fluvioglacial landscapes

As mentioned in chapter 13, fluvioglacial accumulations are occasionally so extensive (particularly in the ice sheet 'esker zone' and 'ice disintegration zone') that wide areas may be referred to as fluvioglacial landscapes. In these areas eskers, kames, delta-moraines and large outwash accumulations may be found in juxtaposition, occasionally partly or completely burying the morainic topography which may be the representative of an earlier phase of glacier deposition. There are good examples from British Columbia (Ryder, 1972), Ireland (Synge, 1970), and eastern Norway (Andersen, 1965). Generally these expanses of fluvioglacial landforms are related to mid-latitude ice caps and ice sheets which have disintegrated rapidly by downwasting. They are less common in glaciated troughs in the uplands. This is partly because valley glaciers often remain active during retreat, inhibiting the survival of subglacial or englacial meltwater forms and allowing fluvioglacial landforms to persist only in ice marginal and pro-glacial situations. Also, meltwater features built on valley floors are often destroyed or buried by pro-glacial fluvial processes, which have to operate on a relatively narrow valley floor.

Nevertheless, there are many fine examples of fluvioglacial landform associations which have managed to survive in outlet glacier troughs (Figure 16.13). In the mid latitudes, the most suitable locations seem to have been troughs and other depressions orientated transversely to the direction of regional ice movement. Ice in such depressions was often cut off from its source of supply, and stagnated *in situ*, providing ideal situations for the concentration of meltwater and hence the formation of suites of fluvioglacial landforms (Sugden, 1970). Fluvioglacial landscapes can also be found in outlet glacier troughs parallel with the direction of regional ice movement, particularly where there has been rapid ice stagnation such as that which follows a surge. The valley of Kaldalon in north west Iceland is a fine example where kame terraces, subglacially engorged eskers, kettles, kames, esker ridges and pitted outwash are all found together inside a Neoglacial terminal moraine (John and Sugden, 1962). Another spectacular fluvioglacial landscape is shown in Figure 16.14.

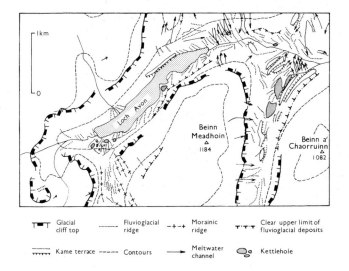

Figure 16.13
Fluvioglacial deposits and other features associated with stagnating ice in two troughs of the central Cairngorm mountains, Scotland.

Bearing in mind what has already been said in chapters 14 and 15, it is worth drawing attention to the regional contrasts which occur in fluvioglacial landscape development. The features described in this chapter are clearly best developed in the northern hemisphere middle latitudes, where deglaciation was generally accompanied by large scale meltwater production, and where the annual melting regime encouraged very high stream power values during the early summer. On the other hand, as pointed out earlier, fluvioglacial depositional landforms are extremely rare in the vicinity of tropical glaciers; much ablation is achieved by direct evaporation, and the more or less regular diurnal melting cycle throughout the year means that there is no build-up of meltwater in any particular season. Hence meltwater plays only a minor role in the transport of glacial detritus, and almost all output from the glacier system is by strictly glacial processes, leading to spectacular moraines.

In polar areas, meltwater production is also minimal, for melting rates are lowered by reduced summer air temperatures. Most meltwater which is produced remains in supraglacial situations, especially on glaciers which are frozen to their beds in their marginal zones. In the sub-Antarctic, the authors have noted that there were no extensive fluvioglacial sedi-

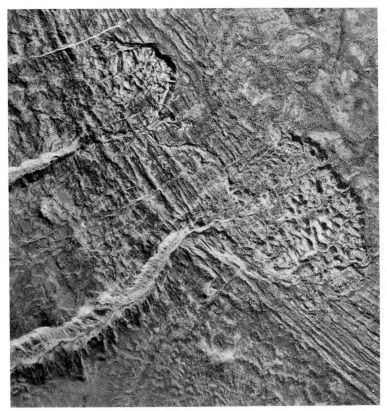

Figure 16.14
Vertical air photograph (scale 1 : 25,000) of two accumulations of fluvioglacial deposits which formed be-neath wasting ice in Herrejaure, Lule Lappmark, Sweden. The main feeder channels are clearly shown, as are the marginal channels marking the former positions of the downwasting ice edge. Both eskers and kames can be seen among the depositional features. *(Copyright Statens Lantmäteriverk, Sweden.)*

ments or landforms in the South Shetland Islands, in spite of meltwater production near glacier margins. And in the Canadian Arctic, Bird (1967) has commented that the relatively poor development of Wisconsin dead-ice landforms in areas such as Keewatin was probably the result of a cold thermal regime, even during ice wastage, which inhibited the production of basal meltwater. For some parts of sub-polar glaciers, the proportions of meltwater which can operate englacially or subglacially are much greater (Rucklidge, 1956; Baranowski, 1973). Maag (1969; 1971) and Iken (1972) have shown that meltwater on White glacier, Axel Heiberg Island, and other glaciers in the vicinity, is both glaciologically and geomor-phologically effective. Fluvioglacial depositional forms are rapidly created, albeit on a small scale compared with mid latitude areas. Most of these forms are supraglacial or marginal, and it is debatable whether extensive esker systems, for example, can form and survive be-neath sub-polar glaciers. As in the other fields of study discussed in this book, there is ample scope here for both the armchair theorist and the intrepid field worker.

Further reading

ALLEN J. R. L. 1970: *Physical processes of sedimentation.* Unwin Univ. Books, London (248 pp).

JOPLING, A. V. and MCDONALD, B. C. (eds), 1975: *Glaciofluvial and glaciolacustrine sedimentation.* Special Publication of Economic Paleontologists and Mineralogists, 23, Tulsa, Oklahoma (320 pp).

LUNDQVIST, J. 1972: Ice-lake types and deglaciation pattern along the Scandinavian mountain range. *Boreas* **1** (1), 27-54.

PRICE, R. J. 1973: *Glacial and fluvioglacial landforms.* Oliver and Boyd, Edinburgh (242 pp).

SAUNDERSON, H. C. 1977: The sliding bed facies in esker sands and gravels: a criterion for full-pipe (tunnel) flow? *Sedimentology* **24**, 623-38.

SHAW, J. 1972: Sedimentation in the ice-contact environment, with examples from Shropshire (England). *Sedimentology* **18**, 23-62.

SHREVE, R. L. 1972: Movement of water in glaciers. *J. Glaciol.* **11** (62), 205-14.

Additional references (March 1982)

Important symposia volumes are:

Symposium on dynamics of large ice masses, Ottawa, 21-25 August, 1978. J. Glaciol. **24** (90), (520 pp), 1979.

Symposium on glacier beds: the ice-rock interface, Ottawa, 15-19 August, 1978. J. Glaciol. **23** (89), (445 pp), 1979.

Proceedings of the symposium on processes of glacial erosion and sedimentation, Geilo, Norway, 25-30 August, 1980. Annals of Glaciology **2** (192 pp), 1981.

SCHLÜCHTER, CH., 1979: *Moraines and varves: origin, genesis, classification.* Balkema, Rotterdam (441 pp).

Others:

BOULTON, G. S. 1978: Boulder shapes and grain size distribution as indicators of transport paths through a glacier, and till genesis. *Sedimentology* **25**, 773-99.

COLBECK, S. C. (Ed) 1980: *Dynamics of snow and ice masses.* Academic Press, New York.

COLLINS, D. N. 1977: Hydrology of an alpine glacier as indicated by the chemical composition of meltwater. *Zeits. für Gletscherkunde und Glazialgeologie* **13**, 219-38.

DENTON, G. H. and HUGHES, T. J. 1981: *The last great ice sheets.* Wiley, New York (484 pp).

EYLES, N. 1979: Facies of supraglacial sedimentation on Icelandic and Alpine temperate glaciers. *Can. J. Earth Sci.* **16**, 1341-61.

GORDON, J. 1981: Ice-scoured topography and its relationship to bedrock structure and ice movement in parts of northern Scotland and West Greenland. *Geogr. Annlr. Stockh.* **63A** (1-2), 55-65.

GROVE, J. 1979: The glacial history of the Holocene. *Progress in Physical Geography* **3** (1), 1-54.

HAMBREY, M. J., MILNES, A. G. and SIEGENTHALER, H. 1980: Dynamics and structure of Griesgletscher, Switzerland. *J. Glaciol.* **25**, 215-28.

LAWSON, D. E. 1979: Sedimentological analysis of the western terminus region of the Matanuska Glacier, Alaska. *U.S. Army Cold Reg. Res. Engng. Lab. Rept.*, 79-9, (112 pp).

LORIUS, C., MERLIVAT, L., JOUZEL, J. and POURCHET, M. 1979: A 30,000-year isotope climatic record from Antarctic ice. *Nature* **280**, 644-8.

MARTINELLI, M., SOMMERFIELD, R. A., THOMAS, R., HARRISON, W., DENHARTOG, S., FRANKENSTEIN, G., RANGO, A. and SHARTRAN, M. 1979: Research on snow and ice. *Rev. Geophys. and Space Phys.* **17**, 1253-88.

MENZIES, J. 1978/79: A review of the literature on the formation and location of drumlins. *Earth-Science Reviews* **14**, 315-59.

MENZIES, J. 1979: The mechanics of drumlin formation with particular reference to the change in pore-water content of the till. *J. Glaciol.* **22** (87), 373-84.

PATERSON, W. S. B. 1981: *The physics of glaciers.* Pergamon, Oxford. (380 pp). Second edition.

SOUCHEZ, R. A. and LORRAIN, R. D. 1978: Origin of the basal ice layer from Alpine glaciers indicated by its chemistry. *J. Glaciol.* **20** (82), 319-28.

VIVIAN, R. 1980: The nature of the ice-rock interface. *J. Glaciol.* **25** (92), 267-77.

WHITTECAR, G. R. and MICKELSON, D. M. 1979: Composition, internal structures, and an hypothesis of formation of drumlins, Waukesha County, Wisconsin, U.S.A. *J. Glaciol.* **22** (87), 357-71.

References

AARTOLAHTI, T. 1972: On deglaciation in southern and western Finland. *Fennia* **114** (84 pp).

ADAM, D. P. 1969: Ice ages and the thermal equilibrium of the earth. *Interim Res. Rept* **15**, Dept Geochronology, Univ. Arizona (26 pp).

1975: Ice ages and the thermal equilibrium of the earth II. *Quat Res.* **5** (2), 161–71.

ADAMS, W. P. 1966: Ablation and runoff on the White glacier. *Axel Heiberg Res. Rept, Glaciology* **1**, McGill Univ., Montreal (77 pp).

ADIE, R. J. 1975: Permo-Carboniferous glaciation of the southern hemisphere. In Wright, A. E. and Moseley, F. (eds) *Ice ages: ancient and modern*, Seel House Press, Liverpool, 287–300.

AHLMANN, H. W:SON 1919: Geomorphological studies in Norway. *Geogr. Annlr* **1**, 1–210.

1937: Vatnajökull in relation to other present day Iceland glaciers. Ch. 4 in Ahlmann, H. W:son and Thorarinsson, S., Vatnajökull, *Geogr. Annlr* **19**, 146–231

1948: Glaciological research on the North Atlantic coasts. *R. geogr. Soc. Res. Ser.* **1** (83 pp).

ALEXANDER, H. S. 1932: Pothole erosion. *J. Geol.* **40**, 305–37.

ALEXANDER, M. and WORSLEY, P. 1973: Stratigraphy of a Neoglacial end moraine in Norway. *Boreas* **2** (3), 117–42.

ALLEN, J. R. L. 1970: *Physical processes of sedimentation.* Unwin Univ. Books, London (248 pp).

1971a: Transverse erosional marks of mud and rock: their physical basis and geological significance. *Sedimentary Geology* **5** (3/4), 1–385.

1971b: A theoretical and experimental study of climbing-ripple cross-lamination, with a field application to the Uppsala esker. *Geogr. Annlr* **53**A, 157–87.

ANDERSEN, B. G. 1965: The Quaternary of Norway. In Rankama, K. (ed) *The Quaternary* **1**, Interscience, New York, 91–138.

ANDERSEN, J. L. and SOLLID, J. L. 1971: Glacial chronology and glacial geomorphology in the marginal zones of the glaciers Midtdalsbreen and Nigardsbreen, south Norway. *Norsk geogr. Tidsskr.* **25**, 1–38.

ANDERSEN, S. A. 1931: Om Aase og Terrasser inden for Susas Vadomraade. *Dan. Geol. Unders.* **2** (54) (201 pp).

ANDREWS, J. T. 1963: The cross-valley moraines of north-central Baffin Island, NWT: a descriptive analysis. *Geogr. Bull.* **19**, 49–77.

1970: A geomorphological study of postglacial uplift with particular reference to Arctic Canada. *Inst. Br. Geogr. Sp. Pub.* **2** (156 pp).

1972a: Glacier power, mass balances, velocities and erosion potential. *Z. Geomorph.* NF **13**, 1–17.

1972b: Englacial debris in glaciers. *J. Glaciol.* **11** (61), 155.

1973: The Wisconsin Laurentide ice sheet: dispersal centers, problems of rates of retreat, and climatic implications. *Arctic and Alpine Res.* **5** (3), 185–99.

1974: Cainozoic glaciations and crustal movements of the Arctic. In Ives, J. D. and Barry, R. G. (eds) *Arctic and alpine environments*, Methuen, London, 277–317.

ANDREWS, J. T. and BARNETT, D. M. 1972: Analysis of strandline tilt directions in relation to ice centres and post-glacial crustal deformation, Laurentide ice sheet. *Geogr. Annlr* **54**A, 1–11.

ANDREWS, J. T., BARRY, R. G., BRADLEY, R. S., MILLER, G. H. and WILLIAMS, L. D. 1972: Past and present glaciological responses to climate in eastern Baffin Island. *Quat. Res.* **2**, 303–14.

ANDREWS, J. T., BARRY, R. G. and DRAPER, L. 1970: An inventory of the present and past glacierization of Home Bay and Okoa Bay, east Baffin Island, NWT, Canada, and some climatic and palaeoclimatic considerations. *J. Glaciol.* **9** (57), 337–62.

ANDREWS, J. T. and DUGDALE, R. E. 1971: Quaternary history of northern Cumberland Peninsula, Baffin Island, NWT, part V: factors affecting corrie glacierization in Okoa Bay. *Quat. Res.* **1** (4), 532–51.

ANDREWS, J. T., FAHEY, B. D. and ALFORD, D. 1971: Note on correlation coefficients derived from cumulative distributions with reference to glaciological studies. *J. Glaciol.* **10** (58), 145–7.

ANDREWS, J. T., FUNDER, S., HJORT, C. and IMBRIE, J. 1974: Comparison of the glacial chronology of eastern Baffin Island, east Greenland, and the Camp Century accumulation record. *Geology* **2** (7), 355–8.

ANDREWS, J. T. and IVES, J. D. 1972: Late- and postglacial events (< 10,000 BP) in the eastern Canadian Arctic with particular reference to the Cockburn moraines and break-up of the Laurentide ice sheet. In Vasari, Y., Hyvärinen, H. and Hicks, S. (eds) *Climatic changes in Arctic areas during the last ten thousand years*, Oulu, 149–74.

ANDREWS, J. T. and KING, C. A. M. 1968: Comparative till fabrics and till fabric variability in a till sheet and a drumlin; a small-scale study. *Proc. Yorks. geol Soc.* **36**, 435–61.

ANDREWS, J. T. and LEMASURIER, W. E. 1973: Rates of Quaternary glacial erosion and corrie formation, Marie Byrd Land, Antarctica. *Geology* October 1973, 75–80.

ANDREWS, J. T. and MAHAFFY, M. A. 1975: Rate of growth of the Laurentide ice sheet based on a three-dimensional numerical ice flow model. In *Abstracts with programs* **7** (7), Geol Soc. Am., 977–978.

ANDREWS, J. T. and SMITHSON, B. B. 1966: Till fabrics of the cross-valley moraines of north-central Baffin Island, NWT, Canada. *Bull: geol Soc. Am.* **77**, 271–90.

ANON. 1966: Mass balance terms. *J. Glaciol.* **8** (52), 3–7.

ARMSTRONG, T. E., ROBERTS, B. and SWITHINBANK, C. 1973: *Illustrated glossary of snow and ice* (2nd edn), Scott Polar Research Institute, Cambridge (60 pp).

ARRHENIUS, G. 1952: Sediment cores from the east Pacific. *Swedish Deep-Sea Exped. Rept* **5**, Elander, Göteborg (227 pp).

AVSIUK, G. A. 1955: Temperaturnoe Sostoianie Lednikov. *Iszestiia Akademii Nauk SSR, Seriia Geograficheskaia* **1**, 14–31.

BAIN, G. W. 1931: Spontaneous rock expansion. *J. Geol.* **39**, 715–35.

BAKKER, J. P. 1965: A forgotten factor in the interpretation of glacial stairways. *Z. Geomorph.* NF **9**, 18–34.

BAKKER, J. P. and LE HEUX, J. W. N. 1952: A remarkable new geomorphological law. *Koninklijke Nederlandsche Akademie van Wetenschappen B* **50**, 959–66 and 1154–62.

BANHAM, P. H. 1966: The significance of till pebble lineations and their relation to folds in two Pleistocene tills at Mundesley, Norfolk. *Proc. Geol Ass.* **79**, 507–12.

    1975: Glacitectonic structures: a general discussion with particular reference to the contorted drift of Norfolk. In Wright, A. E. and Moseley, F. (eds) *Ice ages: ancient and modern*, Seel House Press, Liverpool, 69–94.

BANHAM, P. H. and RANSON, C. E. 1968: Structural study of the contorted drift and disturbed chalk at Weybourne, north Norfolk. *Geol. Mag.* **102**, 167–74.

BANNACEF, A., BEUF, S., BIJU-DUVAL, B., DeCHARPAL, O., GARIEL, O., and ROGNON, P. 1971: Example of cratonic sedimentation—Lower Palaeozoic of Algeria Sahara. *Am. Ass. Petr. Geol. Bull.* **55**, 2225–2245.

BARANOWSKI, S. 1970: The origin of fluted moraine at the fronts of contemporary glaciers. *Geogr. Annlr* **52**, 68–75.

    1973: Geyser-like water spouts at Werenskioldbreen, Spitzbergen. *Symposium on the Hydrology of Glaciers, Cambridge, 9–13 Sept. 1969, Int. Ass. scient. Hydrol. Pub.* **95**, 131–3.

BARNES, H. L. 1956: Cavitation as a geological agent. *Am. J. Sci.* **254**, 493–505.

BARNETT, D. M. and HOLDSWORTH, G. 1974: Origin, morphology, and chronology of sublacustrine moraines, Generator Lake, Baffin Island, NWT, Canada. *Can J. Earth Sci.* **11** (3), 380–408.

BARRY, R. G. 1973: Conditions favouring glacierization and deglacierization in North America from a climatological viewpoint. *Arctic and Alpine Res.* **5** (3), 171–84.

BARRY, R. G. and HARE, F. K. 1974: Arctic climate. In Ives, J. D. and Barry, R. G. (eds) *Arctic and alpine environments*, Methuen, London, 17–54.

BATTEY, M. H. 1960: Geological factors in the development of Veslgjuv-botn and Veslskautbreen. In Lewis, W. V. (ed) *Norwegian cirque glaciers*, R. geogr. Soc. Res. Ser.. **4**, 5–10.

BAUER, A. 1955: Contribution à la connaissance du Vatnajökull-Islande. *Jökull* **5**, 11–22.

BAYROCK, L. A. 1967: Catastrophic advance of the Steele glacier, Yukon, Canada. *Boreal Inst. Univ. Alberta, Edmonton, Occ. Pub.* **3** (35 pp).

BEAUMONT, P. 1971a: Stone orientation and stone count data from the lower till sheet, eastern Durham. *Proc. Yorks. geol Soc.* **38** (3), 343–60.

    1971b: Break of slope in particle-size curves of glacial tills. *Sedimentology* **16**, 125–8.

BENEDICT, J. B. 1973: Origin of rock glaciers. *J. Glaciol.* **12** (66), 520–22.

BERGERSEN, O. F. and FOLLESTAD, B. A. 1971: Evidence of fossil ice wedges in Early Weichselian deposits at Foss-Eikjeland, Jaeren, southwest Norway. *Norsk geogr. Tiddsskr.* **25**, 39–45.

BERGTHORSSON, P. 1969: An estimate of drift ice and temperature in Iceland in 1000 years. *Jökull* **19**, 94–101.

BERNARD, C. 1971a: Les marques sous-glaciaires d'aspect plastique sur la roche en place (p-forms): observations sur la bordure du bouclier canadien et examen de la question – 1. *Rev. Géogr. Montréal* **25** (2), 111–27.

1971b: Les marques sous-glaciaires d'aspect plastique sur la roche en place (p-forms): leur rapport avec l'environnement et avec certaines marques de corrasion – 2. *Rev. Géogr. Montréal* **25** (3), 265–279.

BEZINGE, A. and SCHAFER, F. 1968: Pompes d'accumulation et eaux glaciaires. *Bulletin technique de la Suisse Romande* **20**, *Symp. de l'AIRH, Lausanne, October 1968*, 282–90.

BIRD, J. B. 1967: *The physiography of Arctic Canada.* Johns Hopkins Press, Baltimore (336 pp).

BIRKELAND, P. W. 1974: *Pedology, weathering and geomorphological research.* OUP, London (285 pp).

BIROT, P. 1968a: *The cycle of erosion in different climates.* Batsford, London (144 pp).

1968b: Les développements recents des théories de l'érosion glaciaire. *Ann. Géogr.* **419**, 1–13.

BISHOP, B. C. 1957: Shear moraines in the Thule area, northwest Greenland. *US Army SIPRE Rept* **17** (46 pp).

BJÖRNSSON, H. 1971: Baegisárjökull, north Iceland. Results of glaciological investigations, 1967–8, part 1: mass balance and general meteorology. *Jökull* **3** (21), 1–23.

BLAKE, W. Jr 1970: Studies of glacial history in Arctic Canada. 1: Pumice, radiocarbon dates, and differential postglacial uplift in the eastern Queen Elizabeth Islands. *Can J. Earth Sci.* **7** (2), 634–64.

BOULTON, G. S. 1967: The development of a complex supraglacial moraine at the margin of Sorbreen, Ny Friesland, Vestspitsbergen. *J. Glaciol.* **6** (47), 717–36.

1968: Flow tills and related deposits on some Vestspitsbergen glaciers. *J. Glaciol.* **7** (51), 391–412.

1970a: On the origin and transport of englacial debris in Svalbard glaciers. *J. Glaciol.* **9** (56), 213–229.

1970b: The deposition of subglacial and melt-out tills at the margins of certain Svalbard glaciers. *J. Glaciol.* **9** (56), 231–45.

1971: Till genesis and fabric in Svalbard, Spitzbergen. In Goldthwait, R. P. (ed) *Till, a symposium,* Ohio State Univ. Press, 41–72.

1972a: The role of thermal regime in glacial sedimentation. In Price, R. J. and Sugden, D. E. (eds) *Polar geomorphology, Inst. Br. Geogr. Spec. Pub.* **4**, 1–19.

1972b: Englacial debris in glaciers: reply to the comments of Dr J. T. Andrews. *J. Glaciol.* **11** (61), 155–6.

1972c: Modern Arctic glaciers as depositional models for former ice sheets. *QJ geol Soc. Lond.* **128** (4), 361–93.

1974: Processes and patterns of glacial erosion. In Coates, D. R. (ed) *Glacial geomorphology,* State Univ. of NY, Binghampton, 41–87.

1975: Processes and patterns of sub-glacial sedimentation: a theoretical approach. In Wright, A. E. and Moseley, F. (eds) *Ice ages: ancient and modern,* Seel House Press, Liverpool, 7–42.

BOULTON. G. S. and VIVIAN, R. 1973: Underneath the glaciers. *Geogr. Mag.* **45** (4), 311–19.

BOWEN, D. Q. 1973: The Pleistocene succession of the Irish Sea. *Proc. Geol Ass.* **84** (3), 249–72.

BRADLEY, R. S. 1973a: Recent freezing level changes and climatic deterioration in the Canadian arctic archipelago. *Nature, Lond.* **243**, 398–400.

1973b: Seasonal climatic fluctuations in Baffin Island during the period of instrumental records. *Arctic* **6** (3), 230–43.

BRADLEY, R. S. and MILLER, G. H., 1972: Recent climatic change and increased glacierization in the eastern Canadian Arctic. *Nature, Lond.* **237**, 385–7.

BRAY, J. R. 1967: Variation in atmospheric carbon-14 activity relative to a sunspot-auroral solar index. *Science* **156**, 640–42.

1968: Glaciation and solar activity since the fifth century BC and the solar cycle. *Nature, Lond.* **220**, 672–4.

BRETZ, J. H. 1935: Physiographic studies in east Greenland. In Boyd, L. A. (ed) The fiord region of east Greenland, *Am. geogr. Soc. Spec. Pub.* **18**, 161–266.

BRISTOW, C. R. and COX, F. C. 1973: Eastern England (additional note). In Mitchell, G. F., Penny, L. F., Shotton, F. W. and West, R. G. (eds) A correlation of Quaternary deposits in the British Isles, *Geol Soc. Lond. Spec. Rept* **4** (99 pp).

BROCHU, M. 1954: Lacs d'érosion différentielle glaciaire sur le Bouclier Canadien. *Revue Géomorph. dyn.* **6**, 274–9.

BROOKS, C. E. P. 1949: *Climate through the ages* (2nd edn). Benn, London (395 pp).

BRYSON, R. A., WENDLAND, W. M., IVES, J. D. and ANDREWS, J. T. 1969: Radiocarbon isochrones on the disintegration of the Laurentide ice sheet. *Arctic and Alpine Res.* **1** (1), 1–13.

BUCKLEY, J. T. 1969: Gradients of past and present outlet glaciers. *Geol Surv. Pap. Can.* **69–29** (13 pp).

BUDD, W. 1966: The dynamics of the Amery ice shelf. *J. Glaciol.* **6** (45), 335–58.

1969: The dynamics of ice masses. *Australian Nat. Antarctic Exped. Series A* **4**. *Glaciology Pub.* **108** (216 pp).

BUDD, W. F. 1970: Ice flow over bedrock perturbations. *J. Glaciol.* **9** (55), 29–48.

1975: A first simple model for periodically self-surging glaciers. *J. Glaciol.* **14** (70), 3–22.

BUDD, W. F. and CARTER, D. B. 1971: An analysis of the relation between the surface and bedrock profiles of ice caps. *J. Glaciol.* **10** (59), 197–209.

BUDD, W., JENSSEN, D. and RADOK, U. 1970: The extent of basal melting in Antarctica. *Polarforschung* **6** (39), 293–306.

BUDD, W. F., JENSSEN, D. and RADOK, U. 1971: Derived physical characteristics of the Antarctic ice sheet (Mark 1). *Univ. Melbourne, Meteorol Dept Pub.* **18** (178 pp).

BÜDEL, J. 1948: Die klima-morphologischen Zonen der Polarländer. *Erdkunde* **2**, 22–53.

BUDYKO, M. I. 1956: *The heat balance of the earth's surface*. Gidrometeorologicheskoe Izd., Leningrad (255 pp).

BULL, C. 1963: Glaciological reconnaissance of the Sukkertoppen ice cap, southwest Greenland. *J. Glaciol.* **4** (36), 813–16.

1971: Snow accumulation in Antarctica. In Quam, L. O. (ed) Research in the Antarctic, *AAAS, Washington, Pub.* **93**, 367–421.

BULL, C. and CARNEIN, C. A. 1970: The mass balance of a cold glacier: Meserve glacier, south Victoria Land, Antarctica. In Gow, A. J. *et al.* (eds) *International Symposium on Antarctic Glaciological Exploration (ISAGE)*, *Int. Ass. scient. Hydrol. Pub.* **86**, 429–46.

BULL, C. and WEBB, P. N. 1973: Some recent developments in the investigation of the glacial history and glaciology of Antarctica. In van Zinderen Bakker, E. M. (ed) *Palaeoecology of Africa and of the surrounding islands and Antarctica* **8**, Balkema, Cape Town, 55–84.

CAILLEUX, A. and TRICART, J. 1959: Initiation à l'étude des sables et des galets. *Centre Doc. Univ. Paris* (3 vols: 376 pp, 194 pp, 202 pp).

CALDER, N. 1974: *The weather machine and the threat of ice*. BBC, London (143 pp).

CALKIN, P. E. 1973: Glacial processes in the ice free valleys of southern Victoria Land, Antarctica. In Fahey, B. D. and Thompson, R. D. (eds) Research in polar and alpine geomorphology, *3rd Guelph Symposium on Geomorphology*, Geo Abstracts, Norwich, 167–186.

CALKIN, P. E. and NICHOLS, R. L. 1972: Quaternary studies in Antarctica. In Adie, R. J. (ed) *Antarctic geology and geophysics*, Universitetsforlaget, Oslo. 625–43.

CAREY, S. W. and AHMAD, N. 1961: Glacial marine sedimentation. In Raasch, G. O. (ed) *Geology of the Arctic* **2**, Toronto Univ. Press, 865–94.

CAROL, H. 1947: The formation of *roches mountonnées*. *J. Glaciol.* **1** (1), 57–9.

CARRUTHERS, R. G. 1947, 1948: The secret of the glacial drifts. *Proc. Yorks. geol Soc.* **27**, 43–57 and 129–172.

1953: *Glaciol drifts and the undermelt theory*. Harold Hill, Newcastle-upon-Tyne (42 pp).

CATCHPOLE, A. 1972: Time and temperature: a model of temporal interrelationships. *Can Geogr.* **16** (4), 365–73.

CHAMBERLIN, T. C. 1885: The rock scorings of the great ice invasions. *US geol Surv. 7th Annual Rept*, 147–248.

1894: Proposed genetic classification of Pleistocene glacial formations. *J. Geol.* **2**, 517–38.

CHAPMAN, C. A. and RIOUX, R. L. 1958: Statistical study of topography, sheeting and jointing in granite, Acadia National Park, Maine. *Am. J. Sci.* **256**, 111–27.

CHAPPELL, J. 1973: Astronomical theory of climatic change: status and problem. *Quat. Res.* **3**, 221–236.

CHARLESWORTH, J. K. 1955: Late glacial history of the highlands and islands of Scotland. *Trans R. Soc. Edinb.* **62**, 769–928.

1957: *The Quaternary era*. Edward Arnold, London (2 vols, 1700 pp).

CHORLEY, R. J. 1959: The shape of drumlins. *J. Glaciol.* **3** (25), 339–44.

CHORLEY, R. J. and KENNEDY, B. A. 1971: *Physical geography: a systems approach*. Prentice-Hall International, London (370 pp).

CHORLTON, J. C. and LISTER, H. 1968: Snow accumulation over Antarctica. In Gow, A. J. *et al.* (eds) *International Symposium on Antarctic Glaciological Exploration* (ISAGE), *Int. Ass. scient. Hydrol. Pub.* **86**, 254–63.

1971: Geographical control of glacier budget gradients in Norway. *Norsk geogr. Tidsskr.* **25,** 159–164.

CHURCH, M. 1972: Baffin Island sandurs: a study of Arctic fluvial processes. *Bull. geol Surv. Can.* **216** (208 pp).

CLAPPERTON, C. M. 1968: Channels formed by the superimposition of glacial meltwater streams, with special reference to the east Cheviot hills, northeast England. *Geogr. Annlr* **50,** 207–20.

1971: The location and origin of glacial meltwater phenomena in the eastern Cheviot hills. *Proc. Yorks. geol Soc.* **38,** 361–80.

1972: The Pleistocene moraine stages of west central Peru. *J. Glaciol.* **11** (62), 255–63.

CLAPPERTON, C. M., GUNSON, A. R. and SUGDEN, D. E. 1975: Loch Lomond readvance in the eastern Cairngorms. *Nature* **253** (5494) 710–12.

CLAPPERTON, C. M. and SUGDEN, D. E. 1972: The Aberdeen and Dinnet glacial limits reconsidered. In Clapperton, C. M. (ed) *North east Scotland geographical essays*, Aberdeen, 5–11.

CLAYTON, K. M. 1965: Glacial erosion in the Finger Lakes region (New York State, USA). *Z. Geomorph.* **9,** 50–62.

CLAYTON, K. M. 1974: Zones of glacial erosion. In Brown, E. H. and Waters, R. S. (eds) *Progress in geomorphology, Inst. Brit. Geogr. Spec. Pub.* **7,** 163–76.

CLAYTON, K. M. and LINTON, D. L. 1964: A qualitative scale of intensity of glacial erosion. *20th int. geog. Congr. (London), Abstracts of papers,* 18–19.

CLAYTON, L. 1964: Karst topography on stagnant glaciers. *J. Glaciol.* **5** (37), 107–12.

CLAYTON, L. and MORAN, S. R. 1974: A glacial process form model. In Coates, D. R. (ed) *Glacial geomorphology,* State Univ. NY, Binghampton, 88–119.

COACHMAN, L. K., HEMMINGSEN, E. and SCHOLANDER, P. F. 1956: Gas enclosures in a temperate glacier. *Tellus* **8** (4), 415–23.

COLBECK, S. C. and EVANS, R. J. 1973: A flow law for temperate glacier ice. *J. Glaciol.* **12** (64), 71–86.

COLLINS, S. G. 1972: Survey of the Rusty glacier area, Yukon Territory, Canada 1967–70. *J. Glaciol.* **11** (62), 235–53.

CONOLLY, J. R., and EWING, M. 1965: Pleistocene glacial-marine zones in North Atlantic deep sea sediments. *Nature, Lond,* **208,** 135–8.

COOK, J. H. 1946a: Ice contacts and the melting of ice below a water level. *Am. J. Sci.* **244,** 502–12.

1946b: Kame complexes and perforation deposits. *Am. J. Sci.* **244,** 573–83.

COOKE, H. B. S. 1973: Pleistocene chronology: long or short? *Quat. Res.* **3,** 206–20.

COOKE, R. U., and WARREN, A. 1973: *Geomorphology in deserts.* Batsford, London (374 pp).

COOPE, G. R. 1970: Climatic interpretations of late Weichselian coleoptera from the British Isles. *Rev. Géogr. phys. Géol. dyn* **12** (2), 149–55.

COOPE, G. R., MORGAN, A. and OSBORNE, P. J. 1971: Fossil Coleoptera as indicators of climatic fluctuations during the last glaciation in Britain. *Palaeogeogr. Palaeoclimatol. Palaeoecol.* **10,** 82–101.

COOPE, G. R., and SANDS, C. H. S. 1966: Insect faunas of the last glaciation from the Tame valley, Warwickshire. *Proc. R. Soc.* B, **165,** 389–412.

COURT, A. 1957: The classification of glaciers. *J. Glaciol.* **3** (21), 3–7.

COWAN, W. R. 1968: Ribbed moraine: till fabric analysis and origin. *Can J. Earth Sci.* **5,** 1145–59.

CRAIG, B. G., and FYLES, J. G. 1960: Pleistocene geology of Arctic Canada. *Geol Surv. Pap. Can.* **60–10** (21 pp).

DAHL, R. 1963: Shifting ice culmination, alternating ice covering and ambulent refuge organisms? *Geogr. Annlr* **45,** 122–38.

1965: Plastically sculptured detail forms on rock surfaces in northern Nordland, Norway. *Geogr. Annlr* **47,** 83–140.

DAMON, P. E. 1971: The relationship between Late Cenozoic volcanism and tectonism and orogenic-epeirogenic periodicity. In Turekian, K. K. (ed) *Late Cenozoic glacial ages,* Yale Univ. Press, 15–35.

DANSGAARD, W., JOHNSEN, S. J., CLAUSEN, H. B., and LANGWAY, C. C. 1971: Climatic record revealed by the Camp Century ice core. In Turekian, K. K. (ed) *Late Cenozoic glacial ages,* Yale Univ. Press, 37–56.

DAVIES, J. L. 1969: *Landforms of cold climates.* MIT Press, Massachusetts (200 pp).

DAVIS, J. R. and NICHOLS, R. L. 1968: The quantity of meltwater in the Marble Point–Gneiss Point Area, McMurdo Sound, Antarctica. *J. Glaciol.* **7** (50), 313–20.

DAVIS, P. T. and WRIGHT, C. 1975: Extent of Little Ice Age snowcover in the eastern Canadian Arctic. In *Abstracts with programs* **7** (7), Geol Soc. Am. 1046–7.

DAVIS, W. M. 1900: Glacial erosion in France, Switzerland and Norway. *Proc. Boston Soc. Nat. Hist.* **29,** 273–321.

1909: Glacial erosion in north Wales. *QJ geol Soc. Lond.* **65,** 281–350.

1920: A roxen lake in Canada. *Scott. geogr. Mag.* **41,** 65–74.

DE GEER, G., 1897: Om rullstensåsarnas bildningssätt. *Geol. Fören i Stockh. Förh.* **19,** 366–88.

DEMOREST, M. 1938: Ice flowage as revealed by glacial striae. *J. Geol.* **46,** 700–25.

DENTON, G. H., ARMSTRONG, R. L., and STUIVER, M. 1971: The Late Cenozoic glacial history of Antarctica. In Turekian, K. K. (ed) *Late Cenozoic glacial ages,* Yale Univ. Press, 267–306.

DENTON, G. H., and KARLÉN, W. 1973: Holocene climatic variations – their pattern and possible cause. *Quat. Res.* **3,** 155–205.

DENTON, G. H. and PORTER, S. C. 1970: Neoglaciation. *Sci. American* **222,** 101–10.

DERBYSHIRE, E. 1961: Subglacial col gullies and the deglaciation of northeast Cheviots. *Trans Inst. Br. Geogr.* **29,** 31–46.

1962: Fluvioglacial erosion near Knob Lake, central Quebec–Labrador, Canada. *Bull. geol Soc. Am.* **73,** 1111–26.

1968: Cirques. In Fairbridge, R. W. (ed) *The encyclopedia of geomorphology* **1,** St Martin's Press, New York (608 pp).

DERBYSHIRE, E., and EVANS, I. S. 1976: The climatic factor in cirque variation. In Derbyshire, E. (ed) *Geomorphology and Climate,* Wiley, New York.

DEWART, G. 1966: Moulins on Kaskawulsh glacier, Yukon Territory. *J. Glaciol.* **6** (44), 320–21.

DICKSON AND POSEY, 1967: Maps of snow-cover probability for the northern hemisphere. *Monthly Weather Review* **95,** 347–53.

DOLGUSHIN, L. D. 1961: The main features of modern glaciation of the Urals. *Int. Ass. scient. Hydrol* **54,** 335–48.

DONN, W. L. *et al.* 1962: Pleistocene ice volumes and sea-level lowering. *J. Geol.* **70,** 206–14.

DONNER, J. J. 1965: The Quaternary of Finland. In Rankama, K. (ed) *The Quaternary* **1,** Interscience, New York, 199–272.

DOORNKAMP, J. C., and KING, C. A. M. 1971: *Numerical analysis in geomorphology.* Edward Arnold, London (372 pp).

DORT, W. Jr 1967: Internal studies of Sandy glacier, Southern Victoria Land, Antarctica. *J. Glaciol.* **6** (46), 524–40.

DRAKE, L. D. 1968: Till studies in New Hampshire. *Unpub. Ph.D. diss. Ohio State Univ.* (106 pp).

1971: Evidence for ablation and basal till in east central New Hampshire. In Goldthwait, R. P. (ed) *Till, a symposium,* Ohio State Univ. Press, 73–91.

DREIMANIS, A. 1971: Procedures of till investigations in North America: a general review. In Goldthwait, R. P. (ed) *Till, a symposium,* Ohio State Univ. Press, 27–37.

DREIMANIS, A., and VAGNERS, U. J. 1971: Bimodal distribution of rock and mineral fragments in basal tills. In Goldthwait, R. P. (ed) *Till, a symposium,* Ohio State Univ. Press, 237–50.

DREWES, H., FRASER, G. D., SNYDER, G. L. and BARNETT, H. F. 1961: Geology of Unalaska Island and adjacent insular shelf, Aleutian Islands, Alaska. *US geol Surv. Bull.* **1028–5,** 583–676.

DREWRY, D. J. 1972: The contribution of radio echo sounding to the investigation of Cenozoic tectonics and glaciation in Antarctica. In Price, R. J. and Sugden, D. E. (eds) *Polar geomorphology, Inst. Brit. Geogr. Spec. Pub.* **4,** 43–57.

1975: Initiation and growth of the east Antarctic ice sheet. *J. geol Soc. Lond.* **131,** 255–73.

DUGDALE, R. E. 1972: A statistical analysis of some measures of the state of a glacier's 'health'. *J. Glaciol.* **11** (61), 73–9.

DUNN, P. R., THOMSON, B. P. and RANKAMA, K. 1971: Late Precambrian glaciation in Australia as a stratigraphic boundary. *Nature, Lond.* **231,** 498–502.

DURY, G. H. 1953: A glacial breach in the northwestern highlands. *Scott. geogr. Mag.* **69,** 106–17.

DYSON, J. L. 1952: Ice-ridged moraines and their relation to glaciers. *Am. J. Sci.* **350,** 204–12.

EASTERBROOK, D. J. 1964: Void ratios and bulk densities as means of identifying Pleistocene till. *Bull. geol Soc. Am.* **75** (8), 745–50.

EBERS, E. 1926: Dir bisherigen Ergebrisse der Drumlinforschung. Eine Monographie der Drumlins. *Neues Jahrbuch für Min., Geol. und Palaönt* **53** (A), 153–270.

EDELMAN, N. 1972: Meandrande glacialrännor. *Terra* **84** (3), 104–7.

EINARSSON, T., HOPKINS, D. M., and DOELL, R. R. 1967: The stratigraphy of Tjörnes, northern Ireland, and the history of the Bering Land Bridge. In Hopkins, D. M. (ed) *The Bering land bridge,* Stanford Univ. Press, Stanford, 312–25.

ELLISTON, G. R. 1973: Water movement through the Gornergletscher. *Symposium on the Hydrology of Glaciers, Cambridge, 9–13 Sept. 1969. Int. Ass. scient. Hydrol. Pub.* **95,** 79–84.

ELSON, J. A. 1957: Origin of washboard moraines. *Bull. geol Soc. Am.* **68,** 1721 (Abstract).

1961: The geology of tills. *Proc. 14th Canadian Soil Mechanics Conf.*, Nat. Res Council, Ottawa, 5–17.

1969a: Washboard moraines and other minor moraine types. In Fairbridge, R. W. (ed) *Encyclopedia of geomorphology* **3,** Reinhold, New York, 1213–19.

1969b: Late Quaternary marine submergence of Quebec. *Rev. Géogr. Montréal* **23,** 247–58.

EMBLETON, C. 1961: The geomorphology of the Vale of Conway, north Wales, with particular reference to its deglaciation. *Trans Inst. Br. Geogr.* **29,** 47–70.

1964: Subglacial drainage and supposed ice-dammed lakes in northeast Wales. *Proc. Geol. Ass.* **75** (1), 31–8.

EMBLETON, C., and KING, C. A. M. 1975a: *Glacial geomorphology.* Edward Arnold, London; Halsted, New York (583 pp)

1975b: *Periglacial geomorphology.* Edward Arnold, London; Halsted, New York (203 pp).

EMILIANI, C. 1955: Pleistocene temperatures. *J. Geol.* **63,** 538–78.

1966: Palaeotemperature analysis of Caribbean Cores P6304–8 and P6304–9 and a generalized temperature curve for the past 425,000 years. *J. Geol.* **74,** 109–26.

1969: Interglacial high sea levels and the control of Greenland ice by the procession of the equinoxes. *Science* **166,** 1503–04.

ENGELN, O.·D. von 1912: Phenomena associated with glacier drainage and wastage, with special reference to observations in the Yakutat Bay region, Alaska. *Z. Gletscherk.* **6,** 104–50.

1918: Transportation of debris by icebergs. *J. Geol.* **26,** 74–81.

1937: Rock sculpture by glaciers: a review. *Geog. Rev.* **27,** 478–82.

EPSTEIN, S. *et al.* 1970: Antarctic ice sheet: stable isotope analysis of Byrd Station cores and interhemispheric climatic implications. *Science* **168,** 1570–72.

EVANS, I. S. 1969: The geomorphology and morphometry of glacial and nival areas. In Chorley, R. J. (ed) *Water, earth and man,* Methuen, London, 369–80.

1972: Inferring process from form: the asymmetry of glaciated mountains. In Adams, W. P. and Helleiner, F. M. (eds) *International geography* **1,** Univ. Toronto Press, 17–19.

EVANS, I. S., and COX, N. 1974: Geomorphometry and the operational definition of cirques. *Area* **6** (2), 150–53.

EVENSON, E. B. 1971: The relationship of macro and micro-fabric of till and the genesis of glacial landforms in Jefferson County, Wisconsin. In Goldthwait, R. P. (ed) *Till, a symposium,* Ohio State Univ. Press, 345–64.

EVTEEV, S. A. 1960: Determination of the rate of erosive activity of the east Antarctic ice sheet. *Akad. Nauk SSSR, Mezhduved, Komit Proved. MGG, Sbornik Statei, IX Razdel Programmy MGG (Gliatsiologiia)* **5,** 88–94.

EWING, M. and DONN, W. L. 1959: A theory of ice ages. *Science* **129.** 463–5.

EYTHORSSON, J. 1935: On the variations of glaciers in Iceland. *Geogr. Annlr* **17,** 121–37.

1949: Variations of glaciers in Iceland 1930–1947. *J. Glaciol.* **1** (5), 250–52.

EYTHORSSON, J., and SIGTRYGGSSON, H. 1971: The climate and weather of Iceland. *The zoology of Iceland* **1** (3), 1–62.

FAIRBAIRN, H. W., HURLEY, P. M., CARD, K. D. and KNIGHT, C. J. 1969: Correlation of radiometric ages of Nipissing diabase and Huronian metasediments with Proterozoic orogenic events in Ontario. *Can J. Earth Sci.* **6,** 489–97.

FAIRBRIDGE. R. W. (ed) 1967: *The encyclopedia of atmospheric sciences and astrogeology.* Reinhold, New York (1200 pp).

1970: South Pole reaches the Sahara. *Science* **168,** 878–81.

1971: Upper Ordovician glaciation in northwest Africa? Reply. *Bull. geol Soc. Am.* **82,** 269–74.

1972: Climatology of a glacial cycle. *Quat. Res.* **2,** 283–302.

FALCONER, A. 1969: Processes acting to produce glacial detritus. *Earth Sci. J.* **3** (1), 40–43.

FEININGER, T. 1971: Chemical weathering and glacial erosion of crystalline rocks and the origin of till. *US geol Surv. Prof. Pap.* **750**–C, 65–81.

FIDALGO, F. and RIGGI, J. C. 1965: Los rodados Patagonicos en la Meseta del Guenguel y alrededores (Santa Cruz). *Asoc. geol. Argentina Rev.* **20,** 273–325.

FILLON, R. H. 1975: Deglaciation of the Labrador continental shelf. *Nature, Lond.* **253** (5491), 429–31.

FISHER, D. 1973: Subglacial leakage of Summit Lake, British Columbia by dye determinations. *Symposium on the Hydrology of Glaciers, Cambridge, 9–13 Sept. 1969, Int. Ass. scient. Hydrol. Pub.* **95,** 111–116.

FISHER, D. A. and JONES, S. J. 1971: The possible future behaviour of Berendon glacier, Canada—a further study. *J. Glaciol.* **10** (58), 85–92.

FISHER, J. E. 1962: Ogives of the Forbes type on Alpine glaciers and a study of their origins. *J. Glaciol.* **4** (31), 53–61.

1963: Two tunnels in cold ice at 4,000 m on the Breithorn. *J. Glaciol.* **4** (34), 513–20.

FLINT, R. F. 1928: Eskers and crevasse fillings. *Am. J. Sci.* **235,** 410–16.

1929: The stagnation and dissipation of the last ice sheet. *Geog. Rev.* **19,** 256–289.

1930: The origin of the Irish eskers. *Geog. Rev.* **20,** 615–630.

1971: *Glacial and Quaternary Geology.* Wiley, New York (892 pp).

FOGELBERG, P. 1970: Geomorphology and deglaciation at the second Salpausselkä between Vääksy and Vierumäki, southern Finland. *Comm. Phys-Math. Helsinki* **39** (90 pp).

FOSTER, H. 1970: Sarn Badrig, a submarine moraine in Cardigan Bay, north Wales. *Z. Geomorph.* NF **4** (4), 475–86.

FRAKES, L. A., MATTHEWS, J. L. and CROWELL, J. C. 1971: Late Palaeozoic glaciation: part III, Antarctica. *Bull. geol Soc. Am.* **82** (6), 1581–604.

FRANCIS, E. A. 1970: Quaternary. In Johnson, G. A. L. and Hickling, G. (eds) Geology of County Durham, *Trans Nat. Hist. Soc. Northumb.* **41,** 134–52.

1975: Glacial sediments: a selective review. In Wright, A. E. and Moseley, F. (eds) *Ice Ages, ancient and modern,* Seel House Press, Liverpool, 43–68.

FREDÉN, C. 1967: A historical review of the Ancylus Lake and the Svea River. *Geol. Fören. i Stockh. Förh.* **89,** 239–67.

FREDRIKSSON, S. 1969: The effects of sea ice on flora, fauna and agriculture. *Jökull* **19,** 146–57.

FRISTRUP, B. 1966: *The Greenland ice cap* (transl. D. Stoner). Univ. Washington Press, Seattle (312 pp).

FRITTS, H. C. 1965: Dendrochronology. In Wright, H. E. and Frey, D. G. (eds) *The Quaternary of the United States,* Princeton Univ. Press, NJ, 871–80.

FROMM, E. 1949: Datering av den senglaciala utvecklingen i Norrbottens kustland. *Geol. Fören. i Stockh. Förh.* **71,** 313–27.

FRYE, J. C. 1973: Pleistocene succession of the central interior United States. *Quat. Res.* **3,** 275–83.

FRYE, J. C. and WILLMAN, H. B. 1973: Wisconsinan climatic history interpreted from Lake Michigan lobe deposits and soils. *Mem. geol Soc. Am.* **136,** 135–52.

FUNDER, S. 1972: Deglaciation of the Scoresby Sund fjord region, northeast Greenland. In Price, R. J. and Sugden, D. E. (eds) Polar geomorphology, *Inst. Br. Geogr. Spec. Pub.* **4,** 33–42.

FUNDER, S. and HJORT, C. 1973: Aspects of the Weichselian chronology in central east Greenland. *Boreas* **2** (2), 69–84.

GALIBERT, M. G. 1962: Recherches sur les processus d'érosion glaciaires de la Haute Montagne Alpine. *Bull. Ass. Géogr. Francais* **303/4,** 8–46.

GALIBERT, G. 1965: *La haute montagne alpine.* Thèse Lettres, Toulouse (406 pp).

GALLOWAY, R. W. 1956: The structure of moraines in Lyngdalen, north Norway. *J. Glaciol.* **2,** 730–33.

GARWOOD, E. J. and GREGORY, J. W. 1898: Contributions to the glacial geology of Spitsbergen. *QJ geol Soc. Lond.* **54,** 197–227.

GATES, W. L. 1974: Global climate simulated by the 2-level MINTZ-ARAKAWA model: comparison with observation. In Mapping the atmospheric and oceanic circulations and other climatic parameters at the time of the Last Glacial Maximum about 17,000 years ago, *Clim. Res. Unit Res. Pub* **2,** Univ. East Anglia, Norwich, 111–12.

GAUDIN, A. M. 1926: An investigation of crushing phenomena. *Trans Am. Inst. Min. and Metallurg. Engineers* **73,** 253–316.

GJESSING, J. 1960: Isavsmeltingstidens drenering, dens forløp og formdannende virkning i Nordre Atnedalen med sammenlignende studier fra Nordre Gudbransdalen og Nordre Østerdalen. *Ad Novas* **3,** Norw. geogr. Soc. Spec. Pub (492 pp).

1965: On 'plastic scouring' and 'subglacial erosion'. *Norsk geogr. Tidsskr.* **20,** 1–37.

1966: Some effects of ice erosion on the development of Norwegian valleys and fjords. *Norsk geogr. Tidsskr.* **20** (8), 273–99.

1967: Potholes in connection with plastic scouring forms. *Geogr. Annlr* **49**A, 178–87.

GLEN, J. W. 1954: The stability of ice-dammed lakes and other water-filled holes in glaciers. *J. Glaciol.* **2** (15), 316–18.

1955: The creep of polycrystalline ice. *Proc. R. Soc.* A. **228** (1175), 519–38.

1956: Measurement of the deformation of ice in a tunnel at the foot of an ice fall. *J. Glaciol.* **2** (20), 735–45.

GLEN, J. W. and LEWIS, W. V. 1961: Measurements of side-slip at Austerdalsbreen, 1959. *J. Glaciol.* **3** (30), 1109–122.

GLÜCKERT, G. 1973: Two large drumlin fields in central Finland. *Fennia* **120** (37 pp).

GODARD, A. 1965: *Recherches de géomorphologie en Ecosse du nordouest*. Les Belles Lettres, Paris (702 pp).
GOLDTHWAIT, R. P. 1951: Development of end moraines in east-central Baffin Island. *J. Geol.* **59**, 567–577.
1958: Wisconsin age forests in western Ohio, 1: age and glacial events. *Ohio J. Sci.* **58**, 209–30.
1960: Study of ice cliff in Nunatarssuaq, Greenland. *Tech. Rept Snow Ice Permafrost Res. Establ.* **39**, 1–103.
1966: Evidence from Alaskan glaciers of major climatic changes. In Sawyer, J. S. (ed) *World climate from 8,000 to 0 BC*, R. met. Soc. London, 40–53.
1968: Surficial geology of the Wolfeboro-Winnipesaukee area, New Hampshire. *New Hampshire Dept Resources and Econ. Development* (60 pp).
1971: Introduction to till, today. In Goldthwait, R. P. (ed) *Till, a symposium*, Ohio State Univ. Press, 3–26.
1973: Jerky glacier motion and meltwater. *Symposium on the Hydrology of glaciers, Cambridge, 9–13 Sept. 1969, Int. Ass. scient. Hydrol. Pub.* **95**, 183–8.
GOLUBEV, G. N. 1973: Analysis of the runoff and flow routing for a mountain glacier basin. *Symposium on the Hydrology of Glaciers, Cambridge, 9–13 Sept. 1969, Int. Ass. scient. Hydrol. Pub.* **95**, 41–50.
GOODELL, H. G. *et al.* 1968: The Antarctic glacial history recorded in sediments of the Southern Ocean. *Palaeogeogr. Palaeoclimatol. Palaeoecol.* **5**, 41–62.
GORDON, J. E. 1975: Morphometry of Scottish cirques. *Paper presented at annual meeting of Institute of British Geographers, Mimeo* (23 pp).
GOW, A. J. 1969: On the rates of growth of grains and crystals in South Polar firn. *J. Glaciol.* **8** (53), 241–52.
1970: Preliminary results of studies of ice cores from the 2164 m deep drill hole, Byrd Station, Antarctica. In Gow, A. J. *et al.* (eds) *International Symposium on Antarctic Glaciological Exploration (ISAGE)*, *Int. Ass. scient. Hydrol. Pub.* **86**, 78–90.
1971: Depth-time-temperature relationships of ice crystal growth in polar glaciers. *US Army Cold Reg. Res. Engng. Lab. Res. Rept* **300** (18 pp).
GRABERT, H. 1967: Ergebnis und Ausdeutung radiometrischer Untersuchungen an Graniten des Brasilianischen Schildes. *Neues Jahrb. Geol. u. Paläont. Monatsch.* **5**, 268–81.
GRAF, W. L. 1970: The geomorphology of the glacial valley cross-section. *Arctic and Alpine Res* **2**, 303–312.
GRANÖ, O. 1958: The Vessö esker of south Finland and its economic importance. *Fennia* **82**, 3–33.
GRAVENOR, C. P. 1953: The origin of drumlins. *Am. J. Sci.* **251**, 674–81.
1974: The Yarmouth drumlin field, Nova Scotia, Canada. *J. Glaciol.* **13** (67), 45–54.
GRAVENOR, C. P. and KUPSCH, W. O. 1959: Ice disintegration features in western Canada. *J. Geol.* **67**, 48–64.
GRAVENOR, C. P. and MENELEY, W. A. 1958: Glacial flutings in central and northern Alberta. *Am. J. Sci.* **256**, 715–28.
GRINDLEY, G. W. 1967: The geomorphology of the Miller Range, Tramantanta Mountains, with notes on the glacial history and neotectonics of east Antarctica. *NZ J. Geol. Geophys.* **10** (2) 557–98.
GRIPP, K. 1927: Beiträge zur Geologie von Spitzbergen. *Abh. Geb. Naturw.*, Hamburg **21**, 1–38.
1929: Glaciologische und geologische Ergebnisse der Hamburgischen Spitzbergen-Expedition 1927. *Naturwiss. Verein in Hamburg. Abh. Geb. Naturw.* **22** (2–4), 146–249.
1938: Endmoränen. *Int. Geogr. Congr. Abstracts* IIA.
GROSVAL'D, M. G. and KOTLYAKOV, V. M. 1969: Present-day glaciers in the USSR and some data on their mass balance. *J. Glaciol.* **8** (52), 9–22.
GUDMANDSEN, P. 1975: Layer echoes in polar ice sheets. *J. Glaciol.* **15** (73), 95–101.
GUDMUNDSSON, G. and SIGBJARNARSON, G. 1972: Analysis of glacier runoff and meteorological observations. *J. Glaciol.* **11** (63), 303–18.
GUILCHER, A. 1969: Pleistocene and Holocene sea-level changes. *Earth-Sci. Rev.* **5**, 69–97.
GUNN, B. M. 1964: Flow rates and secondary structures of Fox and Franz Josef glaciers, New Zealand. *J. Glaciol.* **5** (38), 173–90.
HAEFELI, R. 1961: Contribution to the movement and the form of ice sheets in the Arctic and the Antarctic. *J. Glaciol.* **3** (30), 1133–151.
1968: Gedanken zum Problem der glazialen Erosion. *Felsmech. Ingenieurgeol.* **4**, 31–51.
1970: Changes in the behaviour of the Unteraargletscher in the last 125 years. *J. Glaciol.* **9** (56), 195–212.
HAGGETT, P. and CHORLEY, R. J. 1969: *Network analysis in geography*. Edward Arnold, London (348 pp).

HALLIDAY, W. R. and ANDERSON, C. H. 1970: Glacier caves; a new field of speleology. *Studies in Speleology* **2,** 53–9.

HAMILTON, W. and KRINSLEY, D. 1967: Upper Paleozoic glacial deposits of South Africa and southern Australia. *Bull. geol Soc. Am.* **78,** 783–800.

HAMMEN, T. van der, WIJMSTRA, T. A. and ZAGWIJN, W. H. 1971: The floral record of the Late Cenozoic of Europe. In Turekian, K. K. (ed) *Late Cenozoic glacial ages,* Yale Univ. Press, 391–424.

HANSEN, B. L. and LANGWAY, C. C. 1966: Deep-core drilling and ice core analysis, Camp Century, Greenland, 1961–1966. *Antarctic J. US* **1,** 207–08.

HANSEN, S. 1965: The Quaternary of Denmark. In Rankama, K. (ed) *The Quaternary* **1,** Interscience, New York, 1–90.

HANWELL, J. D. and NEWSON, M. D. 1973: *Techniques in physical geography.* Macmillan, London (230 pp).

HARE, F. K. 1968: The Arctic. *J.R. met. Soc.* **94,** 439–59.

HARLAND, W. B. and HEROD, K. N. 1975: Glaciations through time. In Wright, A. E. and Moseley, F. (eds) *Ice ages: ancient and modern,* Seel House Press, Liverpool, 189–216.

HARRISON, A. E. 1964: Ice surges on the Muldrow glacier. *J. Glaciol.* **5** (39), 265–368.

HARRISON, P. W. 1957: A clay-till fabric: its character and origin. *J. Geol.* **65,** 275-308.

1960: Original bedrock composition of Wisconsin till in central Indiana. *J. Sed. Petrol.* **30,** 432-446.

HARRISON, W. D. 1972: Temperature of a temperate glacier. *J. Glaciol.* **11** (61), 15–29.

1975: Temperature measurements in a temperate glacier. *J. Glaciol.* **14** (70), 23–30.

HARTSHORN, J. H. 1958: Flowtill in southeastern Massachusetts. *Bull. geol Soc. Am.* **69,** 477–82.

HARTSHORN, J. H. and ASHLEY, G. M.: 1972: Glacial environment and processes in southeastern Alkaska. *Coastal Res. Center Tech. Rept* **4** -CRC, Univ. Massachussetts (69 pp).

HASTENRATH, S. L. 1971: On the Pleistocene snow-line depression in the arid regions of the South American Andes. *J. Glaciol.* **10** (59), 255–67.

HATHERTON, T. (ed) 1965: *Antarctica.* Methuen, London (511 pp).

HATTERSLEY-SMITH, G. 1974: Present arctic ice cover. In Ives, J. D. and Barry, R. G. (eds) *Arctic and alpine environments,* Methuen, London, 195–223.

HATTERSLEY-SMITH, G., FUZESY, A. and EVANS, S. 1969: Glacier depths in northern Ellesmere Island: airborne radio echo sounding in 1966. *Can Dept of Nat. Defence DREO Tech. Note* **69–6** (23 pp).

HATTERSLEY-SMITH, G. and SERSON, H. 1973: Reconnaissance of a small ice cap near St Patrick Bay, Robeson Channel, Northern Ellesmere Island, Canada. *J. Glaciol.* **12** (66), 417–21.

HAYNES, V. M. 1968a: Nature of glaciated landforms. *Nature, Lond.* **217** (5133), 1035–6.

1968b: The influence of glacial erosion and rock structure on corries in Scotland. *Geogr. Annlr* **50**A, 221–34.

1972: The relationship between the drainage areas and sizes of outlet troughs of the Sukkertoppen ice cap, west Greenland. *Geogr. Annlr* **54**A (2), 66–75.

HAYS, J. D. and OPDYKE, N. D. 1967: Antarctic radiolaria, magnetic reversals, and climatic change. *Science* **158,** 1001–11.

HELLAAKOSKI, A. 1931: On the transportation of materials in the esker of Laitila. *Fennia* **52,** 1–41.

HENOCH, W. E. S. 1971: Estimate of glacier's secular (1948–66) volumetric change and its contribution to the discharge in the Upper North Saskatchewan river basin. *J. Hydrol.* **12,** 145–60.

HEUBERGER, H. 1974: Alpine Quaternary glaciation. In Ives, J. D. and Barry, R. G. (eds) *Arctic and alpine environments,* Methuen, London, 319–38.

HEUBERGER, H. and BESCHEL, R. 1958: Beitrage zur Datierung alter Gletscherstande in Hochstubai (Tirol). *Schlern-Schriften* **90,** 73–100.

HEUSSER, C. J. 1960: Late Pleistocene environments of North Pacific North America. *Geogr. Soc. Am. Spec. Pub.* 35 (308 pp).

HILL, A. 1970: The relationship of drumlins to the directions of ice movement in north Co. Down. In Stephens, N. and Glasscock, R. E. (eds) *Irish Geographical Studies,* Queen's Univ., Belfast, 53–9.

1971: The internal composition and structure of drumlins in North Down and South Antrim, Northern Ireland. *Geogr. Annlr* **53**A, 14–31.

HILLEFORS, Å. 1974: The stratigraphy and genesis of stoss- and lee-side moraines. *Bull. geol Inst. Univ. Upsala,* NS **5,** 139–54.

HJULSTRÖM, F. 1935: Studies of the morphological activities of rivers as illustrated by the river Fyris. *Bull. geol Inst. Univ. Upsala* **25,** 221–527.

1954: Geomorphology of the area surrounding the Hoffellssandur. *Geogr. Annlr* **1–2**

HOBBS, W. H. 1926: *Earth features and their meaning.* Macmillan, New York, (506 pp).

HODGE, S. M. 1974: Variations in the sliding of a temperate glacier. *J. Glaciol.* **13** (69), 349–69.

HOEL, A. and WERENSKIOLD, C. 1962: Glaciers and snowfields in Norway. *Norsk Polarinstitutt Skrifter* **114,** (291 pp).

HOLDSWORTH, G. 1973a: Barnes ice cap and englacial debris in glaciers. *J. Glaciol.* **12** (64), 147–8.

   1973b: Ice deformation and moraine formation at the margin of an ice cap adjacent to a proglacial lake. In Fahey, B. D. and Thompson, R. D. (eds) Research in polar and alpine geomorphology, *3rd Guelph Symposium on Geomorphology*, Geo Abstracts, Norwich, 187–99.

HOLDSWORTH, G. and BULL, C. 1970: The flow law of cold ice; investigations on Meserve glacier, Antarctica. In Gow, A. J. *et al.* (eds) *International Symposium on Antarctic Glaciological Exploration (ISAGE)*, *Int. Ass. scient. Hydrol. Pub.* **86,** 204–16.

HOLLIN, J. T. 1962: On the glacial history of Antarctica. *J. Glaciol.* **4** (32), 173–95.

   1964: Origin of ice ages: an ice shelf theory for Pleistocene glaciation. *Nature, Lond.* **202** (4937), 1099-100.

   1965: Wilson's theory of ice ages. *Nature, Lond.* **208,** 12–16.

   1969: Ice-sheet surges and the geological record. *Can J. Earth Sci.* **6,** 903–10.

HOLLINGWORTH, S. E. 1931: The glaciation of western Edenside and the Solway basin. *QJ geol Soc. Lond.* **87,** 281–359.

HOLMES, C. D. 1941: Till fabric. *Bull. geol Soc. Am.* **52,** 1299–354.

   1947: Kames. *Am. J. Sci.* **245,** 240–49.

   1960: Evolution of till-stone shapes, central New York. *Bull. geol Soc. Am.* **71,** 1645–60.

HOLMES, G. W. 1955: Morphology and hydrology of the Mint Julep area, southwest Greenland. *Mint Julep Reports, Part II, Arctic Desert Topic Information Center U.S. Air University Pub.* A–**104**–B.

HOLTEDAHL, H. 1958: Some remarks on the geomorphology of continental shelves off Norway, Labrador and southeast Alaska. *J. Geol.* **66,** 461–71.

   1959: Geology and paleontology of Norwegian sea bottom cores. *J. Sed. Petrol.* **29,** 16–29.

   1967: Notes on the formation of fjords and fjord valleys. *Geogr. Annlr* **49**A, 188–203.

HOLTEDAHL, H. and SELLEVOLL, M. 1972: Notes on the influence of glaciation on the Norwegian continental shelf bordering on the Norwegian sea. In Dahl, E., Strömberg, J–O. and Tandberg, O. G. (eds) The Norwegian Sea region: its hydrography, glacial and biological history, *Ambio Spec. Rept* **2,** 31–8.

HOLTEDAHL, O. 1929: On the geology and physiography of some Antarctic and sub-Antarctic islands. *Scient. Results Norw. Antarct Exped.* **3** (172 pp).

   1970: On the morphology of the west Greenland shelf with general remarks on the 'marginal channel' problem. *Marine Geol.* **8,** 155–72.

HOPE, R., LISTER, H. and WHITEHOUSE, R. 1972: The wear of sandstone by cold, sliding ice. In Price, R. J. and Sugden, D. E. (eds) Polar geomorphology, *Inst. Br. Geogr. Spec. Pub.* **4,** 21–31.

HOPKINS, W. 1862: On the theory of the motion of glaciers. *Phil. Trans R. Soc.* **152** (2), 677–745.

HOPPE, G. 1948: Isrecessionen från Norrbottens Kustland. *Geographica* **20** (112 pp).

   1951: Drumlins in nordöstra Norrbotten. *Geogr. Annlr* **33,** 157–65.

   1952: Hummocky moraine regions, with special reference to the interior of Norrbotten. *Geogr. Annlr* **34,** 1–72.

   1957: Problems of glacial morphology and the ice age. *Geogr. Annlr* **39,** 1–17.

   1959: Glacial morphology and inland ice recession in northern Sweden. *Geogr. Annlr* **41,** 193–212.

HOPPE, G., KINDBLOM, B. O., KLEIN, K., KLINGSTRÖM, A. and E. 1959: Glacialmorfologi och isrörelser i ett lappländskt fjällområde. *Geogr. Annlr* **41,** 1–14.

HOPPE, G. and SCHYTT, V. 1953: Some observations on fluted moraine surfaces. *Geogr. Annlr* **35,** 105–115.

HUGHES, O. L. 1964: Surficial geology, Nichicun-Kaniapiskau map area, Quebec. *Bull. geol. Surv. Can.* **106** (20 pp).

HUGHES, T. 1970: Convection in the Antarctic ice sheet leading to a surge of the ice sheet and possibly to a new ice age. *Science* **170** (3958), 630–33.

   1972: Is the west Antarctic ice sheet disintegrating? *ISCAP (Ice Streamline Co-operative Antarctic Project) Bull.* **1** (77 pp).

   1975: The west Antarctic ice sheet: instability, disintegration and initiation of Ice Ages. *Rev. Geophys. and Space Phys.* **13** (4), 502–26.

HYVÄRINEN, H. 1973: The deglaciation history of eastern Fennoscandia – recent data from Finland. *Boreas* **2** (2), 85–102.

IKEN, A. 1972: Measurements of water pressure in moulins as part of a movement study of the White glacier, Axel Heiberg Island, NWT, Canada. *J. Glaciol.* **11** (61), 53–8.

IMBRIE, J. 1974: The CLIMAP game plan: design for a global atmosphere–ocean modelling experiment.

In Mapping the atmospheric and oceanic circulations and other climatic parameters at the time of the Last Glacial Maximum about 17,000 years ago, *Clim. Res. Unit Res. Pub.* **2**, Univ. East Anglia, Norwich, 109–10.

IMBRIE, J., VAN DONK, J. and KIPP, N. G. 1973: Palaeoclimatic investigation of a late Pleistocene Caribbean deep-sea core: comparison of isotopic and faunal methods. *Quat. Res.* **3**, 10–38.

INTERNATIONAL ASSOCIATION OF SCIENTIFIC HYDROLOGY 1954: *The international classification for snow.* Commission on Snow and Ice.

IVES, J. D. 1958: Glacial drainage channels as indicators of late-glacial conditions in Laborador–Ungava: a discussion. *Cahiers de Géogr. Québec* **5**, 57–72.

1960: Glaciation and deglaciation of the Helluva Lake area, central Labrador-Ungava. *Geogr. Bull.* **15**, 46–64.

1962: Indication of recent extensive glacierization in north central Baffin Island, NWT. *J. Glaciol.* **4** (32), 197–206.

IVES, J. D., ANDREWS, J. T. and BARRY, R. G. 1975: Growth and decay of the Laurentide ice sheet and comparisons with Fenno-Scandinavia. *Die Naturwiss.* **62**, 118–25.

IVES, J. D. and KING, C. A. M. 1955: Glaciological observations on Morsarjökull, southwest Vatnajökull, Iceland. *J. Glaciol.* **2** (17), 477–82.

IVES, J. D. and KIRBY, R. P. 1964: Fluvioglacial erosion near Knob Lake, central Quebec–Labrador, Canada: discussion. *Bull. geol. Soc. Amer.* **75**, 917–22.

IWAN, W. 1936: Beobachtungen am Drangajökull, N.W. Island. *Z. de. Gesellschaft für Erdkunde* 3–4, 102–114.

JAHNS, R. H. 1943: Sheet structure in granite: its origin and use as a measure of glacial erosion in New England. *J. Geol.* **51** (2), 71–98.

JÄRNEFORS, B. 1952: A sediment-petrographic study of glacial till from Pajala district, northern Sweden. *Geol. Fören i Stockh. Förh.* **74**, 185–211.

JOHN, B. S. 1970a: Pembrokeshire. In Lewis, C. A. (ed) *The glaciations of Wales*, ch. 10, Longmans, London, 229–65.

1970b: The Pleistocene drift succession at Porth-clais, Pembrokeshire. *Geol. Mag.* **107**, 439–57.

1972a: Evidence from the South Shetland Islands towards a glacial history of West Antarctica. In Price, R. J. and Sugden, D. E. (eds) Polar Geomorphology, *Inst. Br. Geogr. Spec. Pub.* **4**, 75–92.

1972b: A Late Weichselian kame terrace at Mullock Bridge, Pembrokeshire. *Proc. Geol Ass.* **83**, 213–229.

1973: Vistulian periglacial phenomena in southwest Wales. *Biul. Peryglac.* **22**, 185–212.

1974: Northwest Iceland reconnaissance 1973 (Durham Univ. Vestfirðir Project). *Dept Geog., Durham Univ. Spec. Pub.* (54 pp).

JOHN, B. S. and SUGDEN, D. E. 1962: The morphology of Kaldalon, a recently deglaciated valley in Iceland. *Geogr. Annlr* **44**, 347–65.

1971: Raised marine features and phases of glaciation in the South Shetland Islands. *Brit. Ant. Surv. Bull.* **24**, 45–111.

1975: Coastal geomorphology of high latitudes. *Progress in geography* **7**, Edward Arnold, London, 53–132.

JOHNSEN, S. J., DANSGAARD, W. and CLAUSEN, H. B. 1970: Climatic oscillations 1200–2000 AD. *Nature, Lond.* **227**, 482–3.

JOHNSEN, S. J., DANSGAARD, W., CLAUSEN, H. B. and LANGWAY, C. C. 1972: Oxygen isotope profiles through the Antarctic and Greenland ice sheets. *Nature, Lond.* **235**, 429–34.

JOHNSON, P. G. 1972: The morphological effects of surges of the Donjek glacier, St Elias Mountains, Yukon Territory, Canada. *J. Glaciol.* **11** (62), 227–34.

KALIN, M. 1971: The active push moraine of the Thompson glacier. *Axel Heiberg Island Res. Rept, Glaciology* **4** (68 pp).

KAMB, B. 1964: Glacier geophysics. *Science* **146**, 353-65.

1970: Sliding motion of glaciers: theory and observation. *Rev. Geophys. and Space Phys.* **8** (4), 673–728.

KAMB, B. and LACHAPELLE, E. 1964: Direct observation of the mechanism of glacier sliding over bedrock. *J. Glaciol.* **5** (38), 159–72.

KARLÉN, W. 1973: Holocene glacier and climatic variations, Kebnekaise mountains, Swedish Lapland. *Geogr. Annlr* **55A**, 29–63.

KASSER, P. 1967: *Fluctuations of glaciers 1959–1965.* IASH/UNESCO, Belgium (100 pp).

KELLY, M. R. 1968: Floras of middle and upper Pleistocene age from Brandon, Warwickshire. *Phil. Trans B* **254**, 401–15.

KENDALL, P. F. 1902: A system of glacier-lakes in the Cleveland hills. *QJ geol Soc. Lond.* **58,** 471–571.

KENNETT, J. P. 1970: Pleistocene palaeoclimates and foraminiferal biostratigraphy in sub-antarctic deep sea cores. *Deep Sea Res.* **17,** 125–40.

KING, C. A. M. 1959: Geomorphology in Austerdalen, Norway. *Geogr. J.* **125,** 357–69.

1970: Feedback relationships in geomorphology. *Geogr. Annlr* 52A (3–4), 147–59.

KING, C. A. M. and BUCKLEY, J. 1968: The analysis of stone size and shape in Arctic environments. *J. Sed. Petrol.* **38,** 200–14.

KING, C. A. M. and LEWIS, W. V. 1961: A tentative theory of ogive formation. *J. Glaciol.* **3** (29), 913–939.

KING, H. G. R. 1969: *The Antarctic.* Blandford, London (276 pp).

KIRBY, R. P. 1969. Till fabric analyses from the Lothians, central Scotland. *Geogr. Annlr* 51A, 48–60.

KIRCHNER, G. 1963: Observations at bore holes sunk through the Schuchert Gletscher in northeast Greenland. *J. Glaciol.* **4** (35), 817–18.

KIRWAN, L. P. 1962: *The white road: a history of polar exploration.* Penguin, London (408 pp).

KLIMASZEWSKI, M. 1964: On the effect of the preglacial relief on the course and the magnitude of glacial erosion in the Tatra mountains. *Geogr. Polonica* **2,** 11–21.

KLOKOV, V. D. 1973: Magnitude of meltwater runoff. *Soviet Ant. Exped. Info. Bull.* **8** (3), 154–7.

KLUTE, F. 1921: Über die Ursachen der letzten Eiszeit. *Geog. Z.* **27,** 199–203.

KNIGHTON, D. 1972: Meandering habit of supraglacial streams. *Bull. geol Soc. Amer.* **83,** 201–04.

KOCH, L. 1945: The east Greenland ice. *Medd. om. Grønland* **130,** 1–375.

KOERNER, R. M. 1961a: Glaciological observations in Trinity Peninsula and the islands in Prince Gustav channel, Graham Land, 1958–59. *Falkland Islands Dependencies Surv. Preliminary Glaciological Rept* **2** (44 pp).

1961b: Glaciological observations in Trinity Peninsula, Graham Land, Antarctica. *J. Glaciol.* **3** (30), 1063–74.

1968: Fabric analysis of a core from the Meighen ice cap, NWT, Canada. *J. Glaciol.* **7** (51), 421–430.

1970: The mass balance of the Devon Island ice cap, NWT, Canada, 1961–66. *J. Glaciol.* **9** (57), 325–36.

KORPELA, K. 1969: Die Weischsel-Eiszeit und ihr Interstadial in Perapohjola (nordliches Nordfinnland) im Licht von Submoranen Sedimenten. *Fin. Acad. Sci. Ann.* A3 (84) (109 pp).

KRIMMEL, R. M. and MEIER, M. F. 1975: Glacier applications of ERTS images. *J. Glaciol.* **15** (73), 391–402.

KUENEN, P. H. 1951: Mechanics of wave formation and the action of turbidity currents. *Geol. För. Stockh. Förh.* **6,** 149–62.

KUJANSUU, R. 1967: On the deglaciation of western Finnish Lapland. *Bull. Comm. géol. Finl.* **232** (98 pp).

KUKLA, G. J. 1972: Insolation and glacials. *Boreas* **1,** 63–96.

KUKLA, G. J. and KUKLA, H. J. 1972: Insolation regime of interglacials. *Quat. Res.* **2,** 412–24.

KUKLA, G. J., MATTHEWS, R. K. and MITCHELL, J. M. Jr 1972: The end of the present interglacial. *Quat. Res.* **2** (3), 261–9.

KUPSCH, W. O. 1962: Ice-thrust ridges in western Canada. *J. Geol.* **70,** 582-94.

LA CHAPELLE, E. 1965: The mass budget of Blue glacier, Washington. *J. Glaciol.* **5** (41), 609–23.

LADURIE, E. LE ROY, 1971: *Times of feast, times of famine.* Doubleday, New York, (428 pp).

LAMB, H. H. 1959: The southern westerlies. *QJ R. met. Soc.* **123,** 287–97.

1963: On the nature of certain climatic epochs which differed from the modern (1900–1939) normal. In Changes of climate, *Unesco Arid Zone Res. Ser.* **20,** 125–50.

1964: The role of atmosphere and oceans in relation to climatic changes and the growth of ice sheets on land. In Nairn, A. E. M. (ed) *Problems of palaeoclimatology,* John Wiley, London, 332–48.

1966: *The changing climate.* Methuen, London (236 pp).

1969: The new look of climatology. *Nature, Lond.* **223** (5212), 1209–15.

1971: Climates and circulation regimes developed over the northern hemisphere during and since the last ice age. *Palaeogr. Palaeoclimatol. Palaeoecol.* **10,** 125–62.

1972: *Climate: present, past and future* **1,** *Fundamentals and climate now.* Methuen, London (613 pp).

1974: The data available and course established for the development of the Little Ice Age in recent centuries in Europe and other parts of the world. In Mapping the atmospheric and oceanic circulations and other climatic parameters at the time of the Last Glacial Maximum about 17,000 years ago, *Climatic Res. Unit Res. Pub.* **2,** Univ. East Anglia, Norwich, 97–100.

LAMB, H. H. and JOHNSON, A. I., 1959, 1961: Climatic variation and observed changes in the general circulation. *Geogr. Annlr* **41,** 94–134; **43,** 363–400.

LAMB, H. H., LEWIS, R. P. W. and WOODROFFE, A. 1966: Atmospheric circulation and the main climatic variables between 8000 and 0 BC: meteorological evidence. In Sawyer, J. S. (ed) *World climate from 8,000 to 0 BC*, R. met. Soc., London, 174–217.

LAMB, H. H., PROBERT-JONES, J. R. and SHEARD, J. W. 1962. A new advance of the Jan Maven glaciers and a remarkable increase in precipitation. *J. Glaciol.* **4** (33), 355–66.

LAMB, H. H. and WOODROFFE, A., 1970: Atmospheric circulation during the last ice age. *Quat. Res.* **1** (1), 29–58.

LAMPLUGH, G. W. 1911: On the shelly moraines of the Sefström glacier. *Proc. Yorks. geol Soc.* **17**, 216–241.

LANG, H. 1968: Relations between glacier runoff and meteorological factors observed on and outside the glacier. *Int. Ass. scient. Hydrol.* **79**, 429–39.

LARSEN, J. A. and BARRY, R. G. 1974: Palaeoclimatology. In Ives, J. D. and Barry R. G. (eds) *Arctic and alpine environments*, Methuen, London, 253–76.

LEE, H. A. 1959: Surficial geology of southern district of Keewatin and the Keewatin ice divide, NWT. *Bull. geol Surv. Can* **51** (42 pp).

1963: Glacial fans in till from the Kirkland Lake fault: a method of gold exploration. *Geol Surv. Pap. Can.* **63–45** (36 pp).

1965: Investigation of eskers for mineral exploration. *Geol Surv. Pap. Can.* **68–22** (16 pp).

LEMKE, R. W. 1958: Narrow linear drumlins near Velva, North Dakota. *Am. J. Sci.* **256**, 270–83.

LEOPOLD, L. B., WOLMAN, M. G. and MILLER, J. P. 1964: *Fluvial processes in geomorphology*. Freeman, San Francisco (522 pp).

LEWIS, W. V. 1940: The function of meltwater in cirque formation. *Geogr. Rev.* **30**, 64–83.

1947: The formation of *roches moutonnées* – some comments on Dr H. Carol's article. *J. Glaciol.* **1** (2), 60–63.

1949: An esker in process of formation: Böverbreen, Jotunheimen. *J. Glaciol.* **1** (6), 314–19.

1954: Pressure release and glacial erosion. *J. Glaciol.* **2** (16), 417–22.

(ed) 1960: Norwegian cirque glaciers. *R. geogr. Soc. Res. Ser.* **4** (104 pp).

LIESTØL, O. 1967: Storbreen glacier in Jotunheimen, Norway. *Norsk Polarinst, Skrifter* **141** (63 pp).

LINDSTRÖM, E. 1973: Deglaciation, sediment och högsta kustlinje i Nordvästra Ångermanland. *UNGI Rapport* **26**, Uppsala (372 pp).

LINTON, D. L. 1955: The problem of tors. *Geogr. J.* **121**, 470–87.

1959: Morphological contrasts between eastern and western Scotland. In Miller, R. and Watson, J. W. (eds) *Geographical essays in memory of Alan G. Ogilvie*, Edinburgh, 16–45.

1962: Glacial erosion on soft-rock outcrops in central Scotland. *Biul. Peryglac.* **11**, 247–57.

1963: The forms of glacial erosion. *Trans Inst. Br. Geogr.* **33**, 1–28.

1964: Landscape evolution. In Priestley, R., Adie, R. J. and Robin, G. De Q. (eds) *Antarctic research*, Butterworths, London, 85–94.

1968: Divide elimination by glacial erosion. In Wright, H. E. and Osburn, W. H. (eds) Arctic and alpine environments, *Proc. 7th INQUA Congr.*, 241–8.

LINTON, D. L. and MOISLEY, H. A. 1960: The origin of Loch Lomond. *Scott. geogr. Mag.* **76**, 26–37.

LISTER, H. 1973: Glacial origin of pro-glacial boulders. *Symposium on the Hydrology of Glaciers, Cambridge, 9–13 Sept. 1969, Int. Assoc. scient. Hydrol. Pub.* **95**, 151–6.

LISTER, H., PENDLINGTON, A. and CHORLTON, J. 1968: Laboratory experiments on abrasion of sandstones by ice. *Int. Ass. scient. Hydrol.* **79**, 98–106.

LJUNGNER, E. 1930: Spaltektonik und Morphologie der Schwedischen Skagerrak-Küsle, 3: Die Erosionsformen. *Bull. Geol. Univ. Uppsala* **21**, 255–478.

1948: East-west balance of the Quaternary ice caps in Patagonia and Scandinavia. *Bull. Geol. Inst. Univ. Uppsala* **33**, 12–96.

LLIBOUTRY, L. 1953: More about advancing and retreating glaciers in Patagonia. *J. Glaciol.* **2** (13), 168–72.

1957: Banding and volcanic ash on Patagonian glaciers. *J. Glaciol.* **3** (21), 20–25.

1958a: La dynamique de la mer de glace et la vague de 1891–95 d'après les mesures de Joseph Vallot. *Int. Ass. scient. Hydrol.* **47**, 125–238.

1958b: Studies of the shrinkage after a sudden advance, blue bands and wave ogives on Glacier Universidad (central Chilean Andes). *J. Glaciol.* **3** (24), 261–8.

1964, 1965: *Traité de glaciologie*. Masson, Paris (2 vols, 428 pp and 162 pp).

1968: General theory of subglacial cavitation and sliding of temperate glaciers. *J. Glaciol.* **7** (49), 21–58.

1969a: How ice sheets move. *Science J.* **5** (4), 50–55.

1969b: Contribution à la théorie des ondes glaciaires. *Can J. Earth Sci.* **6** (4), 943–53.

1970a: Ice flow from ice sheet dynamics. In Gow, A. J. *et al.* (eds) *International Symposium on Antarctic Glaciological Exploration (ISAGE), Int. Ass. scient. Hydrol. Pub.* **86**, 217–28.

1970b: Current trends in glaciology. *Earth-Science Rev.* **6**, 141–67.

1971: The glacier theory. In Ven Te Chow (ed) *Advances in hydroscience* **7**, Academic Press, New York, 81–167.

LOEWE, F. 1970a: Screen temperatures and 10 m temperatures. *J. Glaciol.* **9** (56), 263–8.

1970b: The transport of snow on ice sheets by the wind. *Univ. Melbourne Meteorol. Dept Pub.* **13** (69 pp).

LØKEN, O. H., and HODGSON, D. A. 1971: On the submarine geomorphology along the east coast of Baffin Island. *Can J. Earth Sci.* **8** (2), 185–95.

LOOMIS, S. R. 1970: Morphology and structure of an ice-cored medial moraine, Kaskawulsh glacier, Yukon. *Arctic Inst. N. Am. Res. Pap.* **57**, 1–51.

LORENZO, J. L. 1959: *Los Glaciares de Mexico.* Inst. de Geofisica, Cuidad Universitaria, Mexico, DF (114 pp).

LUNDQVIST, G. 1937: Sjösediment från Rogenområdet i Härjedalen. *Sv. geol. Unders.* C **408**,

1958: Beskrivning till jordartskarta över Sverige. *Sv. geol. Unders.* Ba **17** (106 pp).

LUNDQVIST, J. 1960: Aspects of Quaternary geology in middle Sweden. *Guide to excursion C15, 21st Int. geol. Congress*, Stockholm (46 pp).

1965: The Quaternary of Sweden. In Rankama, K. (ed) *The Quaternary* **1**, Interscience, New York, 139–198

1967: Submoräna sediment i Jämtlands Län. *Sv. geol. Unders.* C **618** (267 pp).

1969: Problems of the so-called Rogen moraine. *Sv. geol. Unders.* C **648**, 1–32.

1972: Ice-lake types and deglaciation pattern along the Scandinavian mountain range. *Boreas* **1** (1), 27–54.

1973: Isavsmältningens förlopp i Jämtlands Län. *Sv. geol. Unders.* **66** (12) (187 pp).

LYONS, J. B., SAVIN, S. M. and TAMBURI, A. J. 1971: Basement ice, Ward Hunt ice shelf, Ellesmere Island, Canada. *J. Glaciol.* **10** (58), 93–100.

MAAG, H. 1969: Ice dammed lakes and marginal glacial drainage on Axel Heiberg Island. *Axel Heiberg Island Res. Rept*, McGill Univ., Montreal (147 pp).

1972: Ice-dammed lakes on Axel Heiberg Island, with special reference to the geomorphological effect of the outflowing lake water. In Müller, F. (ed) *Miscellaneous papers, Axel Heiberg Island Res. Rept, IGU Field Tour Ea* **2**, 39–48.

MACCLINTOCK, P. and DREIMANIS, A. 1964: Reorientation of till fabric by overriding glacier ice in the St Laurence valley. *Am. J. Sci.* **262**, 133–42.

MACKAY, J. R. 1959: Glacier ice-thrust features on the Yukon coast. *Geogr. Bull.* **13**, 5–21.

MALAURIE, J. *et al.* 1972: Preliminary remarks on Holocene palaeoclimates in the regions of Thule and Inglefield Land, above all since the beginning of our own era. In Vasari, Y., Hyvärinen, H. and Hicks, S. (eds) *Climatic changes in Arctic areas during the last ten thousand years*, Oulu, 105–36.

MANGERUD, J. 1970a: Interglacial sediments at Fjøsanger, near Bergen, with the first Eemian pollen spectra from Norway. *Norsk geol. Tidsskr.* **50**, 167–81.

1970b: Late Weichselian vegetation and ice-front oscillations in the Bergen district, western Norway. *Norsk geogr. Tidsskr.* **24** (3), 121–48.

MANGERUD, J., ANDERSEN, S. T., BERGLUND, B. E. and DONNER, J. J. 1974: Quarternary stratigraphy of Norden, a proposal for terminology and classification. *Boreas* **3** (3), 109–28.

MANLEY, G. 1955: On the occurrence of ice domes and permanently snow-covered summits. *J. Glaciol.* **2** (16), 453–6.

1966: Problems of the climatic optimum: the contributions of glaciology. In Sawyer J. S. (ed) *World climate from 8,000 to 0 BC*, R. met. Soc., London, 34–39.

1971: Interpreting the meteorology of the late and post-glacial. *Palaeogeogr. Palaeoclimatol. Palaeoecol.* **10**, 163–75.

MANNERFELT, C. M. 1945: Några glacialmorfologiska Formelement. *Geogr. Annlr* **27**, 1–239.

1949: Marginal drainage channels as indicators of the gradients of Quaternary ice caps. *Geogr. Annlr* **31**, 194–9.

MARCUS, M. G. and RAGLE, R. H. 1970: Snow accummulation in the Icefield Ranges, St Elias Mountains, Yukon. *Arctic and Alpine Res.* **2** (4), 277–92.

MARCUSSON, I. 1973: Studies on flow till in Denmark. *Boreas* **2** (4), 213–31.

MARKOV, K. K., BARDIN, V. I., LEBEDEV, V. L., ORLOV, A. I. and SUETOVA, I. A. 1970: The geography of Antarctica. *Israel Progr. Sci. Transl.* Jerusalem (370 pp).

DE MARTONNE, E. 1957: *Traité de geographie physique.* Armand Colin, Paris (482 pp).

MATHEWS, W. H. 1964: Water pressure under a glacier. *J. Glaciol.* **5** (38), 235–40.

MATTHES, F. E. 1930: Geologic history of the Yosemite valley. *US geol Surv. Prof. Pap.* **160** (137 pp).

MATTHEWS, R. K. 1972: Dynamics of the ocean-cryosphere system: Barbados data. *Quat. Res.* **2**, 368–373.

MATHEWS, W. H., KELLOGG, W. W. and ROBINSON, G. D. (eds) 1971: *Man's impact on climate.* MIT Press, Massachusetts.

MATHEWS, W. H. and MACKAY, J. R. 1960: Deformation of soils by glacier ice and the influence of pore pressure and permafrost. *R. Soc. Can. Trans* **54**, 3 (4), 27–36.

MAYO, L. R., MEIER, M. F. and TANGBORN, W. V. 1972: A system to combine stratigraphic and annual mass-balance systems: a contribution to the International Hydrological Decade. *J. Glaciol.* **11** (61), 3–14.

MAYR, R. 1964: Untersuchungen über ausmass und folgen der klima – und gletscherschwankungen seit dem beginn der post glacialen wärm-zeit. *Z. Geomorph.* **9**, 257-85.

MCCALL, J. G. 1960: The flow characteristics of a cirque glacier and their effect on glacial structure and cirque formation. In Lewis, W. V. (ed) Norwegian cirque glaciers, *R. geogr. Soc. Res. Ser.* **4**, 39–62.

M'CORMICK, R. 1884: *Voyages of discovery in the Arctic and Antarctic seas and round the world* **1**, Sampson Low, Marston, Searle and Rivington, London (432 pp).

MCGREGOR, V. R. 1967: Holocene moraines and rock glaciers in the central Ben Ohau range, South Canterbury, New Zealand. *J. Glaciol.* **6** (47), 737–48.

MCKENZIE, G. D. 1969: Observations on a collapsing kame terrace in Glacier Bay National Monument, southeastern Alaska. *J. Glaciol.* **8** (54), 413–25.

MCSAVENEY, M. J. 1973: Folding of cold ice. *Antarctic J. US* **8** (6), 344–6.

MCSAVENEY, M. J. and GAGE, M. 1968: Ice flow measurements on Franz Josef glacier, New Zealand, in 1966. *NZ J. Geol. Geophys.* **11**, 564–92.

MEIER, M. F. 1951: Recent eskers in the Wind River mountains of Wyoming. *Iowa Acad. Sci.* **58**, 291–294.

1960a: Mode of flow of Saskatchewan glacier, Alberta, Canada. *US geol Surv. Prof. Pap.* **351** (70 pp).

1960b: Distribution and variations of glaciers in the United States exclusive of Alaska. *Int. Ass. scient. Hydrol.* **54**, *Gen. Assembly of Helsinki, 1960*, 420–29.

1961: Mass budget of South Cascade glacier 1957-60. *US geol Surv. Prof. Pap.* **424**-B, 206–11.

1964: The recent history of advance-retreat and net budget of South Cascade glacier. *Amer. Geophys. Union Trans* **45**, 608-

1965: Glaciers and climate. In Wright, H. E. and Frey, D. G. (eds) *The Quaternary of the United States*, Princeton Univ. Press, NJ, 795–805.

1967: Why study glaciers? *Am Geophys. Un. Trans* **48**, 798–802.

MEIER, M. F. and JOHNSON, A. 1962: The kinematic wave on Nisqually glacier, Washington. *J. geophys. Res.* **67**, 886.

MEIER, M. F. and POST, A. 1969: What are glacier surges? *Can J. Earth Sci.* **6** (4), 807–17.

MEIER, M. F. and TANGBORN, W. V. 1961: Distinctive characteristics of glacier runoff. *US geol Surv. Prof. Pap.* **424**-B, 14–16.

MEIER, M. F., TANGBORN, W. V., MAYO, L. R. and POST, A. 1971: Combined ice and water balances of Gulkana and Wolverine glaciers, Alaska, and South Cascade glacier, Washington, 1965 and 1966 hydrologic years. *US geol Surv. Prof. Pap.* **715**-A (23 pp).

MELLOR, M. 1964: Snow and ice on the earth's surface. *US Army Cold Reg. Res. Engng Lab. Res. Rept* II-CL (163 pp).

MENARD, H. W. 1971: The Late Cenozoic history of the Pacific and Indian Ocean basins. In Turekian, K. K. (ed) *Late Cenozoic glacial ages*, Yale Univ. Press, 1–14.

MENELEY, W. A. 1964: Geology of the Melfort area (72-A), Saskatchewan. *Unpub. PhD thesis, Univ. Illinois*, 33–9.

MERCER, J. H. 1961: The response of fjord glaciers to changes in the firn limit. *J. Glaciol.* **3** (29), 850–58 1967: Glacier resurgence at the Atlantic/sub-Boreal boundary. *QJ R. Met. Soc.* **93**, 528–34.

1968a: Antarctic ice and Sangamon sea-level. *Int. Ass. scient. Hydrol. Bull.* **79**, 217–25.

1968b: Glacial geology of the Reedy glacier area, Antarctica. *Bull. Geol Soc. Am.* **79** (4), 471–83.

1969: The Alleröd Oscillation: a European climatic anomaly? *Arctic and Alpine Res.* **1**, 227–34.

1970: A former ice sheet in the Arctic Ocean. *Palaeogeogr. Palaeoclimatol. Palaeoecol.* **8**, 19–27.

1971: Cold glaciers in the central Transantarctic mountains, Antarctica: dry ablation areas and subglacial erosion. *J. Glaciol.* **10** (59), 319–21.

1973: Cainozoic temperature trends in the southern hemisphere: Antarctic and Andean glacial evidence. In van Zinderen Bakker E. M. (ed) *Palaeoecology of Africa and of the surrounding islands and Antarctica* **8,** Balkema, Cape Town, 87–114.

MICKELSON, D. M. 1971: Glacial geology of the Burroughs glacier area, southeast Alkaska. *Inst. of Polar Studies Rept* **40,** Ohio State Univ. (149 pp).

MILANKOVICH, M. 1969: *Canon of insolation and the ice age theory.* Israel Progr. Sci. Transl., Jerusalem (484 pp).

MILLER, C. D. 1969: Chronology of Neoglacial moraines in the Dome Peak area, North Cascade range, Washington. *Arctic and Alpine Res.* **1** (1), 49–66.

MILLER, G. H. 1973: Late Quaternary glacial and climatic history of northern Cumberland Peninsula, Baffin Island, NWT Canada. *Quat. Res.* **3** (4), 561–83.

MILLER, G. H. and HARE, P. E. 1975: Use of amino acid reactions in some Arctic Quaternary marine fossils as stratigraphic and geochronological indicators. In *Abstracts with programs* **7** (7), Geol Soc. Am., 1200.

MILLER, M. M. 1952: Glacier tunnel observations in Alaska. *J. Glaciol.* **2** (11), 69–70.

1961: A distribution study of abandoned cirques in the Alaska–Canada Boundary range. In Raasch, G. O. (ed) *Geology of the Arctic* **2,** 833–47.

1974: Entropy and self-regulation of glaciers in Arctic and Alpine regions. In Fahey, B. D. and Thompson, R. D. (eds) *Research in Polar and alpine geomorphology, 3rd Guelph Symposium on geomorphology, 1973,* Geo Abstracts, Norwich 136–58.

MITCHELL, G. F. 1960: The Pleistocene history of the Irish Sea. *Advmt Sci., Lond.* **17,** 313–25.

MITCHELL, G. F., PENNY, L. F., SHOTTON, F. W. and WEST, R. G. 1973: A correlation of Quaternary deposits in the British Isles. *Geol Soc. Lond. Spec. Rept* **4** (99 pp).

MITCHELL, J. M. 1972: The natural breakdown of the present interglacial and its possible intervention by human activities. *Quat. Res.* **2,** 436–45.

MORAN, S. 1971: Glaciotectonic structures in drift. In Goldwait, R. P. (ed) *Till, a symposium,* Ohio State Univ. Press, 127–48.

MORGAN, A. 1973: Late Pleistocene environmental changes indicated by fossil insect faunas of the English midlands. *Boreas* **2** (4), 173–212.

MORGAN, V. I. and BUDD, W. F. 1975: Radio echo sounding of the Lambert glacier basin. *J. Glaciol.* **15** (73), 103–11.

MORNER, N. A. 1972: When will the present interglacial end? *Quat. Res.* **2,** 341–9.

MULLER, E. H. 1974: Origins of drumlins. In Coates, D. R. (ed) *Glacial geomorphology,* State Univ. NY, Binghampton, 187–204.

MÜLLER, F. and IKEN, A. 1972: Velocity fluctuations and water regime of Arctic valley glaciers. *Symposium on the Hydrology of Glaciers, Cambridge, 9–13 Sept. 1969, Int. Ass. scient. Hydrol. Pub.* **95,** 165–82.

MURRAY, R. C. 1953: The petrology of the Carey and Valders tills of northeastern Wisconsin, *Am. J. Sci.* **251,** 140–55.

NACE, R. L. 1969: World water inventory and control. In Chorley, R. J. (ed) *Water, earth and man,* Methuen, London, 31–42.

NAMIAS, J. 1969: Seasonal interactions between the North Pacific Ocean and the atmosphere during the 1960s. *Monthly Weather Review* **97,** 173–92.

NIELSEN, D. N. 1970: Washboard moraines in northeastern North Dakota. *Compass* **47** (3), 154–62.

NIELSEN, L. E. 1969: The ice-dam, powder-flow theory of glacier surges. *Can J. Earth Sci.* **6,** (4), 955–961.

NIERIAROWSKI, W. 1963: Some problems concerning deglaciation by stagnation and wastage of large portions of the ice sheet within the area of the last glaciation in Poland. *Rept 6th Conf. int. Ass. quatern. Res. (Warsaw, 1961),* Lodz **3,** 245–56.

NILSSON, E. 1960: The recession of the land-ice in Sweden during the Alleröd and the Younger Dryas ages. *Int. Geol. Cong. 21st Session (Norden) Rept* **4,** 98–107.

1968: The Late Quaternary history of southern Sweden. *Kungl. Sv. Vetensk. Akad. Handle. Ser. 4.* **12** (1) (117 pp).

NOBLES, L. H. and WEERTMAN, J. 1971: Influence of irregularities of the bed of an ice sheet on deposition rate of till. In Goldthwait, R. P. (ed) *Till, a symposium,* Ohio State Univ. Press, 117–126.

NORLUND, P. 1924: Buried Norsemen at Herfoljness. *Medd. om Grønland* **67,** 228–59.

NORRIS, S. E. and WHITE, G. W. 1961: Hydrologic significance of buried valleys in glacial drift. *US geol Surv. Prof. Pap.* **424**-B, 34–5.

NOUGIER, J. 1972: Aspects de morpho-tectonique glaciaire aux Iles Kerguelen. *Rev. Géogr. Phys. et Géol. dyn.* **14** (5), 499–505.

NYE, J. F. 1952a: A method of calculating the thickness of ice sheets. *Nature, Lond.* **169** (4300), 529–530.

1952b: The mechanics of glacier flow. *J. Glaciol.* **2** (12), 82–93.

1957: The distribution of stress and velocity in glaciers and ice sheets. *Proc. R. Soc.* A **239**, 113–133.

1960: The response of glaciers and ice sheets to seasonal and climatic changes. *Proc. R. Soc.* A **256**, 559–84.

1961: The influence of climatic variations on glaciers. *Int. Ass. scient. Hydrol.* **54**, 397–404.

1965a: The flow of a glacier in a channel of rectangular, elliptic or parabolic cross-section. *J. Glaciol.* **5** (41), 661–90.

1965b: The frequency response of glaciers. *J. Glaciol.* **5** (41), 567–87.

1965c: Stability of a circular cylindrical hole in a glacier. *J. Glaciol.* **5** (40), 505–07.

1970: Glacier sliding without cavitation in a linear viscous approximation. *Proc. R. Soc.* A **315**, 381–403.

1973a: The motion of ice past obstacles. In Whalley, E., Jones, S. J. and Gold, L. W. (eds) *Physics and chemistry of ice*, R. Soc. Can., Ottowa, 387–95.

1973b: Water at the bed of a glacier. *Symposium on the Hydrology of Glaciers, Cambridge, 9–13 Sept. 1969, Int. Ass. scient. Hydrol. Pub.* **95**, 189–94.

NYE, J. F. and FRANK, F. C. 1973: Hydrology of the intergranular veins in a temperate glacier. *Symposium on the Hydrology of Glaciers, Cambridge, 9–13 Sept. 1969. Int. Ass. scient. Hydrol. Pub.* **95**, 157–61.

NYE, J. F. and MARTIN, P. C. S. 1968: Glacial erosion. *Int Ass. scient. Hydrol. Pub.* **79**, 78–86.

OKKO, M. 1965: M. Sauramo's Baltic ice lake B IV–BV–BVI; a re-evaluation. *Ann. Acad. Scient. Fenn.* AIII, **84** (63 pp).

OKKO, V. 1955: Glacial drift in Iceland, its origin and morphology. *Bull. Comm. géol. Finl.* **170**, 1–133.

OLAUSSON, E. 1972: Oceanographic aspects of the Pleistocene of the Arctic Ocean. *Inter-Nord* **12**, 151–170.

OLLIER, C. D. 1969: *Weathering.* Oliver and Boyd, Edinburgh (304 pp).

OLSSON, I. U. (ed) 1970: Radiocarbon variations and absolute chronology. *Nobel Symposium* **12**, Wiley, New York.

OMMANNEY, C. S. L. 1969: A study in glacier inventory: the ice masses of Axel Heiberg Island, Canadian Arctic Archipelago. *Axel Heiberg Res. Rept, Glaciology* **3**, McGill Univ. (105 pp).

ORHEIM, O. 1970: Glaciological investigations of Store Supphellebre, west Norway. *Norsk Polarinstitutt Skr.* **151** (48 pp).

1971: Glaciological studies of Deception and Livingston Island. *Antarct. J. US* **6** (4), 85.

ORVIG, S. 1951: The climate of the ablation period on the Barnes ice cap in 1950. *Geogr. Annlr* **33**, 166–209.

ØSTREM, G. 1959: Ice melting under a thin layer of moraine, and the existence of ice cores in moraine ridges. *Geogr. Annlr* **41**, 228–30.

1964: Ice-cored moraines in Scandinavia. *Geogr. Annlr* **46**, 282–337.

1965: Problems of dating ice-cored moraines. *Geogr. Annlr* **47**, 1–38.

1966: The height of the glacial limit in southern British Columbia and Alberta. *Geogr. Annlr* **48**A (3), 126–38.

1974a: The use of ERTS data to monitor glacier behaviour and snow cover – practical implications for water power production. *Norges Vassdrags-og Elektrisitetsvesen* **24** (12 pp).

1974b: Present alpine ice cover. In Ives, J. D. and Barry, R. G. (eds) *Arctic and alpine environments*, Methuen, London, 226–50.

ØSTREM, G., BRIDGE, C. W. and RANNIE, W. F. 1967: Glacio-hydrology, discharge and sediment transport in the Decade glacier area, Baffin Island, NWT. *Geogr. Annlr* **49**A, 268–82.

ØSTREM, G., HAAKENSEN, N. and MELANDER, O. 1973: *Atlas over Breer i Nord-Skandinavia.* Norges Vassdrags og Elektrisitetsvesen og Stockholm Univ. (315 pp).

ØSTREM, G. *et al.* 1969: *Atlas over breer i Sør-Norge.* Norges Vassdrags-og Electrisitetsvesen, Oslo.

OSWALD, G. K. A. and ROBIN, G. de Q. 1973: Lakes beneath the Antarctic ice sheet. *Nature, Lond.* **245** (5423), 251–4.

OUTCALT, S. I. and MACPHAIL, D. 1965: A survey of neoglaciation in the Front Range of Colorado. *Univ. Colorado Studies, Earth Sci. Ser.* **4** (124 pp).

PAGE, N. R. 1971: Subglacial limestone deposits in the Canadian Rocky mountains. *Nature, Lond.* **229** (5279), 42–3.

PALMER, A. C. 1972: A kinematic wave model of glacier surges. *J. Glaciol.* **11** (61), 65–72.

PARIZEK, R. R. 1969: Glacial ice-contact rings and ridges. *Geol. Soc. Am. Spec. Pap.* **123**, 49–102.

PASCHINGER, V. 1912: Die Schneegrenze in verschiedenen Klimaten. *Peterm. Geogr. Mitt.* **37** (173) (93 pp).

PATERSON, W. S. B. 1964: Variations in velocity of Athabasca glacier with time. *J. Glaciol.* **5,** (39), 277–85.

— 1969: *The physics of glaciers.* Pergamon, Oxford (250 pp).

— 1972: Temperature distribution in the upper layers of the ablation area of the Athabasca glacier, Alberta, Canada. *J. Glaciol.* **11** (61), 31–41.

PATERSON, W. S. B. and SAVAGE, J. C. 1970: Excess pressure observed in a water-filled cavity in Athabasca glacier, Canada. *J. Glaciol.* **9** (55), 103–07.

PELTIER, L. C. 1950: The geographic cycle in periglacial regions as it is related to climatic geomorphology. *Ann. Ass. Amer. Geogr.* **40,** 214–36.

PENCK, A. 1905: Glacial features in the surface of the Alps. *J. Geol.* **13,** 1–17.

PENCK, A. and BRÜCKNER, E. 1909: *Die Alpen im Eiszeitalter.* Leipzig (1199 pp).

PENNINGTON, W. A. 1969: *The history of British vegetation.* EUP, London (152 pp).

PENNY, L. F. and CATT, J. A. 1967: Stone orientation and other structural features of tills in east Yorkshire. *Geol. Mag.* **104,** 344–60.

PENNY, L. F. *et al.* 1969: Age and insect fauna of the Dimlington silts, east Yorkshire. *Nature, Lond.* **224,** 65–7.

PENTTILÄ, S. 1963: The deglaciation of the Laanilä area, Finnish Lapland. *Bull. Comm. géol. Fin.* **203,** 7–71.

PERRY, W. J. and ROBERTS, H. G. 1968: Late Precambrian glaciated pavements in the Kimberley region, Western Australia. *J. geol Soc. Aust.* **15,** 51–6.

PETERSON, D. N. 1970: Glaciological studies on the Casement glacier, southeast Alaska. *Rept Ohio State Univ. Inst. Polar Stud.* **36** (161 pp).

PIPPAN, T. 1965: Morphological studies concerning glaciated areas in Norway's high mountains with special reference to alpine landforms. *Die Erde (Z. der Gesellschaft für Erdkunde, Berlin)* **96** (2), 105–121.

POPPER, K. R. 1972: *The logic of scientific discovery.* Hutchinson, London (480 pp).

PORTER, S. C 1964: Late Pleistocene glacial chronology of north central Brooks range, Alaska. *Am. J. Sci.* **262,** 446–60.

PORTER, S. C. 1975: Glaciation limit in New Zealand's southern alps. *Arctic and Alpine Res.* **7** (1), 33–37.

PORTER, S. C. and DENTON, G. H. 1967: Chronology of neoglaciation in the North American cordillera. *Am. J. Sci.* **265,** 177–210.

POST, A. S. 1960: The exceptional advances of the Muldrow, Black Rapids and Susitna glaciers. *J. geophys. Res.* **65,** 3703–12.

— 1967: Effects of the March 1964 Alaska earthquake on glaciers. *US geol Surv. Prof. Pap.* **544**-D (42 pp).

POST, A. S. and LACHAPELLE, E. R. 1971: *Glacier ice.* Univ. Washington Press (100 pp).

PREST, V. K. 1968: Nomenclature of moraines and ice-flow features as applied to the glacial map of Canada. *Geol Surv. Pap. Can.* **67–57** (32 pp).

— 1969: Retreat of Wisconsin and recent ice in North America. *Geol Surv. Can.* Map 1257A.

PREST, V. K. *et al.* 1968: Glacial map of Canada. Scale 1 : 5 million. *Geol Surv. Can.* Map 1253A.

PRICE, R. J. 1960: Glacial meltwater channels in the upper Tweed drainage basin. *Geogr. J.* **126,** 483–489.

— 1966: Eskers near the Casement glacier, Alaska. *Geogr. Annlr* **48,** 111–25.

— 1969: Moraines, sandar, kames and eskers near Breiðamerkurjökull, Iceland. *Trans Inst. Brit. Geogr.* **46,** 17–43.

— 1970: Moraines at Fjällsjökull, Iceland. *Arctic and Alpine Res.* **2,** 27–42.

— 1973: *Glacial and fluvioglacial landforms.* Oliver and Boyd, Edinburgh (242 pp).

PYTTE, R. (ed) 1969: Glasiologiske undersökelser i Norge 1968. *Norges Vassdrags-og Elektrisitetsvesen Rapport* **5/69,** Oslo (149 pp).

— (ed) 1970: Glasiologiske undersökelser i Norge, 1969. *Norges Vassdragsdirektoratet Hydrologisk Avdeling,* Oslo (96 pp).

RAMBERG, H. 1964: Notes on model studies of folding of moraines in piedmont glaciers. *J. Glaciol.* **5** (38), 207–18.

RAPP, A. 1960: Talus slopes and mountain walls at Tempelfjorden, Spitzbergen. *Norsk Polarinst. Skr.* **119** (96 pp).

RAYMOND, C. F. 1971: Flow in a transverse section of Athabasca glacier, Alberta, Canada. *J. Glaciol.* **10** (58), 55–84.

RAYNAUD, D. and LORIUS, C. 1973: Climatic implications of total gas content in ice at Camp Century. *Nature, Lond.* **243** (5405), 283–4.

READING, H. G. and WALKER, R. G. 1966: Sedimentation of Eocambrian tillites and associated sediments in Finnmark, northern Norway. *Palaeogeogr. Palaeoclimatol. Palaeoecol.* **2,** 177–212.

REED, B., GALVIN, C. J. and MILLER, J. P. 1962: Some aspects of drumlin geometry. *Am. J. Sci.* **260**, 200–10.

REID, J. R. 1970: Geomorphology and glacial geology of the Martin River glacier, Alaska. *Arctic* **23** (4), 254–67.

REYNAUD, L. 1973: Flow of a valley glacier with a solid friction law. *J. Glaciol.* **12** (65), 251–8.

RICH, J. L. 1943: Buried stagnant ice as a normal product of a progressively retreating glacier in a hilly region. *Am. J. Sci.* **241**, 95–9.

RICHTER, K. 1936: Refugestudien im Engabre, Fondalsbre, und ihren Vorlandsedimenten. *Z. Gletscher.* **24**, 22–30.

ROBIN, G. de Q. 1955: Ice movement and temperature distribution in glaciers and ice sheets. *J. Glaciol.* **2** (18), 523–32.

1958: Glaciology III. Seismic shooting and related investigations. *Norwegian-British-Swedish Antarctic Exped. 1040–52. Scientific Results*, Oslo **5**, 126–7.

1969: Initiation of glacier surges. *Can J. Earth Sci.* **6** (4), 919–28.

1972: Polar ice sheets: a review. *Polar Record* **16** (100), 5–22.

ROBIN, G. de Q. and BARNES, P. 1969: Propagation of glacier surges. *Can J. Earth Sci.* **6** (4), 969–77.

ROBIN, G. de Q., SWITHINBANK, C. W. M. and SMITH, B. M. E. 1969: Interpretation of radio echo sounding in polar ice sheets. *Phil. Trans R. Soc.* A **265**, 437–505.

RÖTHLISBERGER, H. 1968: Erosive processes which are likely to accentuate or reduce the bottom relief of valley glaciers. *Int. Ass. scient. Hydrol.* **79**, 87–97.

1972: Water pressure in intra- and subglacial channels. *J. Glaciol.* **11** (62), 177–203.

1974: Möglichkeiten und Grenzen der Gletscherüberwachung. *Neuen Zürcher Zeitung* **196**, 2–15.

ROWLANDS, B. M. 1971: Radiocarbon evidence of the age of an Irish Sea glaciation in the Vale of Clwyd. *Nature, Lond.* **230** (9), 9–11.

RUCKLIDGE, M. A. 1956: A glacier water-spout in Spitsbergen. *J. Glaciol.* **2** (19), 637–9.

RUDBERG, S. 1954: Västerbottens berggrundsmorfologi. Ett försök till rekonstruktion av preglaciala erosionsgenerationer i Sverige. *Geographica* **25** (457 pp).

1970: The areas of bare rock in Scandinavia. *Acta Geographica Lodziensia* **24**, 389–97.

1973: Glacial erosion forms of medium size – a discussion based on four Swedish case studies. *Z. Geomorph.* **17**, 33–48.

RUNDLE, A. S. 1970: Snow accumulation and ice movement on the Anvers Island ice cap, Antarctica: a study of mass balance. In Gow, A. J. *et al.* (eds) *International Symposium on Antarctic Glaciological Exploration (ISAGE), Int. Ass. scient. Hydrol. Pub.* **86**, 377–90.

RUSSELL, I. C. 1893: Malaspina glacier. *J. Geol.* **1**, 219–45.

RUTFORD, R. H. 1972: Glacial geomorphology of the Ellsworth mountains. In Adie, R. J. (ed) *Antarctic geology and geophysics*, Universitetsforlaget, Oslo 225–32.

RUTFORD, R. H. *et al.* 1972: Tertiary glaciation in the Jones Mountains. In Adie, R. J. (ed) *Antarctic geology and geophysics*, Universitetsforlaget, Oslo, 239–43.

RUTKIS, J. 1971: Tables on relative relief in middle and western Europe. *UNGI Rapport* **9**, Uppsala (94 pp).

RUTTEN, M. G. 1960: Ice-pushed ridges, permafrost and drainage. *Am. J. Sci.* **258** (4), 293–7.

RUTTER, N. W. 1965: Foliation pattern of Gulkana glacier, Alaska range, Alaska. *J. Glaciol.* **5** (41), 711–18.

RYDER, J. M. 1972: Pleistocene chronology and glacial geomorphology: studies in southwestern British Columbia. In Slaymaker, O. and McPherson, H. J. (eds) *Mountain geomorphology: geomorphological processes in the Canadian Cordillera*, BC Geog. Ser. **14**, 63–72.

SACHS, H. M. 1973: North Pacific radiolarian assemblages and their relationship to oceanographic parameters. *Quat. Res.* **3**, 73–88.

SANCETTA, C. A., IMBRIE, J., KIPP, N. G., MCINTYRE, A. and RUDDIMAN, W. F. 1972: Climatic record in North Atlantic deep-sea core V23-82: comparison of the last and present interglacials based on quantitative time series. *Quat. Res.* **2**, 363–7.

SAURAMO, M. 1929: The Quaternary geology of Finland. *Bull. Comm. géol. Fin.* **86** (110 pp).

SAVAGE, J. C. and PATERSON, W. S. B. 1963: Borehole measurements in the Athabasca glacier. *J. geophys. Res.* **68**, 4521–36.

SAWYER, J. S. (ed) 1966: *World climate from 8,000 to 0 BC*. R. met. Soc., London (229 pp).

SCHOU, A. 1949: The landscapes. In Nielsen, N. *Atlas of Denmark*, Copenhagen.

SCHOVE, D. J. 1955: The sunspot cycle, 649 BC to AD 2000. *J. Geophys. Res.* **60**, 127–46.

SCHUMM, S. A. and LICHTY, R. W. 1965: Time, space and causality in geomorphology. *Am. J. Sci* **263**, 110–119.

SCHUMM, S. A. and SHEPHERD, R. G. 1973: Valley floor morphology: evidence of subglacial erosion. *Area* **5** (1), 5–9.

SCHYTT, V. 1964: Scientific results of the Swedish glaciological expedition to Nordaustlandet, Spitzbergen, 1957 and 1958. Parts I and II. *Geogr. Annlr* **46** (3), 243–81.

1967: A study of 'ablation gradient'. *Geogr. Annlr* **49**A (2–4), 327–32.

1969: Some comments on glacier surges in eastern Svalbard. *Can J. Earth Sci.* **6** (4), 867–73.

SCOTT POLAR RESEARCH INSTITUTE, 1974: *Ice sheet surface and sub-ice relief*: *90° E–180°*. Antarctica, radio echo sounding Map Series A. Scale 1:5 million. Cambridge.

SEDDON, B. 1957: Late-glacial cwm glaciers in Wales. *J. Glaciol.* **3** (22) 94–9.

SELLERS, W. D., 1965: *Physical climatology*. Univ. of Chicago Press, Chicago (272 pp).

SHACKLETON, N. J. and OPDYKE, N. O. 1973: Oxygen isotope and palaeomagnetic stratigraphy of equatorial Pacific core V28–238: oxygen isotope temperatures and ice volumes on a 10^5 year and 10^6 year scale. *Quat. Res.* **3,** 39–55.

SHARP, R. P. 1949: Studies of superglacial debris on valley glaciers. *Am. J. Sci.* **247,** 289–315.

1951: Thermal regime of firn on Upper Seward glacier, Yukon Territory, Canada. *J. Glaciol.* **1** (9), 461–5.

1960: *Glaciers*. Condon Lectures, Oregon State System of Higher Education, Oregon (78 pp).

1974: Ice on Mars. *J. Glaciol.* **13** (68), 173–86.

SHAW, D. M. and DONN, W. L. 1971: A thermodynamic study of Arctic palaeoclimatology. *Quat. Res.* **1,** 175–87.

SHAW, J. 1971: Mechanism of till deposition related to thermal conditions in a Pleistocene glacier. *J. Glaciol.* **10** (60), 363–73.

1972: Sedimentation in the ice-contact environment, with examples from Shopshire (England). *Sedimentology* **18,** 23-62.

SHREVE, R. L. 1972: Movement of water in glaciers. *J. Glaciol.* **11** (62), 205–14.

SHUMSKII, P. A. 1950: The energy of glaciation and the life of glaciers. *SIPRE Translation* **7,** *Corps of Engineers, US Army* (27 pp).

SIMPSON, G. C. 1957: Further studies in world climate. *QJ R. met. Soc.* **83,** 459–81.

SISSONS, J. B. 1958: Supposed ice-dammed lakes in Britain, with particular reference to the Eddleston valley, southern Scotland. *Geogr. Annlr* **40,** 159–87.

1960a: Some aspects of glacial drainage channels in Britain, part I, *Scott. geogr. Mag.* **76,** (3) 131–146.

1960b: Subglacial, marginal and other glacial drainage in the Syracuse-Oneida area, New York. *Bull. geol Soc. Am.* **71,** 1575–88.

1961a: Some aspects of glacial drainage channels in Britain, Part II. *Scott. geogr. Mag.* **77,** 15–36.

1961b: A subglacial drainage system by the Tinto hills, Lanarkshire. *Trans Edinb. geol Soc.* **18,** 175–193.

1963: The glacial drainage system around Carlops, Peebleshire. *Trans Inst. Brit. Geogr.* **32,** 95–111.

1964a: The Perth Readvance in central Scotland. *Scott. geogr. Mag.* **80,** 28–36.

1964b: The glacial period. In Watson, J. W. and Sissons, J. B. (eds) *The British Isles, a systematic geography*. Nelson, London, 131–51.

1967: *The evolution of Scotland's scenery*. Oliver and Boyd, Edinburgh (259 pp).

SKINNER, R. G. 1973: Quaternary stratigraphy of the Moose river basin, Ontario. *Bull. Geol Surv. Canada* **225** (77 pp).

SLATER, G. 1926: Glacial tectonics as reflected in disturbed drift deposits. *Proc. geol Assoc.* **37,** 392–400.

1927: The structure of the disturbed deposits of Møene Klint, Denmark. *Trans R. Soc. Edinb.* **55,** 289–302.

1929: The structure of drumlins exposed on the south shore of Lake Ontario. *NY St. Mus. Bull.* **281,** 3–19.

SMALL, R. J. and CLARK, M. J. 1974: The medial moraines of the lower Glacier de Tsidjiore Nouve, Valais, Switzerland. *J. Glaciol.* **13** (68), 255–64.

SMALLEY, I. J. and UNWIN, D. J. 1968: The formation and shape of drumlins and their distribution and orientation in drumlin fields. *J. Glaciol.* **7** (51), 377–90.

SMITH, D. I. 1972: The solution of limestone in an Arctic environment. In Price, R. J. and Sugden, D. E. (eds) Polar Geomorphology, *Inst. Brit. Geogr. Spec. Pub.* **4,** 187–200.

SMITH, H. T. U. 1948: Giant glacial grooves in northwest Canada. *Am. J. Sci.* **246,** 503–14.

SMITH, J. 1960: Glacier problems in South Georgia. *J. Glaciol.* **3** (28), 705–14.

SOLLID, J. L. 1963/64: Isavsmeltingsforløpet langs hovedvasskillet mellom Hjerkinn og Kviknesskogen. *Norsk Geogr. Tidsskr.* **19**, 51–76.

SOONS, J. M. 1960: Glacial retreat stages in Kinrosshire. *Scott. geogr. Mag.* **76**, 46–57.

1971: New Zealand's capricious Franz Josef. *Geogr. Mag.* **43** (7), 490–94.

SÖRENSEN, R. 1974: Ice recession in the Oslofjord area. In Aario, R. and Piispanen, R. (eds) *XI Nordiska Geologiska Vintermötet, Oulu, Jan. 3–5, 1974*, **B**, 53.

SOUCHEZ, R. 1966: Sur les mécanismes de l'érosion en Antarctique. *Bull. Soc. belge Étude Géogr.* **35** (1), 25–34.

1967: Le recult des verrous-gradins et les rapports glaciaire-périglaciaire en Antarctique. *Revue Géomorph. dyn.* **17** (2), 49–54.

SPARKS, B. W. and WEST, R. G. 1972: *The Ice Age in Britain*. Methuen, London (302 pp).

STACEY, F. D. 1969: *Physics of the earth*. Wiley, New York, (324 pp).

STALKER, A. M. S. 1960: Ice-pressed drift forms and associated deposits in Alberta. *Bull. geol Surv. Can.* **57** (38 pp).

STANLEY, A. D. 1969: Observations of the surge of Steele glacier, Yukon Territory, Canada. *Can J. Earth Sci.* **6** (4), 819–30.

STEINER, J. and GRILLMAIR, E. 1973: Possible galactic causes for periodic and episodic glaciations. *Bull. geol Soc. Am.* **84** (3), 1003–18.

STENBORG, T. 1965: Problems concerning winter runoff from glaciers. *Geogr. Annlr* **47**A (3), 141–84.

1968: Glacier drainage connected with ice structures. *Geogr. Annlr* **50**A (1), 25–53.

1969: Studies of the internal drainage of glaciers. *Geogr. Annlr* **51**A (1–2), 13–41.

1970: Delay of runoff from a glacier basin. *Geogr. Annlr* **52**A (1), 1–30.

1973: Some viewpoints on the internal drainage of glaciers. *Symposium on the Hydrology of Glaciers, Cambridge, 9–13 Sept. 1969, Int. Ass. scient. Hydrol. Pub.* **95**, 117–29.

STOKES, J. C. 1958: An esker-like ridge in process of formation, Flåtisen, Norway. *J. Glaciol.* **3** (24), 286–90.

STRÖMBERG, B. 1962: *Studier av isrecessionen i nordöstra Uppland och trakterna kring Ålands hav*. Naturgeogr. Inst. Stockholms Univ. (103 pp).

1965: Mappings and geochronological investigations in some moraine areas of south central Sweden. *Geogr. Annlr* **47**A, 73–82.

SUESS, H. E. 1970: The three causes of secular C¹⁴ fluctuations, their amplitudes and time constants. In Olsson, I.U. (ed) Radiocarbon variations and absolute chronology, *12th Nobel Symp*, Stockholm, 595–604.

SUGDEN, D. E. 1968: The selectivity of glacial erosion in the Cairngorm mountains, Scotland. *Trans Inst. Brit. Geogr.* **45**, 79–92.

1969: The age and form of corries in the Cairngorm mountains, Scotland. *Scott. geogr. Mag.* **85** (1), 34–46.

1970: Landforms of deglaciation in the Cairngorm mountains, Scotland. *Trans Inst. Brit. Geogr.* **51**, 201–19.

1972: Deglaciation and isostasy in the Sukkertoppen ice cap area, west Greenland. *Arctic and Alpine Res.* **4** (2) 97–117.

1974: Landscapes of glacial erosion in Greenland and their relationship to ice, topographic and bedrock conditions. *Inst. Brit. Geogr. Spec. Pub.* **7**, 177–95.

SUGDEN, D. E. and JOHN, B. S. 1965: The raised marine features of Kjove Land, east Greenland. *Geogr. J.* **131**, 235–47.

1973: The ages of glacier fluctuations in the South Shetland Islands, Antarctica. In van Zinderen Bakker, E. M. (ed) *Palaeoecology of Africa and of the surrounding islands and Antarctica* **8**, Balkema, Cape Town, 139–59.

SUGGATE, R. P. 1950: Franz Josef and other glaciers of the Southern Alps, New Zealand. *J. Glaciol.* **1** (8), 422–9.

SVENSSON, H. 1959: Is the cross-section of a glacial valley a parabola? *J. Glaciol.* **3** (25), 362–3.

SWITHINBANK, C. W. M. 1964: To the valley glaciers that feed the Ross ice shelf. *Geogr. J.* **130** (1), 32–48.

1966: A year with the Russians in Antarctica. *Geogr. J.* **132**, 463–75.

1970: Ice movement in the McMurdo Sound area of Antarctica. In Gow, A. J. *et al.* (eds) *International Symposium on Antarctic Glaciological Exploration (ISAGE), Int. Ass. scient. Hydrol. Pub.* **86**, 472–87.

SWITHINBANK, C. W. M. and ZUMBERGE, J. H. 1965: The ice shelves. In Hatherton, T. (ed) *Antarctica*, Methuen, London, 199–220.

SYNGE, F. M. 1950: The glacial deposits around Trim, Co. Meath. *Proc. R. Irish Acad.* **53**B (10), 99–110.

1970: The Irish Quaternary: current views 1969. In Stephens, N. and Glasscock, R. E. (eds) *Irish geographical studies*, Queen's Univ., Belfast, 34–48.

SYNGE, F. M. and STEPHENS, N. 1960: The Quaternary period in Ireland – an assessment. *Ir. Geogr.* **4** (2), 121–30.

TABOR, D. and WALKER, J. C. F. 1970: Creep and the friction of ice. *Nature, London.* **228** (5267), 137–139.

TANNER, V. 1932: The problems of the eskers. *Fennia* **55,** 1–13.

1934: The problems of the eskers. *Fennia* **58** (1).

1937: The problems of the eskers. *Fennia* **63** (1) (31 pp).

1938: *Die Oberflächengestaltung Finnlands.* Soc. Sci. Fennica, Helsinki.

TARLING, D. H. and TARLING, M. P. 1971: *Continental drift.* G. Bell, London (112 pp).

TARR, R. S. 1909: Some phenomena of the glacier margins in the Yakutat Bay region, Alaska. *Z. Gletscherk.* **3**, 81–110.

TARR, R. S. and BUTLER, B. S. 1909: The Yakutat Bay region, Alaska. *US geol Surv. Prof. Pap.* **64,** 1–178.

TARR, R. S. and MARTIN, L. 1914: *Alaskan glacier studies.* Nat. Geogr. Soc., Washington (498 pp).

TAUBER, H. 1970: The Scandinavian varve chronology and C[14] dating. In Olsson, I. U. (ed) Radiocarbon variations and absolute chronology, *12th Nobel Symposium*, Stockholm, 173–96.

TAYLOR, L. D. 1963: Structure and fabric on the Burroughs glacier, south-east Alaska. *J. Glaciol.* **4** (36), 731–52.

THEAKSTONE, W. H. 1965: Recent changes in the glaciers of Svartisen. *J. Glaciol.* **5** (40), 411–31.

1967: Basal sliding and movement near the margin of the glacier Østerdalsisen, Norway. *J. Glaciol.* **6** (48), 805–16.

THOMAS, M. F. 1974: *Tropical geomorphology.* Macmillan, London (332 pp).

THOMAS, R. H. 1973: The creep of ice shelves. *J. Glaciol.* **12** (64), 45–70.

THOMAS, R. H. and COSLETT, P. H. 1970: Bottom melting of ice shelves and the mass balance of Antarctica. *Nature, Lond.* **228** (5266), 47–9.

THORARINSSON, S. 1943: Oscillations of the Icelandic glaciers in the last 250 years. In Ahlmann, H. W:son and Thorarinsson, S., Vatnajökull, *Geogr. Annlr* **25,** 1–54.

1944: Present glacier shrinkage and eustatic changes of sea-level. *Geogr. Annlr* **26,** 131–59.

1953: Some new aspects of the Grímsvötn problem. *J. Glaciol.* **2** (14), 267–74.

1956: *The thousand years' struggle against ice and fire.* Mus. of Nat. Hist., Reykjavik (52 pp).

1956: On the variations of Svinafellsjökull, Skaftafellsjökull and Kviarjökull in Öraefi. *Jökull* **6,** 1–15.

1969: Glacier surges in Iceland, with special reference to the surges of Brúarjökull. *Can J. Earth Sci.* **6** (4), 875–82.

THORÉN, R. 1969: *Picture atlas of the Arctic.* Elsevier, Amsterdam (449 pp).

TODTMANN, E. M. 1932: Glazialgeologische Studien am Südrand des Vatna-Jökull (Sommer 1931). *Forsch. und Fortschritte* **8** (26).

TOLSTIKOV, Y. I. (ed) 1966: *Atlas Antarktiki* **1,** Glavnoye Upravleniye Geodezii Kartografii, Moscow.

TÓMASSON, H. and VILMUNDARDÓTTIR, E. G. 1967: The lakes Stórisjór and Langisjór. *Jökull* **17,** 280–99.

TONINI, M. and ROSSI, G. 1968: Le glacier de la Marmolada – variations depuis 1951. *Int. Ass. scient. Hydrol.* **79,** 193–206.

TRAINER, F. W. 1973: Formation of joints in bedrock by moving glacial ice. *US geol Surv. J. Res.* **1** (2), 229–36.

TRICART, J. 1965: *Principes et méthodes de la géomorphologie.* Masson, Paris (496 pp).

1970: *Geomorphology of cold environments.* Transl. Watson, E., Macmillan, London (320 pp).

TRICART, J. and CAILLEUX, A. 1962: *Le modelé glaciaire et nival.* Sedes, Paris (508 pp).

1965: *Introduction à la géomorphologie climatique.* Sedes, Paris (306 pp).

TUREKIAN, K. K. (ed) 1971: *Late Cenozoic glacial ages.* Yale Univ. Press (606 pp).

UEDA, H. T. and GARFIELD, D. E. 1968: Deep-core drilling program at Byrd Station (1967–1968). *Antarctic J. US* **3** (4), 111–12.

UNTERSTEINER, N. and NYE, J. F. 1968: Computations of the possible future behaviour of Berendon glacier, Canada. *J. Glaciol.* **7** (50), 205–13.

VANNI, M. 1968: Les variations des glaciers italiens au cours des dix dernières années 1957–1966. *Int. Ass. scient. Hydrol.* **79,** 182–92.

VANWORMER, D. and BERG, E. 1973: Seismic evidence for glacier motion. *J. Glaciol.* **12** (65), 259–65.

VERNON, P. 1966: Drumlins and Pleistocene ice flow over the Ards peninsula – Strangford Lough area, County Down, Ireland. *J. Glaciol.* **6** (45), 401–9.

VEYRET, P. 1955: Le lit glaciaire: contradiction apparente des formes et logique réelle des processus d'érosion. *Revue Géogr. alp.* **43**, 495–509.

1971: Observations recentes sur deux glaciers du massif du Mont Blanc. *Ann. der Meteorol.* **5**, 219–220.

VIALOV, S. S. 1958: Regularities of glacial shield movements and the theory of plastic viscous flow. *Int. Ass. scient. Hydrol.* **47**, 266–75.

VIRKKALA, K. 1952: On the bed structure of till in eastern Finland. *Bull. Com. géol. Fin.* **157**, 97–109.

1963: On ice-marginal features in southwestern Finland. *Bull. Com. géol. Fin.* **210** (76 pp).

VIVIAN, R. 1970: Hydrologie et érosion sous-glaciaires. *Revue. Géogr. alp.* **58** (2), 241–64.

VIVIAN, R. 1975: *Les glaciers des Alpes Occidentales.* Allier, Grenoble (513 pp).

VIVIAN, R. and BOCQUET, G. 1973: Subglacial cavitation phenomena under the Glacier d'Argentière, Mont Blanc, France. *J. Glaciol.* **12** (66), 439–51.

VIVIAN, R. and ZUMSTEIN, J. 1973: Hydrologie sous-glaciaire au glacier d'Argentière (Mont Blanc, France). *Symposium on the Hydrology of glaciers, Cambridge, 9–13 Sept. 1969, Int. Ass. Scient. Hydrol. Pub.* **95**, 53–64.

VORONOV, P. S. 1968: Oblaka–razrushiteli gor Antarktidy. *Chelovek i stikh* (141 pp).

WARDLE, P. 1973: Variations of the glaciers of Westland National Park and the Hooker range, New Zealand. *NZ J. Bot.* **11**, 349–88.

WARNKE, D. A. 1970: Glacial erosion, ice rafting and glacial-marine sediments: Antarctica and the southern ocean. *Am. J. Sci.* **269** (3), 276–94.

WASTENSON, L. 1969: Blockstudier i flygbilder. En metodundersökning av möjligheten att kartera markytans blockhalt från flygbilder. *Sveriges geol. Unders.* C **638** (95 pp).

WATERS, R. S. 1954: Pseudobedding in the Dartmoor granite. *Trans R. Geol Soc. Cornwall* **18**, 456–62.

WEERTMAN, J. 1957: On the sliding of glaciers. *J. Glaciol.* **3** (21), 33–8.

1961: Equilibrium profile of ice caps. *J. Glaciol.* **3** (30), 953–64.

1962: Stability of ice-age ice caps. *US Army Cold Reg. Res. Engng Lab. Res. Rept.* **97** (12 pp).

1964a: Rate of growth or shrinkage of non-equilibrium ice sheets. *US Army Cold Reg. Res. Engng Lab. Res. Rept* **145** (16 pp).

1964b: The theory of glacier sliding. *J. Glaciol.* **5** (39), 287–303.

1966: Effect of a basal water layer on the dimensions of ice sheets. *J. Glaciol.* **6** (44), 191–207.

1969: Water lubrication mechanism of glacier surges. *Can J. Earth Sci.* **6** (4), 929–42.

1970: A method for setting a lower limit on the water layer thickness at the bottom of an ice sheet from the time required for upwelling of water into a borehole. In Gow, A. J. *et al.* (eds) *International Symposium on Antarctic Glaciological Exploration (ISAGE), Int. Ass. scient. Hydrol. Pub.* **86**, 69–73.

1972: General theory of water flow at the base of a glacier or ice sheet. *Reviews of Geophysics and Space Physics* **10** (1), 287–333.

1973: Position of ice divides and ice centers on ice sheets. *J. Glaciol.* **12** (66), 353–60.

WEIDICK, A. 1968: Observations on some Holocene glacier fluctuations in west Greenland. *Medd. om Grønland* **165** (6) (202 pp).

1972: Notes on Holocene events in Greenland. In Vasari, Y., Hyvärinen, H. and Hicks, S. (eds) *Climatic change in Arctic areas during the last ten thousand years*, Oulu, 177–204.

WELLMAN, P., MCELHINNY, M. W. and MCDOUGALL, I. 1969: On the polar-wander path for Australia during the Cenozoic. *R. Astron. Soc. Geophys. J.* **18**, 371–95.

WEST, R. G. and DONNER, J. J. 1956: The glaciations of East Anglia and the east midlands: a differentiation based on stone-orientation measurements of the tills. *Q J geol Soc. Lond.* **112**, 69–91.

WHALLEY, W. B. 1971: Observations of the drainage of an ice-dammed lake – Strupvatnet, Troms, Norway. *Norsk geogr. Tidsskr.* **25**, 165–74.

WHITE, G. W. 1974: Buried glacial geomorphology. In Coates, D. R. (ed) *Glacial geomorphology*, State Univ. NY, Binghampton, 331–50.

WHITE, W. A. 1970: Erosion of cirques. *J. Geol.* **78**, 123–26.

1972: Deep erosion by continental ice sheets. *Bull. geol Soc. Am.* **83** (4), 1037–56.

WILLIAMS, J., BARRY, R. G. and WASHINGTON, W. M. 1974: Simulation of the atmospheric circulation using the NCAR global circulation model with Ice Age boundary conditions. *J App. Met.* **13** (3), 305–17.

WILSON, A. T. 1964: Origin of the Ice Ages: an ice shelf theory for Pleistocene glaciation. *Nature, Lond.* **201** (4915), 147–9.

1970: The McMurdo Dry Valleys. In Holdgate, M. W. (ed) *Antarctic ecology* **1**, SCAR, Academic Press, New York, 21–30.

1973: The great antiquity of some Antarctic landforms – evidence for an Eocene temperate glaciation in the McMurdo region. In van Zinderen Bakker, E. M. (ed) *Palaeoecology of Africa and of the surrounding islands and Antarctica* **8**, Balkema, Cape Town, 23–35.

WINTERHALTER, B. 1972: On the geology of the Bothnian Sea, an epeiric sea that has undergone Pleistocene glaciation. *Fin. Geol. Surv. Bull.* **258** (66 pp).

WINTERS, W. A. 1961: Landforms associated with stagnant ice. *Prof. Geogr.* **13**, 19–23.

WISNIEWSKI, E. 1973: Genesis of the Lammi esker (southern Finland). *Fennia* **122** (30 pp).

WOLDSTEDT, P. 1969: Quartär. *Handbuch der stratigraphischen Geologie* **2**, Stuttgart (263 pp).

WOLLIN, G., ERICSON, D. B. and EWING, M. 1971: Late Pleistocene climates recorded in Atlantic and Pacific deep-sea sediments. In Turekian, K. K. (ed) *Late Cenozoic glacial ages*, Yale Univ. Press, 199–214.

WRIGHT, H. E. 1957: Stone orientation in Wadena drumlin field, Minnesota. *Geogr. Annlr* **39**, 19–31.

1972: Interglacial and postglacial climates: the pollen record. *Quat. Res.* **2**, 274–82.

WRIGHT, H. E. and FREY, D. G. (eds) 1965: *The Quaternary of the United States*. Princeton Univ. Press, NJ (922 pp).

WYLLIE, P. J. 1965: Water-spouts on the Britannia Gletscher, northeast Greenland. *J. Glaciol.* **5** (40), 521–3.

YATSU, E. 1966: *Rock control in geomorphology*. Sozosha, Tokyo, (135 pp).

YOUNG, G. J. 1969: Snow cover variations, Axel Heiberg Island, NWT. *Unpub. MSc. thesis, Geogr. Dept, McGill Univ.*, Montreal (87 pp).

YOUNG, J. A. T. 1972: Ice margins of the 19th and 20th centuries in the Venidigergruppe, Hohe Tauern, Austria. *Arctic and Alpine Res.* **4** (1), 73–83.

ZIEGLER, T. (ed) 1972: Slamtransportundersøkelser i Norske breelver, 1970. *Norges Vassdrags-og Elektrisitetsvesen Rapport* **1**/72, Oslo (133 pp).

ZINDEREN BAKKER, E. M. van, 1969: Quaternary pollen analytical studies in the southern hemisphere with special reference to the sub-antarctic. *Palaeoecol. of Africa* **5**, 175–212.

ZUMBERGE, J. H. 1952: The lakes of Minnesota, their origin and classification. *Minn. Geol Surv. Bull.* **35** (99 pp).

1955: Glacial erosion in tilted rock layers. *J. Geol* **63**, 149–58.

Additional references (March 1977)

ANDREWS, J. T., DAVIS, P. T. and WRIGHT, C. 1976: Little Ice Age permanent snowcover in the eastern Canadian Arctic: extent mapped from LANDSAT-1 satellite imagery. *Geogr. Annlr. Stockh.* **58** (A), 71–81.

ANDREWS, J. T. and MAHAFFY, M. A. W. 1976: Growth rate of the Laurentide Ice Sheet and sea level lowering (with emphasis on the 115,000 B.P. sea level low). *Quat. Res.* **6**, 167–83.

BOULTON, G. S. 1976: The origin of glacially fluted surfaces – observations and theory. *J. Glaciol.* **17** (76), 287–309.

BOULTON, G. S. and PAUL, M. A. 1976: The influence of genetic processes on some geotechnical properties of glacial tills. *J. Engng. Geol.* **9**, 159–94.

BUDD, W. F., JENSSEN, D. and YOUNG, N. W. 1975: Computation of the temperature profile from drill data at Vostok Station. *Soviet Antarctic Expedition Information Bulletin* **8** (12), 668–73.

CLARKE, G. K. C. 1976: Thermal regulation of glacier surging. *J. Glaciol.* **16** (74), 231–50.

CLAPPERTON, C. M. 1975: The debris content of surging glaciers in Svalbard and Iceland. *J. Glaciol.* **14** (72), 395–406.

DEWART, G. 1976: Seismic evidence of a wet zone under the West Antarctic ice sheet. *J. Glaciol.* **16** (74), 73–88.

HALLET, B. 1976a: Deposits formed by subglacial precipitation of $CaCO_3$. *Bull. Geol. Soc. Am.* **87**, 1003–1015.

1976b: The effect of subglacial chemical processes on glacier sliding. *J. Glaciol.* **17** (76) 209–21.

HAMBREY, M. J. 1976a: Structure of the glacier Charles Rabot Bre, Norway. *Bull. Geol. Soc. Am.* **87** (11), 1627–37.

1976b: Debris, bubble and crystal fabric characteristics of foliated glacier ice, Charles Rabots Bre, Okstindan, Norway. *Arct. and Alpine Res.* **8** (1), 49–60.

HAMBREY, M. J. and MILNES, A. G. 1975: Boudinage in glacier ice – some examples. *J. Glaciol.* **14** (72), 383–93.

HAYS, J. D., IMBRIE, J. and SHACKLETON, N. J. 1976: Variations in the Earth's orbit: pacemaker of the Ice Ages. *Science* **194** (4270), 1121–32.

HODGE, S. M. 1976: Direct measurement of basal water pressures: a pilot study. *J. Glaciol.* **16** (74), 205–16.

JOHNSON, R. G. and MCCLURE, B. T. 1976: A model for Northern Hemisphere continental ice sheet variation. *Quat. Res.* **6,** 325–53.

LACHAPELLE, E. R. 1973: *Field guide to snow crystals.* Univ. of Washington Press, Seattle and London (101 pp).

MAHAFFY, M. W. 1976: A three-dimensional numerical model of ice sheets: tests on the Barnes Ice Cap, Northwest Territories. *J. Geophys. Res.* **81** (6), 1059–66.

MORRIS, E. M. and MORLAND, L. W. 1976: A theoretical analysis of the formation of glacial flutes. *J. Glaciol.* **17** (76), 311–23.

POST, A. and STREVELER, G. 1976: The tilted forest: glaciological–geologic implications of vegetated neoglacial ice at Lituya Bay, Alaska. *Quat. Res.* **6** (1), 111–17.

REHEIS, M. J. 1975: Source, transportation and deposition of debris on Arapaho glacier, Front Range, Colorado, U.S.A. *J. Glaciol.* **14** (72), 407–20.

ROBIN, G. DE Q. 1976: Is the basal ice of a temperate glacier at the pressure melting point? *J. Glaciol.* **16** (74), 183–96.

SISSONS, J. B. and SUTHERLAND, D. G. 1976: Climatic inferences from former glaciers in the southeast Grampian Highlands, Scotland. *J. Glaciol.* **17** (76), 325–46.

SUGDEN, D. E. 1976: A case against deep erosion of shields by ice sheets. *Geology* **4,** 580–82.
    1977: Reconstruction of the morphology, dynamics and thermal characteristics of the Laurentide ice sheet at its maximum. *Arctic and Alpine Research* **9** (1), 21–47.

THOMAS, R. J. 1976: Thickening of the Ross Ice Shelf and equilibrium state of the West Antarctic Ice Sheet. *Nature* **259** (5540), 180–83.

VALLON, M., PETIT, J.-R. and FABRE, B. 1976: Study of an ice core to the bedrock in the accumulation zone of an Alpine glacier. *J. Glaciol.* **17** (75), 13–28.

WEERTMAN, J. 1976: Sliding – no sliding effect and age determination of ice cores. *Quat. Res.* **6,** 203–207.

WEIDICK, A. 1975: Estimates on the mass balance changes of the Inland Ice since Wisconsin – Weichsel. *Grønlands Geologiske Undersøgelse Rapport* **68** (21 pp).

Additional references (April 1979)

DRAKE, L. D. 1977: Depositional fabrics in basal till reflect alignments during transportation. *Earth Surface Processes* **2,** 309–17.

EVENSON, E. B., DREIMANIS, A. and NEWSOME, J. W. 1977: Subaquatic flow tills: a new interpretation for the genesis of some laminated till deposits. *Boreas* **6,** 115–33.

HAMBREY, M. J. 1977: Structures in ice cliffs at the snouts of three Swiss glaciers. *Journal of Glaciol.* **18,** 407–14.

HOOKE, R. LE B. 1977: Basal temperatures in polar ice sheets. *Quat. Res.* **7,** 1–13.

KUKLA, G. J. 1977: Pleistocene land-sea correlations, I. Europe. *Earth-Science Rev.* **13,** 307–74.

MORLAND, L. W. and MORRIS, E. M. 1977: Stress in an elastic bedrock hump due to glacier flow. *Journal of Glaciol.* **18,** 67–75.

PATERSON, W. S. B., KOERNER, R. M., FISHER, D., JOHNSON, S. J., CLAUSEN, H. B., DANSGAARD, W., BUCHER, P., and OESCHGER, H. 1977: An oxygen isotope climatic record from the Devon Island Ice cap, Arctic Canada. *Nature* **266,** 508–11.

ROBIN, G. de Q. 1977: Ice cores and climatic change. *Phil. Trans. Roy. Soc. London,* B. **280,** 143–68.

ROSE, J. and LETZER, J. M. 1977: Superimposed drumlins. *Journal of Glaciol.* **18,** 471–80.

SHACKLETON, N. J. and OPDYKE, N. D. 1977: Oxygen isotope and palaeomagnetic evidence for early northern hemisphere glaciation. *Nature* **270,** 216–19.

VAN-ZINDEREN BAKKER, E. M. editor 1978: *Antarctic glacial history and world palaeoenvironments.* Balkema, Rotterdam (172 pp.).

WILLIAMS, L. D. 1978: Ice-sheet initiation and climatic influences of expanded snow cover in Arctic Canada. *Quat. Res.* **10** (2), 141–49.

Additional references (March 1982), see p. 336.

Index

The index covers subject matter and place names. An author index is not included in view of the presence of a full bibliography on pages 337–61. **Bold print** indicates significant coverage of a topic.

This book is a study of glacial processes and forms. Not only is it a critical and creative synthesis of recent research, but it also provides significant modifications to accepted theory and approach. In each chapter, the authors have proposed models and hypotheses as an aid to understanding and a stimulus to inquiry. At the same time, they present fundamental data from a wide range of sources, with particular attention to current work in Greenland and Antarctica. Care has been taken at each reprint to give additional references.

'... the authors have produced a *tour de force* in logic, clear discussion and geographical expertise ... this is the most important text on glacial geomorphology.' *Arctic*

'... this book provides the best all-round current introduction to the field.' *Progress in Physical Geography*

'Sur tous ces points, les mêmes qualités d'information, de raisonnement et d'exposition rendent la lecture très profitable. Les auteurs recherchent les explications vraiment scientifiques, mais avec très peu d'appareil mathématique. Des figures suggestives, d'excellentes photographies, une langue claire aident le lecteur. Un remarquable instrument de travail.' *Revue de Géographie Alpine*

'... an unusual and striking textbook about glaciers and their effects on the landscape beneath them ... The book is encyclopaedic, clearly written from a geographical viewpoint, and profusely illustrated with good drawings and photographs.' *Science*

'Few books are better presented for the price and none more captivating to the reader interested in the fascinating phenomenon of ice on the land.' *Nature*

Cover: False-colour ERTS image (no. 8139212185) of the Ejyafjördur region of north Iceland, showing a spectacular landscape of glacial erosion. The white snow-covered mountains stand out clearly and the red colouring emphasizes the dendritic pattern of glacial troughs. The broadest troughs, which are major ice discharge routes, are fed by short tributaries and the main watershed has been breached in two places.

Edward Arnold

ISBN 0 7131 5840 9